Ferry and Brandon's

Cost Planning of
Buildings

Higher Education Library
ea‑ ing Centre
F‑ ‑land College

he last day shov
due items

Multiple Sclerosis Society

A portion of the author's annual royalty payment from sales of this textbook will be donated to the Wakefield and District Branch of the Multiple Sclerosis Society. The branch provides support for families affected by this devastating condition.

Ferry and Brandon's
Cost Planning of Buildings

Eighth Edition

Richard Kirkham

School of the Built Environment
The Liverpool John Moores University

with contributions from

Brian Greenhalgh and Anthony Waterman

Blackwell Publishing

Blackwell Publishing editorial offices:
Blackwell Publishing Ltd, 9600 Garsington Road, Oxford OX4 2DQ, UK
Tel: +44 (0)1865 776868
Blackwell Publishing Inc., 350 Main Street, Malden, MA 02148-5020, USA
Tel: +1 781 388 8250
Blackwell Publishing Asia Pty Ltd, 550 Swanston Street, Carlton, Victoria 3053, Australia
Tel: +61 (0)3 8359 1011

First published in Great Britain by Crosby Lockwood & Sons Ltd 1964; Second edition (metric)
published 1970; Third edition published 1972; Fourth edition published by Granada Publishing 1980;
Fifth edition published 1984; Sixth Edition published by BSP Professional Books 1991; Seventh Edition
published by Blackwell Science Ltd 1999; Eighth edition published by Blackwell Publishing Ltd 2007

3 2009

Library of Congress Cataloging-in-Publication Data

Kirkham, Richard J.
Ferry and Brandon's cost planning of buildings / Richard Kirkham. — 8th ed.
p. cm.
Rev. ed. of: Cost planning of buildings / Douglas J. Ferry, Peter S. Brandon, Jonathan D. Ferry. 7th ed. 1999.
Includes bibliographical references and index.
ISBN 978-1-4051-3070-7 (pbk. : alk. paper)
1. Building–Estimates.   2. Building–Cost control.
I. Brandon, P. S. (Peter S.)   II. Ferry, Douglas J. (Douglas John) Cost planning of buildings.   III. Title.
IV. Title: Cost planning of buildings.

TH435.F36 2007
692—dc22
2007000571

A catalogue record for this title is available from the British Library

Set in 10/12pt Sabon by Graphicraft Limited, Hong Kong
Printed and bound in Singapore by Ho Printing Pte Ltd

For further information on Blackwell Publishing, visit our website:
www.blackwellpublishing.com/construction

Contents

PHASE 3: COST PLANNING AND CONTROL AT PRODUCTION AND OPERATION   287

Chapter 15 Planning and Managing Project Resources and Costs   289

Preface to the Eighth Edition

As an undergraduate student at Liverpool University in the early 1990s, I fondly remember hunting down Ferry and Brandon's text in the dark, narrow corridors of the Harold Cohen Library. I would often use this text to get to grips with the business of estimating, tendering and taking off. I now consider it a great privilege to have been asked to revise and update the eighth edition. Books don't last eight editions simply by chance; the ones that do are always very good ones.

The dilemma for any author updating a long-established textbook is 'how much ought I to change?' I considered a radical approach to the new edition, but on reflection I questioned the wisdom of this and ultimately formed the opinion that the core structure of the book ought not to be changed, as clearly it worked. This edition is therefore still strongly based around the three-phase process advocated by Ferry and Brandon. The layout has been changed slightly to reflect the major changes that have occurred since the seventh edition, not only within the discipline of cost planning, but also the construction industry generally. For example, the treatment of procurement is prominent at the start of the text rather than towards the end. I have also recognised the ever-evolving role of the quantity surveyor, and the prominent role that quantity surveyors now assume throughout the project life-cycle. Moreover, I hope this edition also impresses on the reader the importance of collaborative working between the design team members at the earliest possible stages of the project, and the role that cost planners have in the briefing process. The impact that whole life-cycle costing now has on the cost planning process is also reinforced in this edition.

Many of the principles of elemental cost planning and the techniques used in building up the cost plan etc., have stood the test of time and thus remain unchanged. There is still a great deal of original material in this edition; those readers familiar with the text will no doubt take comfort in this, and I hope that I have struck the right balance between new and old.

I wonder what challenges the construction professions, and in particular the cost planners, will face in the future? The mouth-watering prospect of massive capital investment in built assets around the east of London in time for the 2012 Olympics presents a real opportunity to demonstrate the innovative, dynamic and professional way in which the UK construction industry can deliver prestigious schemes. I hope that this book helps the current crop of undergraduates in quantity surveying, construction management or any other discipline to understand the fundamental importance of effective cost planning, and that they will take this knowledge and be part of something that will change many people's lives for the better. That is the real beauty of the construction industry – you can have a tangible stake in improving people's lives for the better!

Richard Kirkham
Liverpool John Moores University
February 2007

Preface to the First Edition

This book is intended as an introduction to cost planning for practising quantity surveyors and as a textbook for students taking the Final Examination (Quantities Part II) of the Royal Institution of Chartered Surveyors, or the Third Examination of the Institute of Quantity Surveyors. I have therefore assumed that the reader is already familiar with the ordinary processes of quantity surveying, particularly the preparation of bills of quantities, and this is taken for granted in the text; nevertheless I hope that the book may be read with advantage by members of allied professions.

I have tried to present the subject in a way that will be helpful to the surveyor coming to grips with it for the first time, and have concentrated on explaining the basic principles, basic methods and some of the main pitfalls. I have not tried to reprint the masses of tables, charts and detailed examples which have appeared in the technical press, as once the principles have been understood the reader will find that lack of time rather than lack of material will limit his further studies.

I have received a good deal of assistance in the compilation of this work from the various organisations which are mentioned therein, but I would particularly like to mention the help given by the officers of Hertfordshire County Council Architects Department and the Building Cost Advisory Service of the Royal Institution of Chartered Surveyors.

Douglas J. Ferry
Belfast
June 1964

Nomenclature and Acronyms

Σ	the sum of (sigma notation)
P	the principal (in investment terms)
i	the rate of interest
n	time (ordinarily number of years)
r	discount rate or real discount rate (in STPR calculations)
ρ	catastrophe risk and pure time preference (in STPR calculations)
μ	elasticity of the marginal utility of consumption (in STPR calculations)
g	output growth
t	time
$\approx$	approximately equal to
π	pi (3.142 to 3 d.p.)
σ	standard deviation
ACostE	Association of Cost Engineers
AFS	ascertained final sum
AI	artificial intelligence
AIRR	adjusted internal rate of return
ANN	Artificial Neural Network
BCIS	Building Cost Information Service
BMI	Building Maintenance Information Service
BP	back propagation
BQ	bill of quantities
BSI	British Standards Institute
CAAD	Computer Aided Architectural Design
CABE	Commission for Architecture and the Built Environment
CATO	Computer Aided Taking Off System

CBC	Co-ordinated Building Classification
CBS	cost breakdown structure
CDM	Construction Design and Management Regulations 1994
CIBSE	Chartered Institution of Building Services Engineers
CIOB	Chartered Institute of Building
CIRIA	Construction Industry Research and Information Association
CIS	Construction Information Service
CITE	Construction Industry Trading Electronically
CM	construction manager or construction management (in procurement)
CPI	Consumer Price Index
CQS	contractor's quantity surveyor
D&B	design and build
DCF	discounted cash flow
DPM	damp-proof membrane
DCMF	Design, Construct, Manage and Finance
EAC	equivalent annual cost
ECC	Engineering and Construction Contract
E-procurement	electronic procurement
EST	Energy Saving Trust
FV	future value
GDP	Gross Domestic Product
GFA	gross floor area
GMP	guaranteed maximum price
HGCRA	Housing Grants, Construction and Regeneration Act 1996
HMRC	Her Majesty's Revenue and Customs
ICE	Institution of Civil Engineers
IPD	Institute for Professional Development
IRR	internal rate of return
ISO	International Standards Organisation
IT	information technology
ITOCC	Occupiers International Total Occupancy Cost Code
JCT	Joint Contracts Tribunal
KM	knowledge management
LA	local authority
LCC	life-cycle costing
MARR	minimum acceptable rate of return
MC	management contracting
MCDM	multi-criteria decision-making
MLR	minimum lending rate
M&E	mechanical and electrical
MTC	measured term contract
NAO	National Audit Office

NEC	New Engineering Contract
NPS	National Procurement Strategy
NPV	net present value
NR	Network Rail
No	number (of)
NS	net savings
OGC	Office of Government Commerce
PC	practical completion
PFI	Private Finance Initiative
POCA	Property Occupancy Cost Analysis
PPP	Public Private Partnership
PQS	private practice quantity surveyor
PSA	Property Services Agency
PSC	Public Sector Comparator
PV	present value
QS	quantity surveyor
RC	reinforced concrete
RIBA	Royal Institute of British Architects
RICS	Royal Institution of Chartered Surveyors
SBC	Standard Building Contract
SCQS	Society of Construction Quantity Surveyors
SFCA	Standard Form of Cost Analysis
SIR	savings to investment ratio
SMM7	Standard Method of Measurement of Building Works, 7th edition
SPV	Special Purpose Vehicle
STPR	social time preference
TPI	Tender Price Index
TPISH	Tender Price Index of Social Housing
VAT	value added tax (17.5% at time of writing)
WLCC	whole life-cycle costing

Acknowledgements

I would like to express sincere gratitude to Dr Halim Boussabaine (University of Liverpool) for his patience and friendship; updating this text has diverted my time from other activities and had it not been for him, many initiatives we organised would simply not have happened. Similarly, thanks to Mr John Lewis of the same institution, for some useful additions to this text.

Thanks also to the following at Liverpool John Moores University: Mr Bill Atherton for his splendid illustration skills and willingness to cover the odd lecture; Mr John McLoughlin for the production of the diagrams in Chapter 9; Dr Fiona Borthwick and Dr Clare Harris who offered moral support throughout; Anne Roberts and her team in the school office who accommodated my tardiness with the paperwork from time to time, and finally my fellow colleagues on the 'green mile'.

Dr John Schofield, my friend and colleague, chair of the Independent Monitoring Board at HM Prison Altcourse, moved earth and high heaven to accommodate my workload around the duties of the board. His support has been invaluable, and I thank him most sincerely.

Finally to Joanne, Liverpool's finest district nurse, who has had to put up with the kitchen resembling the University Library for quite some time!

About the Authors

Douglas J. Ferry PhD, FRICS formerly Dean of Architecture and Building, New South Wales Institute of Technology and Research Manager with CIRIA. Douglas authored the very first edition of this text in 1964 whilst based at the College of Technology, Belfast where he lectured in quantities and building construction. In that same year he also published *Rationalisation of Measurement* with the Royal Institution of Chartered Surveyors.

Peter S. Brandon DSc, DEng, MSc, FRICS is Director of Salford University 'Think Lab' and Director of Strategic Programmes, School of the Built Environment, University of Salford. Formerly Pro-Vice Chancellor (Research and Graduate College) at the University, he was the inaugurator of several high profile initiatives including Construct IT, the national network for Information Technology in Construction which received the Queen's Anniversary Prize in 2000; SURF, the Centre for Sustainable Urban and Regional Futures; CCI, The Centre for Cnstruction Innovation and the BEQUEST international network.

Jonathan D. Ferry BSc(Hons) is a Manager of Procurement in Projects working for Tube Lines Limited under the London Underground PPP, and was previously a Director of Dearle and Henderson.

Richard J. Kirkham BA(Hons), PhD (Liverpool), MACostE, ICIOB is a Senior Lecturer in Construction Management at the School of the Built Environment, Liverpool John Moores University. His research interests are in whole life-cycle costing, quantitative risk analysis, stochastic modelling techniques and performance measurement for public sector facilities. Prior to this, he worked as a Research Officer at Cranfield University. He has published widely in the field of whole life-cycle cost modelling and is co-author of

Whole Life-Cycle Costing: Risk and Risk Responses. He is a series editor of the RICS Research/Blackwell Construction Series and is Scientific Secretary of CIB-TG62 Complexity and the Built Environment. Richard is a Fellow of the Royal Statistical Society, an Incorporated Member of the Chartered Institution of Building (Vice Chair Liverpool Centre 2006/7 and Chair 2007/8) and was elected a Member of the Association of Cost Engineers in 2004.

About the Contributors

Mr Brian Greenhalgh BSc, MBA, FRICS, FCIOB is currently Head of External Affairs in the School of the Built Environment at Liverpool John Moores University where he specialises in the procurement and management of construction projects. After qualifying as a Chartered Quantity Surveyor, he worked both nationally and internationally before joining the Liverpool Polytechnic. He has served on RICS committees both locally and nationally and has lectured widely on aspects of construction management and contract administration.

Mr Anthony Waterman BA (Hons), MSc (University College London) is Head of Research at Sense Cost Consultancy, a division of Mace. Prior to joining Sense, Anthony worked as a Principal Consultant at the Building Research Establishment after completing his Master's degree in Construction Economics. At BRE he worked on various aspects of whole life-cycle costing and performance modelling, including the development of PSC models for Prime Contracting and PFI schemes and has published several reports and papers.

Introduction

Chapter 1
The Discipline of Cost Planning

1.1 Buildings cost money . . .

Buildings lie at the very heart of our everyday lives. We live and work in them; they provide the very means by which modern civilisations function, but it is because of this that it is easy to underestimate their importance. Buildings and structures facilitate the provision of healthcare, education, commerce and justice. In other words, buildings provide not only enclosure but also *de facto* social capital.

Notwithstanding the value[1] aspects of buildings, the construction industry generally is inextricably linked with money. Simply put, buildings cost money, and usually lots of it. This may seem a rather simplistic contention but history reveals that understanding the costs of construction is a skill that has developed over time. In the seventh edition to this text, Douglas Ferry and Peter Brandon referred back to biblical times in order to trace the origins of cost planning, and the reading they quoted from St Luke (Ch.14) gives a fascinating insight:

> Would any of you think of building a tower without first sitting down and calculating the cost, to see whether he could afford to finish it? Otherwise, if he has laid its foundations and then is not able to complete it, all the onlookers will laugh at him. 'There is the man' they will say 'who started to build and could not finish'.

Whilst there are clearly metaphorical connotations within this reading, the point is pretty clear. To build well you must first plan. Interestingly, the final part of the reading is a harrowing reminder to many clients and builders in today's society who have not taken heed of good budgetary management.

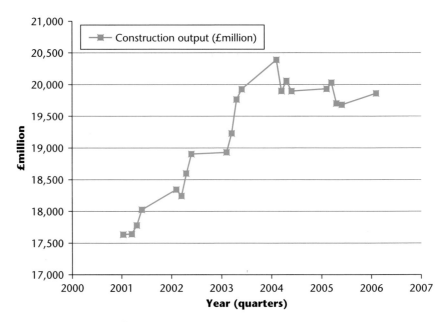

Figure 1.1   Quarterly construction output.

1.2   What happened to the cost plan?

Chapter 7 will consider the relationship between costs of buildings and procurement, procurement being the method by which buildings are delivered to the client. The UK in particular has seen a rapid increase in construction output since the year 2000 (see Figure 1.1), but allied to this has been an increasing focus on project budgets, and moreover the ability to deliver these projects to the projected cost.

Sadly, several high profile construction projects in the UK have been plagued with problems over programme and budget. With public sector construction projects there is a strong emphasis on meeting the budget, so when the project runs into financial difficulties, the taxpayer and media become rather unsympathetic. Some recent examples include the following.

1.2.1   Wembley Stadium, London (2000–2007)

The new national stadium at Wembley is a project that has been mired in controversy with questions over adequate cost planning and budget management. Initially, the cost of the north London stadium was expected to be approximately £200 million. However, in the summer of 2006 the projected cost had increased to some £715 million. The Football Association and Rugby League Challenge Cups, along with a host of other events, were relocated to other stadia as the project rolled on beyond the anticipated completion date.

The project has become the subject of intense media speculation and on 17 July 2006, the Minister of State for Culture, Media and Sport, Richard Caborn, was asked to make a declaration on the target for practical completion. He stated that he was confident that substantial completion of the stadium would be achieved in July 2006, sufficient to enable practical completion (PC) by September 2006. PC was eventually achieved in 2007.

So what happened to the cost plan? The contractors have disputed the projected final cost, arguing that the £715 million figure is the 'cost shown to the banks so that they know that [we] can finance that amount should we need to'.

A combination of factors led to the problems faced by the project team; however, an article in *The Economist* in mid-2005[2] suggested that 'an unanticipated rise in the cost of steel (which doubled in 2004) and the extra labour required to ensure the building is ready for the May FA Cup Final [threw] the management's calculations out of kilter'.

The steelwork issue certainly had a significant impact, not only on the cost plan but also the contractual arrangements between the principal contractor Multiplex and steelwork contractor Cleveland Bridge UK. Early on in the project, Cleveland Bridge entered into a £60 million lump sum contract to design, fabricate, deliver and erect the structural steelwork at Wembley Stadium, including the bowl and huge steel arch[3]. However, in late 2003, Cleveland Bridge and Multiplex entered into formal dispute as the latter argued that it was haemorrhaging cash as a result of market conditions and specific project issues, and it sought significant variation payments or a change to a cost plus contract arrangement. This dispute continued in court as this book went to press.

1.2.2   The Scottish Parliament Building, Edinburgh (1997–2004)

As with Wembley Stadium, the construction of the new Scottish Parliament Building at Holyrood in Edinburgh was shrouded in controversy, resulting in a public enquiry led by Lord Fraser of Carmyllie. In May 1997, the recently elected Labour Government committed to holding a referendum on devolved government in Scotland. In the referendum held on 11 September 1997 almost 75% of those voting agreed that there should be a Scottish Parliament. A new building to house the Parliament was therefore required and in a subsequent white paper it was estimated that the cost of constructing a new Parliament would be between £10 million and £40 million. This estimate was made prior to the identification of a location or a design.

Good cost planning requires critical engagement from all project stakeholders from the outset, and a unified voice of opinion from the client. This was not the case and it could be argued that on this project there was no client. However, one of the most catastrophic decisions to affect the project in terms of cost and progress was the procurement route selected. In his opening speech prior to publication of his report, Lord Fraser said:

> As I have said in my Introduction, while I have a number of sharp criticisms and recommendations to make on matters which ought to have been much better understood, there is no single villain of the piece. There were, however, some catastrophically expensive decisions taken and principal among those was the decision taken – not cleared with Ministers – to follow the procurement route of construction management. I have very real doubts if the extent of the risk remaining with the public purse was properly understood at the time it was adopted and I remain concerned that it was not clearly grasped by the Scottish Parliament for nearly two years after the Project was handed over to the SPCB [Scottish Parliament Corporate Body] when the Parliament gave up trying to have a 'budget' for the building. Any building constructed under the procurement model of construction management costs what it costs.

Lord Fraser's report could not be damning enough of the fact that the project did not have a cost plan. Inadequate briefing was in part responsible for this (some argue that there was no brief at all); the importance of developing a brief and its relationship with the cost plan is covered in this first section of the book. The reader is also encouraged to refer to the further reading sections at the end of the chapters, for further information on briefing.

1.2.3   The British Library at St Pancras, London (1974–1988)

The relationship between cost planning, procurement route selection and the brief was a key feature of the British Library project in London, which by project completion in 1988 had amassed a net increase of £58 million on the original planned cost. Like the other projects described in this chapter, a catalogue of errors occurred which led to the final cost of the project coming in at some £500 million. Principal among these was the decision to adopt the construction management procurement strategy. The National Audit Office (NAO) report heavily criticised this decision, and quite surprisingly Lord Fraser did not allude to this in his Holyrood Enquiry report. Had he done so, the media would no doubt have rallied against the decision-makers in that lessons clearly had not been learned. This unsuitable method of procurement had a fundamental impact on the cost plan, as did the thousands of design changes and variations from the original brief, systematic failures in quality and cost control during production, and failures in the budgetary controls of contracts for the many different works packages undertaken by various subcontractors. The lack of experience in using construction management, allied with the complex nature of the architects' and engineers' contracts and the standard conditions for works contractors, led to a situation that could not possibly sustain the original cost plan. The result was a damning enquiry and an NAO report that was said to be the most critical assessment yet of a public construction project.

1.3   The cost planning process

Cost planning as a process is difficult to define concisely as it involves a variety of procedures and techniques used concurrently by the quantity surveyor (QS) or building economist. Traditional cost planning will usually follow the conventional outline design, detailed design process. In a practical sense, the cost planning process starts with the development of a ballpark figure (or cost bracket) to allow the client to decide whether the project is feasible. More robust techniques for doing this are described in Chapter 5. This feasibility estimate is usually calculated on a unit cost method (e.g. cost per bed for a hospital, cost per student for a school). The estimate is then refined using the elemental method: the building is broken down into its component elements and sub-elements, usually using the Building Cost Information Service (BCIS) cost structure (Appendix B). The elemental method is a system of cost planning and control that enables the cost of a scheme to be monitored during the various stages of design development.

A good cost planning system should:

■ ensure that the tender figure is as close as possible to the first estimate, or that any likely difference between the two is anticipated and within an acceptable range;
■ ensure that the funds available for the projects are allocated effectively and economically to the various elements and sub-elements;
■ always involve the measurement and pricing of approximate quantities at some stage of the process;
■ aim to achieve good value at the desired level of expenditure.

A direct benefit of good cost planning is to reduce project risk. Steps should be taken to ensure that project development budget opportunities and threats are fully identified and assessed.

Ferry and Brandon, in previous editions of this book, described the cost planning process in three phases:

■ *Phase 1*. Defining the brief and setting the budget. In disciplines other than construction project management, this is commonly referred to as scoping or framing.
■ *Phase 2*. The cost planning and control of the design process. This phase is of critical importance since decisions made at design have a direct impact on whole life performance.

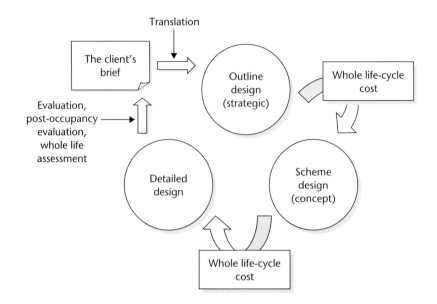

Figure 1.2   The conventional stages of the cost planning process, including whole life-cycle costing.

- *Phase 3*. The cost control of the procurement and construction stages. The construction part of this phase is still as relevant today as it was when Ferry and Brandon first proposed the process. However, procurement has changed and decisions with regard to this are sometimes made prior to design, indeed during Phase 1. It is therefore essential that cost-planning advice recognises the impact that certain procurement decisions will have on design and construction costs. This is addressed in Chapter 8.

Current thinking suggests that cost planning should continue beyond the conventional boundaries and take into account the whole life rather than the period up to PC. Figure 1.2 shows the typical cost planning process, but importantly this is encapsulated within a whole life decision environment; in other words, it should include adequate appraisal of the long-term cost implications of design decisions (such as maintenance, energy and facilities management (FM) costs). This will be explored in depth in Chapter 7.

1.4   Cost planning and the role of the quantity surveyor (QS)

The functions of the quantity surveyor (QS) are broadly concerned with the commercial management of construction projects. This breaks down into two areas of work: the planning and control of project costs, and the management of the terms and conditions of the form of contract agreed by the parties (client and contractor). In this book the first area will be considered.

The planning and control of construction costs cover a range of activities, which may include feasibility studies, cost planning, value engineering, cost–benefit analysis and life-cycle costing, which all take place during the design stage of projects; and the calculation of interim valuations and final accounts, including the cost estimation of variations and changes, which are procedures that take place during the construction stage of the project. QSs can also be known as construction economists, cost engineers or construction commercial managers.

The quantity surveying profession can trace its roots back to the rebuilding of London after the Great Fire of 1666. Before that date, buildings tended to be built on what we now call a design build arrangement, where the client would give the builder an outline of what was wanted, and the master builder would work out the details, arrange all the specialist tradesmen and forward the bills to the client at regular intervals. The difficulty with this arrangement was that the client did not know how much the building was likely to cost before it was finished, and if the client wanted several estimates or quotations[4], each builder would need to calculate the amount of materials, plant and labour required, with the obvious duplication of effort and cost.

With so much rebuilding work required after the Great Fire, a more efficient system of calculating building costs and generating estimates was clearly required. So the independent QS was born, whose role was originally to consider the architect's drawings (and specifications if they were lucky) and to develop a 'Bille of Quantityes', with the purpose of allowing any firm who wished to tender for a project to calculate that tender on the same basis and therefore minimise duplication of effort. This service was originally paid for by the contractors tendering for the work, but over time the role became part of the responsibilities of the client's side, to make sure that all tenderers were issued with identical tender documents.

Up to the beginning of the twentieth century, most large-building construction work was either procured by the government or by private individuals, where cost was not seen as the main criterion. Infrastructure work was slightly different and the considerable amount of canal building in the eighteenth century and railway construction in the nineteenth century was undertaken at considerable expense by corporate organisations. These companies (the railway companies prior to nationalisation) would borrow money from the capital markets to build the permanent way (or P-way as they referred to it) and rolling stock, and they raised revenue through ticket sales to passengers and charging for freight. However, most railway construction projects were grossly over budget, and for all Brunel's image as an icon of railway engineering, he was constantly being sued by construction firms for non-payment of bills on projects where he had lost control of expenditure. Clearly, a further change was required.

The quantity surveying profession therefore took the initiative, spurred by the development of what is now the Royal Institution of Chartered Surveyors (RICS) in 1868, and developed procedures to control construction costs by accurate measurement of the work required and the application of expert knowledge of costs and prices of work, labour, materials and plant. Some time later, they would use their understanding of construction technology to assess the implications of design decisions at an early stage, to ensure good value for money.

The technique of measuring quantities from drawings and specifications prepared by designers, principally architects and engineers, in order to prepare tender/contract documents, is known in the industry as 'taking off'. The quantities of work taken off are used typically to prepare bills of quantities (BQs), which traditionally have been prepared in accordance with one of the published standard methods of measurement as agreed by the quantity surveying profession and representatives of contractor organisations.

Although all QSs will have followed a similar course of education and training (usually to degree level for those entering the profession today), there are many areas of specialisation in which a QS may concentrate. The main distinction between QSs is those who carry out work on behalf of a client organisation, often known as a professional quantity surveyor (private practice QS or PQS), and those who work for construction companies, often known as a contractor's quantity surveyor. The latter is usually responsible for all legal and commercial matters within the contracting organisation and because of this many are now termed commercial managers.

1.5   Public sector building procurement

The innovations in procurement since the mid-1990s have radically changed the professional remit of QSs and other building professionals involved in providing strategic advice to clients on procurement and design. The design and economics of construction are inextricably linked with the procurement process, and this is recognised in the tranche of documentation issued by the Office of Government Commerce (OGC) under the umbrella of 'Achieving Excellence in Construction'[5].

Achieving Excellence in Construction was launched in March 1999, by the Chief Secretary to the Treasury, to improve the performance of central government departments, executive agencies and non-departmental public bodies (NDPBs) as clients of the construction industry. It put in place a strategy for sustained improvement in construction procurement performance and in the value for money achieved by government on construction projects, including those involving maintenance and refurbishment.

Key aspects include the use of partnering and development of long-term relationships, the reduction of financial and decision-making approval chains, improved skills development and empowerment, the adoption of performance measurement indicators and the use of tools such as value and risk management and whole life-cycle costing (WLCC).

With all these changes, QSs are now seen as the financial managers of the construction team who add value by monitoring time and quality as well as the traditional function of cost. The role of QSs has therefore changed significantly from its humble origins and they are now responsible for ascertaining a long-term view of building projects, assessing options and providing clients with comprehensive information on which to base investment decisions.

1.6 International dimensions

The profession of quantity surveying is a peculiarly British institution, owing chiefly to the historical context outlined above. Through emigration in the nineteenth and twentieth centuries, the British construction procurement system has been exported to Commonwealth countries, so that in English speaking countries such as Australia, New Zealand, South Africa and Canada, there are well-established firms of QSs represented by their own national professional bodies. The construction industries in these countries have developed separately, showing that the skills of the QS have been found to be valuable. In Europe, the pre- and post-contract roles are generally split, so the feasibility and cost planning function is taken by the *économiste de la construction* in France or the *baueconomist* in Germany. The post-contract function of valuations and final account preparation is often taken by the resident engineer, assisted by a technician cost engineer. In the USA the situation is substantially the same, although many large firms of QSs are now operating very successfully where clients see the considerable value of core QS skills in technology, law and economics. Traditionally, these skills were held by different professions who did not necessarily have detailed knowledge of the construction industry.

1.7 The future of cost planning

Cost planning, like any other discipline within construction, is continually developing and responding to the ever-changing demands of today's clients. In order to meet this challenge, planners should acquire and continue to develop over their careers a basic set of competences. This book aims to provide the first step, but future cost planners must recognise the importance of:

■ understanding the impact of early stage strategic decisions on the project life-cycle costs and performance;

Introduction

- recognising the importance of sustainability through a wider understanding of how buildings perform in use; environmental and energy issues should act as the catalyst to a wider appreciation of whole life-cycle costing;
- harnessing the benefits of increased interdisciplinary collaborative working with other professionals within the client, design and construction teams;
- engaging with the design team from the outset in order to foster a greater understanding of the impact of design decisions on whole life-cycle cost;
- developing effective risk management strategies with the cost planning process.

Chapter 2 examines some of the recent developments in cost planning discipline and highlights potential future advances.

Notes

(1) The concept of value is based on the relationship between satisfying needs and expectations, and the resources required to achieve them.
(2) 'Project management: Overdue and over budget, over and over again.' *The Economist*, 9 June 2005.
(3) 'Ruthless but lawful.' *QS News*, 14 July 2006.
(4) The difference between an estimate and a quotation is very important. An estimate is only an indication of what the cost of the project will be and may change if, for example, material or labour prices change. A quotation, on the other hand, is a fixed price and will only change if the client varies their instructions.
(5) The full 'Achieving Excellence in Construction' documentation can be downloaded from the Office of Government Commerce website at www.ogc.gov.uk

Further reading

Ashworth, A. (2004) *Cost Studies of Buildings*. Prentice Hall, Harlow.
Ashworth, A. and Hogg, K. (2001) *Willis's Practice and Procedure for the Quantity Surveyor*. Blackwell Science, Oxford.
Brook, M. (2004) *Estimating and Tendering for Construction Work*. Butterworth-Heinemann Ltd, Oxford.
Cartlidge, D. (2006) *New Aspects of Quantity Surveying Practice*. Butterworth-Heinemann Ltd, Oxford.
Jaggar, D., Ross, A., Smith, J. and Love, P. (2002) *Building Design Cost Management*. Blackwell Publishing, Oxford.

Chapter 2
Research and Development in Cost Planning Practice

2.1 Introduction

The nature of construction procurement and the complexity of the relationships between the various actors in the project life-cycle have led to a new paradigm in the use of information technology (IT) supported cost planning systems. The gradual transition from paper-based planning and documentation to e-platforms has allowed the industry to offer better services to clients, whilst empowering construction professionals with the capability to capture knowledge from past projects and translate this into future best practice and service delivery – known as knowledge management (KM).

It would not be fair to say that the construction industry has been slow to take up the new paradigm (although many commentators would argue the contrary). The leading edge of the UK construction industry has always embraced technological change and innovation, whereas the majority of firms, because they are relatively small, have been forced to change either through legislation or supply chain pressures. The innovation of the leading firms is rarely recognised, perhaps indicative of the common view held of the industry. Notwithstanding, the inherent nature of the industry and the environment in which it operates are not overtly conducive to IT/KM intensive systems since construction projects:

- are generally non-repetitive and tend to have significant unique features that are likely to be novel;
- contain multiple stakeholders and teams;
- carry risk and uncertainty;
- are usually in the hands of a temporary team which may be subject to frequent change as the work progresses;
- can be long in duration; events inside and outside (externalities) the project may affect the outcome.

Moreover, the construction process *per se* is heavily dependent on team-work, relationships, organisational strategies, human interactions and communication, more so perhaps than in the classical manufacturing scenario. This is the basis for Sir John Egan's *Rethinking Construction* report (this is touched upon throughout this book). The widespread use of IT-supported systems brings obvious benefits, but the key to success is really in striking the right balance between the abilities of humans and of IT within a structured organisational hierarchy.

This chapter considers the IT paradigm in the context of KM and e-procurement, looking at issues such as tender documentation submission and communication with the various supply chain members.

2.2   Knowledge management in construction

KM is a concept that pervades many industries, such as financial services, commercial enterprise and public service delivery. It is an umbrella term to describe a range of practices utilised by organisations to 'identify, create, represent, and distribute knowledge for reuse, awareness and learning [throughout] the organisation'[1].

The British Standards Institute (BSI) through *PAS 2001 Knowledge Management: A Guide to Good Practice* (the four documents (PD 7500, PD 7501, PD 7502, PD 7503)[2] is working to produce guidance, best practice and policy on the generic implementation of KM. The BSI identifies the following as key benefits of investing in KM within an organisation:

- better service and products for their customers;
- faster generation and application of ideas and innovations;
- access to industry best practices and best-in-class methodologies;
- access to competitor and market intelligence;
- access to internal and external networks of expertise;
- ability to deliver continuous learning;
- likelihood of lower-cost administrative processes;
- more efficient and timely marketing and communications;
- greater (cultural and linguistic) geographic reach;
- reduced loss of knowledge through staff turnover.

The Construction Industry Research and Information Association (CIRIA) has published guidance on the role of KM within construction organisations (see the further reading section at the end of the chapter). The guidance is intended to help senior managers within organisations identify what information and knowledge is important to the organisation, establish methods of eliciting this information from the various sources and utilise the information to positively impact on their operations. This concept has also been called 'organisational learning', i.e. how an organisation ensures that it both captures and *learns from* new data, processes, research and innovation. All of these build up the intangible capital of the organisation and by the continuing improvement in efficiency also have an effect on the bottom-line profits[3].

The complexity[4] of the construction industry is well documented. Typically, large-scale building projects will involve many layers of knowledge, skill and communication, which must engage simultaneously to produce what is often a unique product. Designers, engineers and contractors may use different means of conveying and transmitting the same information and are expected to work together efficiently on complex projects often lasting months or years (PAS 2001). In many cases, these organisations have never worked together before.

Why is KM important for cost planning? The whole essence of KM lies in the feedback of information generated on previous projects. Using past experience and data to provide a platform for informed future decision-making can only increase the quality of service that the industry provides. KM is implicitly recognised in the requirements laid out in BS 7000-4:1996 Design Management Systems, in that 'the client should be made aware of the factors likely to cause the actual construction cost to vary from the cost forecast during design. The client also needs to be informed of the magnitude of any cost estimate inaccuracies that may arise because of a limitation in the amount of design time or information provided to the design team'. As shown in later chapters, cost planning relies on a substantial amount of data, from internal records of the organisation as well as external sources such as supply chain partners, the Building Cost Information Service (BCIS), trade literature and published price books. However, for the cost planner to be most effective, understanding the implications of certain design decisions on the costs of previous projects is essential. KM naturally relies on a good deal of effective communication, ranging from the informal through to the fully automated software-based approaches that are now available. From a cost planning perspective, the benefits of KM include the following:

- It provides a basis for collating cost data that may be used to draw inferences on the effectiveness of certain cost planning strategies through a range of different projects and environments.
- It fosters increased lines of communication that already exist through the supply chain.
- It enables the cost planner to draw the client's attention to likely variations from the forecasted project cost, based on experience of other similar schemes.

A significant part of KM requires the understanding of the external project environment. This makes sense given that while projects can be seen as systems with an easily definable boundary (i.e. cost, time and specification), they also operate within a larger open system (i.e. the economy). Kululanga and McCaffer (2001) identify several mechanisms by which professionals within construction organisations can acquire knowledge from their external business environments, for example by:

- attracting staff from innovative organisations;
- use of experienced practitioners to address their knowledge requirements;
- conducting external benchmarking;
- collaborating with other organisations;
- reviewing innovations in the business environment;
- attending conferences on new developments.

At Loughborough University, research has examined the role that KM assumes in improving business performance. The project, KnowBiz (2000–2003), developed KM principles to facilitate construction companies in collating and managing the data required for input into key business performance measurement models. It addressed how construction organisations, through the implementation of a KM framework, can modify their current practice to adopt a more holistic approach to improved business performance. In Canada, at the Conseil National de Recherches (NRC), research has recognised the need to challenge the automation of existing paper flows and provide the knowledge manager with tools that can be easily used to document previous project experience and track new technological developments, and then incorporate these in the construction strategies for future projects. The authors of the published findings of this project (Udaipurwala and Russell 2002) describe work aimed at creating such tools in the context of a comprehensive decision support system, by developing intelligent representation structures for storing and accessing construction domain knowledge and coupling them with advanced planning tools so as to enable the quick formulation and assessment of initial project plans.

One practical vehicle for integrating KM thinking into the cost planning arena is through the widespread adoption of electronic procurement (e-procurement) systems and e-cost databases. Historically, the industry has not actively sought to collect *and* disseminate potentially sensitive commercial information, probably owing to the adversarial environment in which the industry has operated. E-procurement solutions offer the ideal opportunity for construction professionals to build up comprehensive databases of project information that can be used to improve their processes and service to clients.

2.3   E-procurement and the cost planning function

E-procurement is not a particularly new concept and, as with a good many initiatives within the industry, the catalyst for it has emerged from the government-sponsored public sector. E-procurement is essentially the use of electronic systems to increase efficiency and reduce costs during each stage of the procurement (or purchasing) process. In a typical e-procurement scenario, most if not all tender documentation is issued and communicated via electronic means. This has obvious benefits in terms of cost and time. The OGC is a leading innovator in the field of e-procurement within the UK, publishing in autumn 2002 *eProcurement: Cutting through the hype*, as a guide to e-procurement for the public sector (with applications to construction).

Since autumn 2002 there have been significant developments for e-procurement; legislative changes have encouraged greater use throughout the European Union and new techniques such as electronic reverse auctions[5] (e-auctions) have become common practice, although many in the construction industry view this process with suspicion. The government has launched its drive for greater public sector efficiency following the Treasury's efficiency review in July 2004.

2.3.1   E-collaboration

Throughout this book the importance is emphasised of effective teamwork and communication throughout the cost planning process. Within the UK construction industry, the majority of e-business/IT applications that have been developed have been aimed at increasing the efficiency of project co-operation. The dissipative nature of the project team has often created difficulties in communication and understanding as well as unnecessary organisational complexity, so the ability to work together in real time using project documentation that is available to all team members is particularly welcome. Any system that allows the cost planner to rationalise the processes involved in issuing project documentation (such as measurement, tendering and estimating processes) will inevitably realise considerable cost savings throughout the procurement life-cycle.

Supply chain thinking that has now embedded itself effectively within the industry has provided a suitable catalyst for e-procurement. The efficient and accurate transmission and exchange of cost and design data between members of the supply chain is clearly very important. There is, however, a major difference between how smaller organisations will exchange the critical information necessary for effective construction management and how the major clients, national contractors and the large professional practices will operate on major projects.

2.4   Harnessing digital communication through the supply chain

An interesting survey was conducted by Joe Martin of the BCIS (Martin 2003), in which e-procurement and extranets in the UK construction industry were examined and the conclusions reached that (based on the sample) less than 30% of contractual documents were dispatched in electronic form. There is anecdotal evidence to suggest that one of the biggest problems facing the successful implementation of e-procurement systems lies with contractors' ability to fully integrate into the system (from an IT perspective). For example, simple decisions such as the file format in which documents are digitised within the system (for example, portable document format (PDF)) can sometimes create problems for those in the supply chain who cannot append or mark up drawings digitally in a particular format. There is an assumption that all supply chain members have access to and operate on a common file-sharing protocol, but this is not necessarily the case. These problems pervade software development generally where programmers may develop systems that require a certain standard or operating environment, but the users may not be as technologically up to date as the software and thus are unable to fully utilise the application.

Joe Martin's research also identified the following features of e-procurement within the UK construction industry:

- 51% of bills of quantities (BQs) were prepared by consultants on word processors or spreadsheets, the remainder being prepared on one of a dozen proprietary or bespoke systems. Four proprietary systems dominate the market (CATO, Masterbill, RIPAC and Snape).
- 4% of bills were prepared using digital measurement.
- 29% of BQs were provided to the contractor in digital form: 13% on disk or CD-rom, 15% by e-mail and 1% via a website. This result was at variance with the contractors' view that bills sent electronically accounted for 16% of projects.
- The perceptions were much closer when the question related to the return of priced bills electronically; consultants believed that they receive 6% while contractors believe that they return 8% of projects in this way.

These differences of viewpoint may result from the practice of sending both hard copy and digital formats or making the digital versions available on request only.

Word processed files or spreadsheets are the most common formats for sending bills. Only 10% are sent in Construction Industry Trading Electronically (CITE) format, while 4% are sent as ASCII text files[6]. Contractors reported that on 13% of projects where digital data was available, this could not be imported into their estimating systems; 38% of contractors prefer the information in spreadsheet format while 33% would prefer it in CITE format. Word processing was held to be a useful format by 41% of contractors but 38% classed it as unhelpful. Only 3% of consultants and 4% of contractors had any experience of e-commerce systems.

The research also investigated the specific types of e-documentation from the viewpoint of a large quantity surveying practice. Results revealed that

65% of documentation was presented in PDF format since it could not be tampered with (assuming security is applied), 30% of documentation was in Microsoft Excel format and 5% in CITE. The issue with contractors became apparent when the results identified that they would like to issue documentation in CITE since the data could be re-imported into CATO, but since they rarely received a tender list where all contractors could not use the CITE format, the contractors also stated that they liked spreadsheets which come out of CATO, although they reported that these could not be re-imported.

Finally, the research identified that most e-procurement carried out in construction is for major projects, including schools and hospitals through public private partnerships or private finance initiatives. The capability of construction companies to implement fully functional e-procurement systems was not extensive owing to organisations being apprehensive about implementation of the technology and also the cost. The KM issues were highlighted in that many of the survey respondents identified that the training provision for e-procurement was often limited. Many of the smaller organisations were employing outdated, technically obsolete platforms and this was placing them at variance with several major construction organisations that apply e-procurement principles to differentiate themselves within the market. Smaller organisations were also unable to take advantage of the technology to gain market share owing to cost implications.

2.5   Typical applications used in e-procurement and cost planning

Since the rapid increase in the use of IT-supported cost planning systems, a vast range of applications have emerged on the market. The research described in the previous section identifies four common systems, as follows.

2.5.1   Masterbill Elite and Masterbill Cost Planner

Masterbill is one of the most well-known IT surveying product developers in the UK. Within their suite of products, the key product is Masterbill Elite which comprises:

- a simple word processor for production of Preliminaries, Preambles, Schedules of Work, etc;
- a central database, search and adjustment facilities for use within the cost planning system;
- a standard library, full 'sortation', dimension sheets, dynamic dims and e-tendering for the production of a BQ;
- an estimating system, including the importation, priced library, free-form BQs, sub-contract packaging and adjustment facilities;

■ The reuse, adjustment and monitoring of pre-contract data of a post-contract system.

Masterbill Cost Planner is designed to assist in the monitoring project cost as the design develops from initial sketches through to detailed drawings suitable for tender issue. The Cost Planner interface, which is a feature within the overall system, is designed to enable the surveyor to make full use of the facilities without having to open multiple screens or dialogs. The Explorer provides the user with quick and easy access to all cost plans prepared on the system. The KM aspect of the software lies in the ability to make use of previous cost plans stored in a cost information database. The software uses the standard form of cost analysis items or bespoke cost items as required. The cost plan items are then entered within each elemental section and their associated quantities and rates can be entered direct or calculated using a standard 'dimsheet' or a detailed rate 'build up' sheet. To save time, the system allows an item to be copied between cost plans; the user can check whether it is relevant or not, and if not the surveyor can select an alternative item or adjust the item by:

■ a tender price index
■ a location index
■ a combination of the two.

The software produces a report in the format required by the client or design team.

2.5.2   CATO Cost Planning Module

The CATO (Computer Aided Taking Off System) software toolkit, created by ECL (Elstree Computing Ltd) contains a wide range of modules to cover the various aspects of the project documentation process. The CATO cost planning module, using BCIS data for example, allows the prediction, control and benchmarking of project costs. The tool kit uses the familiar Microsoft operating system including Excel spreadsheets interfaced with explorer style navigation. The system also includes the ability to create both versions and revisions of an estimate or cost plan; a reporting tool with the ability to create report sets including cover sheets, appendices, exclusions and the like; and a comprehensive range of sortation facilities to allow alternative presentation of the cost plan including work group or package analysis for ongoing cost reporting once the project is live.

One leading cost consultancy, Faithful+Gould, has developed a bespoke version of CATO within their organisation. The system was piloted in 2005 within the Glasgow and Edinburgh offices, and rolled out to the rest of the business in 2006. The CATO system is implemented within Faithful+Gould using a central Citrix server, allowing staff throughout the organisation to share and access data and information on all their projects – again recognition of the importance of organisational KM.

2.5.3   RIPAC

RIPAC is produced by CSSP, an international software producer specialising in applications for the construction industry. RIPAC is an integrated system for building, estimating, cost planning, bill preparation and contract administration, using one database from feasibility to final account, similar to the system produced by Masterbill. The system includes an estimating and cost planning module which covers feasibility, estimating, cost planning, approximate quantities, price databases and 'first principles' rate build up. RIPAC also contains an additional whole life-cycle costing (WLCC) module.

2.5.4   Snape

The Snape suite of software includes Snape Contender, essentially an estimating and tendering tool and a detailed estimating system designed for builders, independent estimating consultants, local authorities, etc. Snape Vector is an integrated Microsoft Windows-based software tool that is mainly used by the PQS, as well as those surveyors who carry out similar functions within local authorities.

2.6   E-procurement and IT developments in cost planning/quantity surveying practice

To bring this thinking together, it is perhaps helpful to consider the opinions of two surveyors who have been involved in the development of IT cost planning systems. In IT Showcase Online[7], some interesting points were debated and parts of that are reproduced here.

With the continued development of e-business applications throughout the construction industry, where can the QS really capitalise in terms of streamlining measurement, BQ and tendering processes?

The majority of e-business applications that have been adopted within the construction industry perform a single function – project collaboration, and it is with the increased adoption of these applications that the QS is presented with at least two opportunities to streamline their measurement and tendering processes.

Introduction

The majority of the information a QS receives during a construction pro-ject is design information in drawn format. The increased use of project collaboration systems has led to the drawn information increasingly being made available to the QS in electronic format. When it comes to making use of this drawn information, the QS has two options.

Firstly, he could choose to print the drawings himself in order to continue to use traditional measurement techniques, an option which results in the QS incurring additional costs in hardware and consumables in order to con-tinue to carry out their measurement function. Sticking with the traditional methods means the QS is missing out on an opportunity not just to save on the hardware and consumables costs, but also to radically streamline the amount of time spent on measurement.

Software is now available which supports a second option. It enables the QS to make full use of the electronic format drawings and carry out measure-ments in a fraction of the time by using a similar approach to paper-based drawings 'on screen'. In addition, the software also makes it possible to auto-mate some of the simpler measurement tasks and provides tools to simplify complex measurements such as cut-and-fill and drainage.

The tendering process provides another opportunity for streamlining. The QS can now provide all tendering information in an electronic form to the numerous contractors and sub-contractors that tender on a project and request that tenders are returned electronically, thus enabling the QS to streamline their tender analysis and comparison procedures by once again making maximum use of their software.

However, Masterbill's recent experiences in this area show that while the QS is often enthusiastic about streamlining this process, the contractors are reluctant to adopt alternative ways of working, especially if they perceive it may take longer on the first project whilst they learn the system.

Paul Watkins BSc(Hons), MRICS, Sales & Marketing Director, Masterbill Ltd. Reprinted with permission of the author.

Do fears such as differing data standards, system incompatibility, data accuracy issues or perceived confidentiality breaches threaten to undermine a sharing infrastructure between QSs and their fellow Construction Professionals? Or is seamless interfacing now a reality?

Exchanging data reliably between the different parties of the construction chain is very important today. There is, however, a major difference between how smaller organisations will exchange the critical information necessary for effective construction management and how the major clients, national con-tractors and the large professional practices will operate on major projects. Recently, my company provided a custom software solution for handling the measurement, specification and costing for the building works of a major retailer. All parties in the chain use the software: architects, quantity surveyors, contractors, major subcontractors and project managers. Files are exchanged via the Internet. The results have worked exceptionally well for all parties.

We provided first-rate software for the project, but there was also real clarity of thinking, combined with quality management of the client's particular needs and the operational requirements of the contractors and surveyors. For major clients and contracts it is both desirable and possible to lay down well thought out standards and channels by which information is passed, and insist that all parties conform. With good implementation, all parties gain substantially.

It is just as desirable that the smaller players can exchange information as efficiently as the 'big boys', but the starting point of making this process work is to recognise the world as it really is.

Most contractors/PQSs/project managers work with many different organisations. Inevitably these firms will be using different software that is not directly compatible, and therefore exchanging data using an intermediate file format, commonly CITE, ASCII, Excel, and Microsoft Word/RTF (XML is likely to become important in the future). Our users are nevertheless exchanging files as a matter of routine. The process is usually seamless, but it should be remembered that all these intermediate 'standards' leave room for different interpretation that can cause hiccups.

Take Microsoft Excel as an example. There is a vast range of different ways that bills of quantities can be created. Descriptions can be in a single cell or spread over a number of separate cells. There may or may not be a blank line between each description or there may be lined, blank or hidden columns for internal reference or calculation. Graphic images may be put in a column that is normally used for another purpose, or the length of page, treatment of headers and footers, collections and summaries may vary. The bill may be created on a single sheet, or spread over several.

Our approach with our software programs is to cater for as many standards as possible and make the import and export routines as flexible as possible, so they can accommodate these many different interpretations. Critically, we provide a first-rate technical support service so if we are emailed a file, we can advise what is necessary to make import/export routines work.

The best results are achieved when the software vendor takes the time to explain the issues involved.

John Whitehead, Snape Computers Ltd

The article by Paul Watkins identifies a key issue: the development of IT systems to support project collaboration. This, in essence, encapsulates this chapter and here we shall look at one such application that has been developed to enhance this most important aspect of the construction industry. In order to truly understand the cost planning process it is wise to consider it as part of a wider process within the project life-cycle. Since many projects are dominated by the bottom line of cost, the relative importance of effective cost planning cannot be underestimated.

Certain IT systems have been developed to recognise this need and one such product now widely used in professional practice is Autodesk® Buzzsaw®. This system takes an integrated approach to KM and information sharing; the software is designed to help members of the supply chain within a construction project simplify and manage all project-related documentation.

The system provides an online work environment that integrates a secure project hosting service with CAD-related software, tools and services. The platform provides for a seamless transfer of information between project members through various file formats. The power of the system lies in the ability to mark up drawings and take measurements from these within the platform, and then communicate this either via a revisions list managed online and/or by e-mail. Conventional document management systems have not had the power to cope with real-time project collaboration; this system offers significant improvements in this area. The principal features of Buzzsaw® include:

- multiple project management, online, for internal control and third party access;
- management of company, project and user information;
- managed access control;
- view, review and mark-up and revision of drawings with integrated viewing software;
- supports for over 200 file formats;
- revision management;
- integrated change notification through e-mail;
- integrated discussion threads;
- integration to AutoCAD (2000 onwards).

2.7   Research in cost planning and cost modelling

The future will herald continued levels of growth in the integration of software packages across disciplines within the built environment. Research into cost planning techniques since the early 1990s has largely centred on, or relied on, computer techniques. Historically, researchers transferred manual techniques to automatic computation and subsequently recognised the potential for tools such as sensitivity analysis, regression and simulation. With a good deal of emerging techniques, the problems of data availability and the 'black box' nature of many systems continued (i.e. information is input to the system, output is produced by a transformation model, but the model is hidden from the user). Consequently users were being asked to accept that the model was perfect (which it often wasn't), and at the same time they were barred from bringing their own knowledge and expertise to bear on the problem.

Very few practitioners were, or are, willing to place their faith in such models when their clients' and their own livelihoods depend on the result. Acceptability is a key factor in modelling which does not appear to have been given a high priority among researchers. During the late 1980s and 1990s emphasis swung towards the potential of knowledge-based systems and in turn to the potential for linking with conventional computer models to see whether 'intelligence' can be brought into these systems to relieve routine procedures and improve consistency and performance. The new millennium heralds continued major development in IT that will further change the techniques employed, the nature of the professions and client expectations. It should be expected that greater standardisation and greater integration of systems will continue to erode the traditional boundaries between disciplines. The power of the computer and integrated software may mean that one person or firm can control a wider sphere of activity beyond their normal discipline. This will signal a race between firms to control and manage the technology. It will be interesting to see which professions maintain their status and which firms survive.

2.7.1   Artificial intelligence

Artificial intelligence (AI) techniques are a branch of mathematics and computing that essentially replicate the biological nervous system. AI systems can exhibit a surprising number of the functions of the human brain in that they can learn from experience and generalise from previous examples when evaluating new problems. One of the most common examples of the use of AI in the construction industry is Artificial Neural Networks (ANNs).

The use of AI tools continues to offer exciting developments in the cost modelling arena. Some of the early work conducted on the use of AI for construction cost engineering/estimating purposes was carried out at the Liverpool School of Architecture and Building Engineering in the early 1990s.

Although the use of statistical measures in cost planning is still widespread, AI is now becoming a more widely accepted complementary tool, providing professionals with accurate forecasting techniques that will enable them to make educated and reliable estimates of tender price, construction cost and projects duration (Kirkham, Boussabaine and Grew 1999).

2.7.2   Artificial Neural Networks

ANNs may be described as connectionist models, parallel distributed processing models, neuromorphic systems and possibly a more generalist description, neural computing. ANNs, which simulate neuronal systems of the brain, are useful methods that have attracted the attention of researchers in many disciplinary areas. They have many advantages over traditional cost modelling methods in situations where the input–output relationship of the system under study is not explicitly known (Boussabaine and Kaka 1998).

The five main aspects involved in the neural network process are:

- data acquisition, analysis and problem representation;
- architecture determination;
- learning process determination;
- training of the network;
- testing of the trained network for generalisation evaluation.

Currently in cost modelling, back propagation (BP) algorithms are some of the simplest practical forms of neural networks. BP networks are multi-layered in structure and have transfer functions such as sigmoid or delta. The learning that was described earlier takes place in an iterative fashion. The summation function shown below finds a weight for all inputs into the system, then multiplies the input values by the weights and aggregates them for a weighted sum plus a bias. Typically, the following generalised function for a BP network could be:

$$a_w = \sum_{j=1}^{n} S_i W_{ij} - b_j$$

where S_i = input values, W_{ij} = weight, a_w = weighted sum and b_j = bias.

ANN modelling differs from statistical modelling in the sense that in regression models, the approximated function is assumed and the regression coefficients are calculated, whereas ANN as a function itself is asked to approximate the unknown function that maps the input space to output. In other words, a function is being asked to approximate another function. The model is said to solve the problem if it learns to approximate the function to an arbitrary accuracy.

Neural computing is becoming very fashionable in construction, and is beginning to find applications in WLCC. It is principally used for decision-making, forecasting and optimisation purposes. Some examples of published research in this field include Williams (1994) – back-propagation network for predicting changes in the construction cost index; Li (1999) – ANN for a construction cost estimating model; and Chau and Skibniewski (1994) – ANNs for identifying the key management factors that influence budget performance and construction tender price estimation (Elhag and Boussabaine 2002).

2.7.3   Neuro-fuzzy systems

Neuro-fuzzy is a combination of the explicit knowledge representation of fuzzy logic with the learning capabilities of neural networks that were discussed above. Neuro-fuzzy modelling involves the extraction of rules from a typical data set, and the training of these rules to identify the strength of any pattern within the data set. The system creates membership functions from which linguistic rules can be derived (linguistic rules give descriptions pro rata as opposed to numeric values) (Boussabaine and Elhag 1999). Recent developments in the uses of neuro-fuzzy within construction cost engineering have identified the system to be reliable and to be of value to professionals who require systems that enable them to make more informal decisions on the allocation and management of construction costs. The system tends generally to work better with large data sets, whereas ANN can work with smaller samples as well.

2.7.4   Stochastic modelling

Stochastic models are those that are capable of dealing with probabilities rather than single deterministic 'point' numbers. In AI, stochastic programs work by using probabilistic methods to solve problems, as in simulated annealing, neural networks and genetic algorithms. A problem itself may be stochastic as well, as in planning under uncertainty. A deterministic environment is much simpler to deal with. In many cases in cost planning, the costs that are determined within the system are not certain. As a result it is better to treat these variables as probabilistic values rather than single point estimates. This will be looked at later in Chapter 7 (section 7.7.3). Stochastic models have been used in the prediction of energy consumption in buildings as well as determining service lives and deterioration of building components. Many outline development models for projects also use a simple stochastic model where variables such as revenue and construction cost can be modelled within a simulation such as Monte Carlo analysis, to produce a range of possible outcomes (which are again represented by a probability distribution).

Notes

(1) The definition of knowledge management is taken from Wikipedia. A more comprehensive definition is provided by Royal Dutch/Shell: 'knowledge management practice can be broadly defined as the capabilities by which communities within an organization capture the knowledge that is critical to them, constantly improve it and make it available in the most effective manner to those people who need it, so that they can exploit it creatively to add value as a normal part of their work'.

(2) The BSI standards are aiming to establish the desideratum for a KM framework in construction.

(3) A working paper on knowledge management and organisational learning has been written by the Overseas Development Institute in the UK (see the further reading section at the end of this chapter).

(4) CIB-TG62 (Complexity and the Built Environment) established in 2005 at the University of Liverpool and Liverpool John Moores University engages in the study of complexity theory within construction project environments. More information at http://www.cibworld.nl

(5) Electronic reverse auctions work by enabling suppliers who meet the pre-qualification criteria to submit bids for contracts below a reserve price. Specifications and quantities required are provided to bidders in advance of the auction. The difference between this method and conventional auctions is that the bidders must submit progressively lower bids in order to remain competitive. Each bid is exposed as it is submitted, but the source of the bid is undisclosed. There is a school of thought that would describe this process as contrary to the principles of best value; however, assuming that no deviation from the original performance specification is negotiated, there should not in theory be a problem.

(6) CITE is a collaborative electronic information exchange initiative for the UK construction industry. Exchange formats are provided alongside practical support. CITE avoids the need for companies to set up multiple systems by providing and fully supporting a common solution for the industry. ASCII (American Standard Code for Information Interchange) is a character encoding based on the English alphabet. Like other character representation computer codes, ASCII specifies a correspondence between digital bit patterns and the symbols/glyphs of a written language, thus allowing digital devices to communicate with each other and to process, store, and communicate character-oriented information.

(7) IT Showcase Online is a forum and web-based portal providing information on IT applications in construction, manufacturing, accounting and customer relationship management. http://www.itshowcase.co.uk/pma/521 (Article referred to was accessed on 21 November 2006.)

Further reading

Bowen, A. and Edwards, P. (1996) Interpersonal communication in cost planning during the building design phase. *Construction Management and Economics*, 14, 395–404.

References

Boussabaine, A.H. and Elhag, T.M.S. (1999) Applying fuzzy techniques to cash flow analysis. *Construction Management & Economics*, 17(6), 745–755.

Boussabaine, A.H. and Kaka, A.P. (1998) A neural networks approach for cost flow forecasting. *Construction Management & Economics*, 16(4), 471–479.

Chau, L. and Skibniewski, M.J. (1994) Estimating construction productivity: neural network based approach. *ASCE Journal of Computing in Civil Engineering*, 8(2), 221–233.

CIRIA (2005) Business case for KM in construction (C642). Eds J. Palmer and S. Platt. Construction Industry Research and Information Association, London.

Elhag, T. and Boussabaine, A.H. (2002) Tender price estimation using artificial neural networks: part II. *Journal of Financial Management of Property Construction*, 7(14), 49–64.

Kirkham, R.J., Boussabaine, A.H. and Grew, R.J. (1999) Forecasting the cost of energy in sports centres. Construction and Building Research Conference (COBRA) of the RICS. www.rics.org.uk

Kululanga, G.K. and McCaffer, R. (2001) Measuring knowledge management for construction organizations. *Engineering Construction & Architectural Management*, 8(5/6), 346.

Li, H., Shen, L.Y. and Love, P.E.D. (1999) ANN based mark-up estimation system with self explanatory capacities. *ASCE Journal of Construction Engineering and Management*, **125**(3), 185–189.

Martin, J. (2003) E-Procurement and extranets in the UK Construction industry. Conference paper given at *Federation Internationale des Geometres Working Week*, 13–17 April, Paris.

Overseas Development Institute (2003) *Working Paper 224; Knowledge Management and Organisational Learning: An International Development Perspective, An Annotated Bibliography*. Ingie Hovland, August 2003. Overseas Development Institute, London.

Udaipurwala, A. and Russell, A.D. (2002) Computer-assisted construction methods knowledge management and selection. *Canadian Journal of Civil Engineering*, **29**(3), 499–516, NRC Press, Ottowa.

Williams, T.P. (1994) Predicting changes in construction cost indexes using neural networks. *ASCE Journal of Construction Engineering and Management*, **120**(2), 306–320.

Chapter 3
The Three Stages of Cost Planning

3.1 Introduction

Chapter 1 considered some of the fundamental aspects of cost planning and these concepts were developed further in Chapter 2, which also examined current research and practice in cost planning, with an emphasis on the application of IT infrastructures. Clearly, the discipline is dynamic and ever changing, mainly due to the increasing demands placed on the design team by the client organisation and changes in procurement. The new paradigm of software-centred service delivery has irrevocably changed the way that cost planners work in professional QS and cost consultancy practice but whatever methods or software tools are used to assist the cost planner, the conceptualisation of cost planning as a three-stage event is just as valid today and thus a programme of cost planning and control should comprise:

- Stage 1: The client brief, procurement advice and budget;
- Stage 2: Cost planning and control of the design process;
- Stage 3: Cost control of the production stages.

These stages are looked at briefly here, and through the rest of the book the ideas and concepts that feature within them are developed.

3.2 Stage 1: The outline client brief, procurement strategy and budget

The importance of developing an effective client brief is now well versed, mainly due to the reforms that emerged from the *Rethinking Construction*[1] report. One key recommendation was the exploitation of more formalised approaches to developing the client's brief, and ensuring that this information was translated effectively into the design. The traditional informal arrangements, usually between the client and the architect, were seen to be largely ineffective, particularly for larger and more complex projects with multiple stakeholders.

Today, in most construction projects the brief is a formal statement of need or a document that sets out the client's objectives as well as the functional requirements of the building. It should be in sufficient detail to enable the construction team to execute the detailed design and specification of the work and is therefore an essential reference for the project design team.

Systematic cost planning was first introduced at a time when the national building programme was largely in the public sector. (The first edition of this book provides a fascinating insight into the history of cost planning in the UK.) Budgeting was undertaken on the basis of government formulae for the cost of houses, schools, hospitals, etc., and the skill of the cost planner lay in optimising these in relation to the particular project. Today however, budgeting is concerned much more with the concept of value for money, financing, revenue streams and corporate decision-making. The cost planning process has evolved to be more client-oriented and its importance in the overall project life-cycle has grown accordingly. Later, we shall touch on the concept of measuring value and value management techniques. The latter, and cost planning in general, have in some quarters been perceived as a method of specifying minimum performance standards (or, more cynically, cost cutting). This is not the case; the ultimate aim of value management is to ensure that the project is budgeted correctly for a desired standard and then to ensure that the resulting approved budget is spent effectively.

The traditional method of formulating and reconciling the client brief is through the customary time, cost and quality constraints. This method is common to many other project scenarios and is referred to in BS 6079:2002 Project Management. The requirement to identify the priority that these constraints assume is important; more often than not it will be the cost parameter which dominates, although many developers, for example, will also be focused on time since the project finance may be offset against anticipated future revenues.

At this stage, and depending on the type of project and client, a decision may be made about the method of procurement. Traditionally, this will have been dealt with in Stage 3 (as advocated in previous editions of this book) but now there is strong evidence linking project cost to procurement route; indeed, the examples in Chapter 1 illustrate rather well the consequences of an incorrect procurement route selection. Procurement issues are covered in depth in Chapter 8. Interestingly, many clients (particularly those experienced in dealing with the construction industry) have developed bespoke procurement strategies and these may be specified from the outset; the UK National Health Service ProCure21[2] strategy is an example of such an approach.

It is also useful for the cost planner to have an awareness of the resources being provided by the client to the project. The project cash flow will invariably be affected by the timing of payments by the client. Any potential problems should be identified as early as possible to allow mitigation strategies to be developed within the cost/risk management plan. Finance apart, the involvement of the client in the nuances of the project will also give useful clues as to which procurement route is most appropriate.

3.3   Stage 2: The cost planning and control of the design process

It is at this stage of the process that the effectiveness of the work carried out previously is tested. The greater the quality of preparation work at the pre-design stage, the less likelihood of major revisions and design variations. Ultimately, the effectiveness of the outline brief comes into sharp focus. At this stage, the key tasks are:

- development and preparation of the detailed brief;
- development of the design;
- cost control through the development of the design.

It is unfortunate that in many cases the amount of time and effort spent by the cost planner on each of these three aspects is inversely proportional to their relative importance. Effort is often almost entirely concentrated on the third aspect, with perhaps a little advice given on the second. Sometimes the excessive effort required to keep the design development within cost limits actually stems from unwise decisions made at the earlier stages without the benefit of proper cost investigation. In most cases, if the procurement strategy is right, the design development will usually present few problems. At the early stages, the estimates are mainly based on past experience of a similar building with certain accommodation and of a certain specification level. By adjusting this historical cost, the proposed building ought to be able to be built for a certain amount of money, and it is up to the project team to achieve this. As the design develops, the estimates are increasingly refined to ensure that the designers are kept within parameters, so that the building as finally designed can be built for the original estimate/feasibility sum.

Estimates given at an early stage carry a considerable degree of risk, and as further information becomes available they are almost certain to become subject to amendment. It is important that whenever this happens, the decisions to amend the cost or to amend/abort the scheme can be taken without abortive expenditure being incurred. Therefore the estimates themselves should not be prepared in any more detail than is relevant to the current stage of progress of the design.

Although the establishment of the brief and the investigation of a satisfactory solution are shown as two separate and consecutive functions, there needs to be constant iteration and feedback. Design investigation may suggest modifications to the brief that in turn will need to be investigated. This is all to the good and will probably result in improved performance. Even though it involves a good deal of abortive cost-planning effort, the cost planner's work at this stage is relatively inexpensive compared to the potential benefits and cost savings. The cost control of design development, on the other hand, demands considerable resources and should not be carried out until a satisfactory solution has been defined and agreed. Substantial iteration between investigation of a solution and cost control of the development of design brings nothing but disadvantages. This is the principle involved in decision-tree analysis, where design decisions at the beginning of a project are relatively few but have a large effect on the design and cost; for example, deciding on the foundations or the structural frame. As the design development progresses, the decisions become more numerous but have less individual effect on design and cost, for example the number and specification of internal doors.

In formulating the detailed brief, the cost planner should ensure that they co-operate, preferably on a team basis, with those professionals concerned with the actual building design; representatives of the client's organisation or the client's project managers; valuation surveyors; accountants; and possibly planners. As the investigation of a satisfactory design proceeds further into the realms of building configuration, the cost planner will become increasingly involved with the designer and the design consultants, including perhaps a construction planner, to the gradual exclusion of the other parties.

As previously mentioned, the basic principle to adopt is to move in a series of steps from the ballpark estimating of the outline proposals stage to the detailed costing of production drawings. At each of these steps the process must be monitored. Previous assumptions must be checked in the light of further design development, and any necessary modifications made to the estimates or to the design before proceeding to the next stage. As the brief is developed in more detail, and as the design itself develops, it may become apparent that the building the client wants, with a proper balance between economy and quality, will cost either more or (in theory at any rate) less than has been allowed. It should be possible to reduce or increase the quality of the specification to get back to the original figure, but the cost planner should not automatically assume that the client will want this to be done. The client organisation may be more concerned with getting the building they want than with a potential saving of 5% on the cost; or they may not – but the decision is theirs. The difference between this approach and the situation (without cost planning) of a large and unexpected gap between estimate and tender is that clients make their choice consciously with a knowledge of the amount of money that their decision is costing (or saving) them. In this way the time and expense involved in abortive detailed design work can be avoided.

The above process still assumes that there is a difference between the designers and the constructors and that the constructors do not have any influence on the design process. Clearly, this is changing with the rapid rise of design and build procurement and other forms where there is an overlap between the design and construction roles. Having an early contractor involvement creates massive potential efficiencies in the design development process as the contractor's method statement can be developed concurrently with the design, and the concept of 'buildability' is incorporated. Additionally, the Construction Design and Management Regulations 1994 (CDM) created the role of the planning supervisor, whose function is to develop the Health and Safety Plan during the design stage to establish the potential hazards and risks associated with the site and the design. Having an early contractor involvement would allow many of these hazards and risks to be 'designed out', at no extra cost.

3.4   Stage 3: Cost control of the procurement and construction stages

The original development of cost planning, as already noted, took place largely within the local authority and central government sectors, where because of the system of approvals in force at that time the amount of the tender was the crucial factor and there was less emphasis on the final cost. Today's clients are not satisfied with this procedure and clearly have a significant interest in the final account figure or 'out-turn' cost, as this is what they will actually pay for the facility on completion. Therefore the principle of shifting concentration from estimating the tender figure to estimating the final cost represents a major shift in emphasis.

3.5   The role of the cost planner

The role of the QS has changed significantly since the economic trough of the early 1990s and the rise in alternative forms of procurement and private finance initiative (PFI), and will no doubt continue to evolve in the future. Historically, the QS was often not appointed until the architect (who acts as the surrogate client in the traditional procurement scenario) had prepared the production drawings, or if appointed earlier, would play no part in the project until the production drawing stage was reached and work commenced on the BQs and other formal tender documentation. Today, the QS (or cost planner) is regularly appointed before any other professional consultant and accepts responsibility for the client's financial interests in the project, or in other words assumes the role of project financial manager. When the cost planner is appointed there are a number of issues that will have a bearing on the development of the project budget; it is therefore wise to ensure that these are adequately understood before proceeding further.

Who is responsible for the appointment? If the cost planner has been appointed on the recommendation of the client's architect and is a sub-consultant to them, relations with the client will tend to be conducted through that architect. However, if the appointment has been made by the client as a result of previous experience or outside recommendation, the relationship will usually be more direct. Apart from any other factor, a client who decides to appoint a cost planner directly will probably have an above average interest in costs, which will therefore play a predominant part in the scheme design.

Is the cost planner to be concerned with the total budgeting of the project or limited to particular areas, such as capital expenditure or building and furnishing costs, or merely net building costs, or are they expected to do nothing more than merely give an estimate and then prepare a BQ or other contract documentation?

Who else has been appointed? If other appointments have not been made, the cost planner could assist the cost-oriented client in setting up the team of consultants, and this would obviously be of assistance in the context of total cost control.

What decisions have already been made, or steps taken? Any decisions that have already been taken or implemented will obviously constrain any cost optimisation programme which the cost planner may generate. The cost planner should therefore be hesitant in probing any of these decisions, unless advice is definitely being sought or unless they are so fundamentally wrong as to make it impossible to carry out the task for which the cost planner has been appointed. It would only be in the most extraordinary circumstances, and where the cost planner was in a very strong position, that doubts should be cast on specialist professional advice that the client had already received from valuers, accountants, lawyers, etc.

Is cost control to continue until the completion of the project? In the past, it was not unusual for the cost planner's cost control role to finish at the point where a contract was signed with a contractor. In view of the many major problems that can occur subsequently, this was a very short-sighted policy and is generally no longer the case.

It will therefore be seen that the extent to which the cost planner will be able to use the various techniques described in this book depends not only on the priorities of the client but also on the terms and circumstances of appointment. A cost planner who has been entrusted with the overall cost management of a project must resist the temptation to make firm proposals to the client based on his or her own elementary knowledge of property values, investment, taxation, etc. Cost planners are not experts in these fields (although a large QS/cost consultancy firm may well employ such experts) and the purpose of education in cost planning courses in this area is to make cost planners aware of circumstances in which these matters may be important, and to help them in briefing specialists and assessing specialist advice.

Introduction

3.6   Cost planning practice

The complexity and quantity of work involved in cost planning vary considerably but the professional will aim to reduce the effort to a minimum. QSs intuitively understand which items are of cost significance, where short cuts can be taken, and where detailed and long-drawn-out cost checks can be avoided. Cost significance will be touched on later in this book and forms part of the methodology of WLCC. Experience will also show that the use of computers, standard forms and procedures can hasten the process. For this reason, and also because of the acceleration in working brought about by familiarity with cost and prices, in many larger practices and consultancies the focus on cost planning will be within a separate department rather than being done by the staff who would ordinarily be involved in taking off activities.

The success of good cost planning is correlated to the sufficiency and quality of cost and performance data. There is an argument to suggest that those involved in taking off activities are perhaps more suited to the cost planning role generally since it will give them familiarity with the project and thus make the actual preparation of the BQ easier. Whether it is as valid as it sounds is therefore open to question; certainly a number of public authorities that originally used this method subsequently changed over to separate departments. There is scope for experiment here, but if the cost planner and the taker off are two different people, there must obviously be a close working relationship.

3.7   Key points

- Elemental cost planning should ensure that the tender amount is close to the first estimate, or alternatively that any likely difference between the two is anticipated and is acceptable.
- The cost plan should ensure that the finance available for the project is allocated consciously and economically to the various components and finishes.
- Elemental cost planning does not mean minimum standards and a 'cheap job'; it aims to achieve good value at the desired level of expenditure.
- The measurement and pricing of approximate quantities at some stage is a feature of cost planning.

Notes

(1) *Rethinking Construction*, the report of the Construction Task Force to the Deputy Prime Minister, on the scope for improving the quality and efficiency of

UK construction (commonly referred to as the Egan Report), is now under the auspices of the Constructing Excellence programme.

(2) ProCure21 was originally developed as a procurement system by NHS Estates. Subsequently, NHS Estates was disbanded and reformed as the Estates and Facilities Management Division. ProCure21 uses a framework agreement approach to bring the NHS (client) and construction industry together to build publicly funded capital schemes. It is advocated that the ProCure21scheme enables the NHS to reduce tendering time by a year, whilst ensuring certainty on cost, time and quality constraints.

Further reading

Barrett, P. and Stanley, C. (1999) *Better Construction Briefing*. Blackwell Science, Oxford.

BSI (2002) BS 6079:2002 Project Management. British Standards Institute, London.

Ferry, D.J. (1964) *Cost Planning of Buildings*. Crosby Lockwood & Sons, London.

Introduction

Cost Planning at the Briefing Stage

Chapter 4
Developers' Motivations and Needs

4.1 Developers and development

Phase 1

The property boom that has characterised the UK economy since the mid-1990s led to a significant interest in development activity. This pervades commercial and residential as well as public sector building work and is evidence of the growing underlying economic performance of the UK economy. Notably, the increase in residential development activity has been of significant interest. This has emerged primarily as a result of the growing lack of confidence in the performance of pensions and other financial products. This in turn has led to many individuals (or consortiums of individuals) seeking to invest in built assets; the tangible evidence of this is the proliferation of the 'buy to let' market, where individuals seek to acquire a portfolio of buildings that can be used to provide an alternative revenue stream through letting. Moreover, the critical shortage of affordable housing in the UK, particularly that suited to first-time buyers, has increased the demand for rental accommodation. This provides one example of the motivation for development.

Whilst the term 'developer' (the developer being the person ultimately procuring the building) is now common parlance, it is often the case that this person or organisation is the traditionally defined client or building owner, or in some parts of the world the proprietor. In popular use the term 'developer' is often restricted to those who build for profit *per se*, but this is not correct. Those who build for their own use or for non-profit purposes are equally developers.

All development arises from a consumer demand[1]. Classically, there are two elements of consumer demand: opportunities and preferences. The opportunities element considers variables such as the affordability and consumption possibilities. Preferences, on the other hand, considers the utility function, i.e. what does the consumer like, and by how much does the consumer like this good? In terms of property development then, consumer demand may be either of the following:

- *direct demand*: for commercial property space, for offices, for a town hall or library, etc.
- *indirect demand*: a demand for something which will require building development in order to satisfy it (e.g. demand for motor travel requires further motorway expansion, and increased demand for aviation travel requires development of terminals and airport facilities).

But the demand must always be an economic one. There must be someone (or a group of individuals) willing to invest or who can be induced to do so. Developers, whether individuals or corporate public or private bodies, will be building:

- for profit, when their approach to cost will be governed by the way in which they expect to receive their reward, usually by leasing or by sale; or
- for use, when their approach will be governed by the actual units of accommodation which they require.

The archetypal example of development for profit is the Special Purpose Vehicle (SPV) that is created in Private Finance Initiative (PFI) procurement. The SPV involves members of the PFI consortium committing equity to a PFI scheme. The SPV is a temporary company formed especially for the project. The SPV will then be supported by sub-contractors who deliver the individual elements of the PFI contract (e.g. construction, soft services, training, etc.). The equity holders in the SPV are simply speculative investors seeking the maximum return on investment. Ordinarily they will have no interest in the function, design and layout of the building, but they will of course have an interest in cost.

4.2 Profit development, social development and user development

In considering cost targets we may distinguish between:

- *profit development*, where it is intended that receipts from the disposal or use of the buildings will more than cover costs;
- *social development*, which is usually in the publicly funded sector;
- *user development*, which covers such projects as a private house, or an office building for an insurance company's own occupation.

There is also mixed development, incorporating buildings of more than one of the above types. Because the calculations involved in budgeting for all these types of development are similar, it is easy to forget that the basic situations relating to profit development on the one hand, and social or user development on the other, are fundamentally different.

4.3   Cost targets for profit development

The setting of cost targets for profit development (e.g. office blocks for letting or housing for letting or sale) is related to free enterprise economics, and even where there are grants, subsidies, taxation relief and so on to be taken into account, the criteria are simple. Even if some non-quantifiable element, like preserving the environment, has to be taken into account it will have to be assessed against its effect on profit. A calculation will soon show how much the developer can afford to spend under the heading of building costs after other items of expenditure (see Chapter 8) such as land costs, professional fees, etc., have been taken into account. There is some flexibility because the standard of building will partly determine the rent or selling price that can be asked, but this is often dictated by the environment and by the cost of the land. It would obviously not be economic to erect a high-cost luxury building for sale or rent in an unpopular neighbourhood, nor to try a low-cost, low-income development in a fashionable district where land costs are high and there is good income potential.

There is no room for hard luck stories in carrying out profit development, but on the other hand there is the challenge of working in a real situation where time means money. For example:

- early completion of a retail complex may involve substantial extra profits;
- late completion (e.g. after Christmas instead of before Christmas) could be financially disastrous.

4.4   Cost targets for social or public sector user development

In social or user development, the main object of the exercise is the actual provision of the buildings. It becomes very difficult to set realistic cost targets since there is no definite limit at which an individual building ceases to be possible. In public sector building in particular, it is usually possible (though not always politically expedient) to raise whatever sum of money is required for the purpose. Any constraint is usually at a higher level than the individual project, for example, the proportion of the national budget allocated to education may determine the amount to be spent on school building.

Therefore, to determine a reasonable cost for this type of building it is necessary to set artificial limits based on the cost of similar buildings erected

Phase 1

elsewhere. The gross floor area is too crude for the purposes of this comparison and so various targets based on user requirements have been established, often by the ministries responsible, so as to ensure a nationwide standard. In the case of schools the unit of cost was the number of 'cost places' (a fictitious number of pupils calculated from the teaching space), while for hospitals the number of beds was once the basic yardstick.

These cost targets were usually determined by a set of artificial standards, which had to be adhered to. Although tight in some ways, these usually made exceptions for technical difficulties associated with a particular project. But here we are not in the world of simple profit economics; the early completion of a school or library would usually be a financial burden to the authorities, however welcome it might be to the community.

The cost planning of social development projects, therefore, may resemble the playing of a board game such as Monopoly, where the architect and cost planners try to win according to a set of rules which have little validity in the real world outside. Sometimes the rules may be very crude indeed, for example:

- Money cannot be transferred from one fund to another (so that it is useless trying to save money on furnishings or running costs in order to spend extra on the building).
- Money cannot be spent outside the financial year in which it is allocated.
- No major contract can be let except by competitive tendering on a firm BQ with no allowance for cost fluctuations.
- Any financial considerations other than the contract amount are irrelevant.

In these circumstances the skill in cost planning may consist of loophole designing, to take advantage of the regulations in the same way that a clever accountant takes advantage of the tax laws. An example of this, from outside the UK, occurred when a national system for cost control of flat building gave a greater cost allowance for balconies than the actual cost of providing them. The blocks of flats built during this period can be identified by their lavish provision of unnecessary balconies.

It is recognised by government that unrealistic cost rules lead to bad design, and many far-sighted efforts have been made to do away with the cruder kinds of inconsistency:

- by allowing money saved in one direction to be spent in another;
- by trying to bring the assessment of running costs into the cost comparisons;
- by working out very complex cost criteria for such buildings as hospitals instead of 'so much per bed'.

These are commendable attempts to get nearer to reality, but while they have undoubtedly led to some improvement they also tend to make the 'game' more complicated and the loophole finding more of a challenge to the experts.

It becomes increasingly difficult to avoid the balcony type of inconsistency mentioned above.

The consequence of all this is that everybody can become so preoccupied with trying to meet tight cost targets by clever application of the rules, that they lose sight of the social purpose of the whole exercise. The attempts over the years to get as many houses, flats, schools and hospitals as possible out of the budget, without the consumer checks on satisfactory standards (such as sales or economic rents) which exist in the private sector, have played a large part in creating the massive maintenance programmes with which many public authorities are now faced.

4.5   Cost targets for private user development

This third type of development covers such projects as a private house or an office building for an insurance company's own occupation. User development incorporates some features of both profit and social development. Because of this it is most important to find out the client's real cost priorities and get them defined.

4.6   Cost targets for mixed development

This type of development is a mixture of profit and social development, with perhaps some user development. The cost of some of the buildings, or perhaps a proportion of the total cost, would be met by social funding and the rest is intended to make a profit. A common example of this type is a town-centre development incorporating shops and public amenities. The same remark about clients' priorities applies as in the previous case.

4.7   Cost–benefit analysis (CBA)

In order to overcome some of the difficulties previously outlined, and to justify the public expenditure, the techniques of cost–benefit analysis were developed. Such analyses attempted to quantify all factors, including the various social benefits and disadvantages, and were widely used in connection with traffic and airport schemes and with hospital building. Cost–benefit analysis might be applied, for instance, to a proposal to carry out works to remove a sharp curve and speed restriction on a section of the railway network. The cost of the work could be offset by the saving in fuel, and against the wear and tear on equipment caused by braking and re-acceleration. However, it is possible that on these grounds alone the project might not be feasible. On the other hand, if the curve is removed it might save two minutes

Phase 1

each on a million passenger journeys a year; 30,000 man-hours are worth something, and if they are priced and included among the benefits the scheme might now be justifiable in relation to the national economy.

This is all very well, but the railway company is going to incur the costs and is not going to receive the financial benefit of the 30,000 man-hours, so organisations in their position cannot be expected to take this sort of exercise seriously. In addition, there are the problems of attaching money values to things, that cannot be quantified. As an illustration of this, what is the value of a human life? One approach would be to work this out on the basis of the financial contribution that the individual person is expected to make to the economic life of the community, so that a surgeon might be worth hundreds of thousands of pounds while an unemployed labourer might have a negative value. Even on practical grounds this right-wing approach is obviously unacceptable; if you attempted a cost–benefit analysis of a geriatric hospital on this basis you would find that it would be cheaper to let people die and build a mortuary instead! Or, as another less extreme example, a ring road could be justified that saved a few minutes for 'important' people while wasting the time of humble pedestrians.

It was therefore customary to take a notional figure representing the worth of an 'average' person. There was little wrong with this except that such figures, being notional, were conjured out of thin air and could be used in practice to 'prove' that a politically desired result was the right one. As an example, suppose a traffic improvement scheme was going to cost £100,000 a year and was estimated to reduce road accident deaths by three per year. If you cost a life at £50,000 the scheme would obviously be worthwhile; if you cost a life at £20,000 it wouldn't be. This is not just a theoretical objection. In 1988 the UK Department of Transport arbitrarily doubled the value for a human life used in its calculations, for political purposes; this simple stroke of the pen increased the benefit expected from its road schemes by an average of 4.5%, although nothing in fact had changed. If you remember that the reduced number of deaths will be a guess anyhow, you can see that this sort of exercise is not really worth very much, and the more complicated it gets and the more social benefits that are quantified, the more questionable is the result.

A further difficulty is that by reason of the type of people undertaking these studies, the values assigned to non-quantifiables tend to be those of the cultured middle class; the relative values of preserving the environment as against providing local employment, for instance, might not be those which a working family living in the area would choose.

Cost–benefit analysis in its extreme form is now largely discredited, but its successor in the public domain, option appraisal, draws upon its techniques. Option appraisal can be applied to any proposal for public investment, and involves the appraisal of all possible options (including the do-nothing option). Cost–benefit analysis is used to evaluate those aspects that have a clear money value, both of a capital and recurrent nature, but intangibles are merely assessed and shown separately. It is left to the administrators to make the subjective decisions, knowing the financial outcome of the more tangible parts of each option.

4.8   The client's needs

All building clients will have a set of needs (we shall look at methods for identifying and reconciling these needs in Chapter 5), some of which are more important to them than others. We have to look at these carefully because it is common to be told, for instance, that a low cost is required, or that time is important, without these very basic requirements being defined in more detail – and it is the detail that decides the best way of tackling the project. So we shall now look at the main time and money requirements that the client may have, and the different forms each may take, remembering that some clients may have more than one requirement under each heading.

4.8.1   Time requirements

- *No critical time requirements.* This is quite common, especially in social development.
- *Shortest overall time*, from the inception of the idea to 'turning the key'. This is likely to be the requirement on a simple profit development.
- *Shortest contract period from the time the builder is appointed.* This by itself is not often relevant to the client's needs, but it is surprising how often it is asked for.
- *Shortest contract period from the time that construction actually starts on site.* This is a reasonable requirement where, for example, there is a delay in acquiring the property, or where people have to be moved out of property on the site before demolition can take place, or where the building work will cause inconvenience or disruption.
- *Early start on site.* This may be required where the payment of a grant or subsidy depends on work having been started by a certain date. It is also sometimes asked for by a lazy or incompetent architect to give the client the impression that something is happening at last.
- *Reliable guaranteed completion date stated by contractor.* This may be wanted so that firm arrangements can be made well in advance for commissioning the building.
- *Firm completion date stated by client.* This may apply where the client is under notice to quit existing premises or where there is a particular event that the building must be open for, such as the beginning of the summer season for a hotel.

Phase 1

- *Early completion unwelcome.* It is often wrongly assumed that if a client says that time is critical, early completion will be welcome. This is not necessarily the case; if the client's arrangements are being made on the basis of a particular date, early completion simply means that the client's money has to be wasted watching and maintaining an empty building, and also that the building has to be paid for earlier than anticipated.
- *Phased programme to fit in with plant installation.* This is especially important in the case of sophisticated projects such as TV transmitters or chemical works, where the actual building work is only a small part of the total scheme.
- *Handing over in sections.* This is often very important in alteration works or in rebuilding, where people or processes from one section have to be rehoused elsewhere before work can be carried out in that section.

4.8.2   Cost requirements

- *No critical cost requirements.* This is not very common, but it can occur where the building is only a part of a major development project (e.g. the TV transmitter already mentioned) or where the first consideration is quality.
- *Low total cost of whole project.* By contrast, this is almost always said to be the main priority. However, very often one or more of the following criteria are actually the real ones.
- *Low WLCC* (Chapter 7). Some clients, notably those in the public sector, will either require evidence of low WLCC for compliance with procurement arrangements, or in the case of PFI/Public Private Partnership (PPP) (see Chapter 8) owing to the long concession period. This is also known as the optimum combination of capital and maintenance costs.
- *Low cost of building contract.* The concern here is to keep the lump sum building cost to the minimum, even if this does not minimise financing costs, administrative and supervisory costs, or costs of furnishing and maintenance. It is still quite often required on public sector projects where these different items of expenditure may come out of different funds.
- *Low cost in relation to units of accommodation.* This is a very usual cost requirement, in both the profit and social sectors.
- *Good budgetary control of the project.* In many cases it is important that the final cost should be as close as possible to the initial cost forecasts, even if this is not necessarily the lowest cost that the competitive market might produce at the time the actual orders are placed.
- *Good forecast of cost at contractual commitment.* This is required by clients who want an accurate forecast of final cost before committing themselves to major expenditure.

- *Best combination of capital and maintenance costs.* One would like to think this was more common than it is. There are all sorts of reasons, like taxation, grants, cost yardsticks, etc., which tend to prevent these two types of cost from being weighted equally (see Chapter 7).
- *Low capital cost.* This is a more usual requirement, especially if the building is going to be sold or if running costs come out of a different fund.
- *Low maintenance cost.* This is less usual, but may be required if maintenance is going to be inconvenient, for example by putting the building out of commission while it is going on.
- *Timing of cash flow.* This may be required in order to optimise the cost on a discounted cash flow basis (see Chapter 6) or to phase in with the availability of the client's funds.
- *Minimum capital commitment.* This would be required if the client wanted the contractor to bear most of the cost until the building was handed over.
- *Share in risk of development.* A variation where the contractor is paid by a share in the profits; this has been used on large speculative developments.

4.8.3   Quality requirements

- *No critical quality requirements.* Generally, buildings like factories, distribution depots, etc. that are constructed using simple steel frames with minimal cladding and fenestration will fall into this category. The 'bus shelter', as it were.
- *High quality.* Often, large corporations that are procuring a prestigious building such as a head office will seek to specify the highest levels of quality in order to demonstrate corporate wealth. Hospital buildings will also have specialist areas such as operating theatres where the highest levels of quality and workmanship will have to be attained.
- *Medium quality requirements.* Most buildings are likely to fall into this category.

Under the provisions of ISO 15686-1: Service Life Planning, designers and particularly analysts who are involved in WLCC studies are encouraged to use the factor method for determining the service life of components. One of the factors is workmanship, and depending on the client brief that factor would have to be modified accordingly.

The client, of course, may wish to combine three or four or more requirements from the above list; as a result there are many different sets of possibilities, and the way the project is undertaken should reflect the client's individual priorities and combination of needs. A standard solution should not be adopted simply because it is the one the design team are most comfortable with.

Phase 1

4.9   Key points

Profit development:

- The only purpose of the development is to make a profit, and profit is quantifiable. If an adequate profit cannot be foreseen the development will not be undertaken, however much it may be needed.
- Once a final decision has been made on potential levels of revenues, the cost will have been determined and must not be exceeded. Any variances from the cost plan in certain areas will have to be balanced by a saving elsewhere.
- A misjudgement of either the costs or the expected receipts of a scheme will have exactly the same effect on its profitability; neither is more important than the other.
- If the expected profit is not made the project will be a failure from the client's point of view, however pleasant or useful the resulting development may be.

Social development:

- The cost is not a clear-cut measurement of the effectiveness of the project (as it is with profit development); the benefits of a hospital, clinic, school or police station are largely unquantifiable.
- Therefore nobody really knows whether they ought to be spending twice as much money (or half as much) on buildings of this kind, and no amount of cost planning is going to give them the answer.
- The most that cost planning can do is to help use the total allocated funds more effectively within the current framework of rules, and accept that the basic values will be decided for political reasons.

Note

(1) Consumer demand is considered within the context of the theory of market demand, a well-established branch of economic theory, and underpins a good deal of thinking in the economics discipline generally. An understanding of the properties of market demand is crucial to an understanding of the evolution of prices and the nature of economic equilibrium. The theory and study of consumer demand (in microeconomics) states that individuals use their finite resources to make purposeful choices. The theory assumes that consumers understand their choices (possibilities) and the prices (opportunity costs) associated with each choice. Furthermore, it assumes that consumers consider all the alternatives available and select the one that they like best. This later concept refers to the theory of utility.

Chapter 5
Client Identification and the Briefing Process

5.1 Introduction

Chapter 4 briefly discussed the client's requirements. These can often be numerous and complex, when they should be clearly defined and well scoped. It is the responsibility of the design team, and other professionals who may provide cost advice, to obtain as much information as possible from the client in order to understand not only the scope of the project, but also the budgetary parameters in which it will operate. Therefore, the process of understanding the client through formalised methods is important. This chapter discusses the types of clients that engage with the construction industry, the process of briefing and client identification and how the client's brief should be aligned with the budget. Finally, some simple budgetary examples will be explored, which bring together the content of the first part of this book.

5.2 The client

As early as 1966, the Tavistock Institute was drawing attention to the increasingly intricate nature of client organisations[1], contending:

> ... that they were complex systems[2] of differing interests and that their relationship is seldom with a single member of the building industry ... These client systems are made up of both congruent and competing sets of understandings, values and objectives. Much design and even building work has proved to be abortive because unresolved or unrecognised conflicts of interests or objectives between stakeholders, within client systems, have only come to light after the building process has been initiated.

The almost classic example of this, referring back to the Scottish Parliament project in Chapter 1, is the complexity of the client group involving the project sponsor, the MSPs[3], taxpayers, parliament staff, the general public, etc. – all stakeholders exhibiting competing and ambiguous requirements within the client system. This ultimately led to an untenable situation where the project reached the point of almost critical failure on numerous occasions.

Since the 1960s, clients have demonstrated an increasingly demanding approach to dealing with the construction industry. This emerged through a series of influential reports such as that by the British Property Federation (1983)[4] and more recently, the Egan Report (1998). These reports were in part a variation on a theme, generally expressing dissatisfaction with building industry performance. The industry responded through a variety of innovative mechanisms, one example being radical changes to traditional procurement systems through the use of design and build, management contracting/construction management, partnering, supply chain management and lean construction. These will be explored in greater depth in Chapter 8.

The Egan Report was particularly damning of the industry, contending that clients were highly dissatisfied with the construction industry due to:

- projects not being delivered on time;
- projects not being delivered on cost;
- general dissatisfaction with the quality of buildings;
- poor relationships between client organisation and industry.

Since then, however, a great deal of change has occurred and clients have established efficient systems of working with the industry through such mechanisms as prime contracting and framework agreements. These are discussed further in Chapter 8.

The extent of a client's knowledge and understanding of the construction industry and the processes involved in working relationships has a major effect on the client's ability to be effectively involved in the development process; consequently there is wide scope for possible client involvement. The service that the construction industry provides to a client should be designed to suit the particular needs and expectations of the client. There are two distinct aspects to a client's needs and expectations. First, a client has various requirements of the building:

- its location, size, shape, performance and facilities;
- when it is available;
- its cost etc.

These will depend on the functional requirements which the building is to perform and the objective which it is intended to serve. The second aspect of a client's needs and expectations is the type and level of management service required.

5.3   Type and level of management service required by a client

A client needs a particular type and level of management service to help them define and obtain the building required. Depending on the extent of the client's knowledge, the management service that a client expects may be quite different from the one that is needed. The type of management service that a client needs is determined largely by the extent to which they will be involved in the construction process; this will be either a 'minimum client role' or an 'extended role' taking in a substantial part of the management of a project. Often it is better to consider these two extremes as a dimension (i.e. clients will tend towards or not tend towards high levels of involvement). The extent to which a client will be involved in the construction process depends on both their ability and willingness to be involved.

A client's ability to be involved is dependent on the amount of appropriate knowledge and resources that they possess. To be fully involved a client must:

■   have adequate knowledge of the construction industry and the construction process, relative to the nature of the project;
■   be fully aware of what is involved in the role that they wish to play.

When a client does not have the appropriate level of knowledge, they will be less than fully involved and will require a more extensive service from their professional advisers. A number of previous studies have used the terms sophisticated and naive when referring to the level of a client's knowledge and experience of the construction process. In reality, this is again a dimension issue as few clients can be classified as clearly one or the other of these definitions (i.e. most can be considered as being either relatively sophisticated or relatively naive). Other researchers (Masterman and Gameson 1994) have classified clients by experience and end-use: secondary inexperienced, secondary experienced, primary inexperienced and primary experienced. All except the most sophisticated clients will be unable to inform the construction industry accurately of their own status in this respect. It is therefore mainly a matter of perception, and the client's view of their level of sophistication may be quite different from that of an experienced construction professional.

Knowledge and awareness of the client role are not by themselves sufficient; the client must also have adequate resources of the correct nature to perform their project role successfully. The level of involvement of the client depends on their willingness to be involved, as well as their ability. This is important because clients do display varying levels of willingness to be involved in the project. Chapter 8 gives a useful insight into this area.

In general, a client's willingness to be involved will reflect the knowledge and resources that they possess. However, other factors may also influence their level of involvement:

Phase 1

■ The resources that a sophisticated client possesses may be committed on other projects.

■ Sophisticated clients may recognise that they do not possess the particular expertise or level of experience appropriate to a specific project.

■ The client may prefer to pass responsibility for the project to others, especially where they foresee a high degree of complexity, uncertainty and risk.

■ Conversely, even with the minimum of knowledge and resources, a client will attempt to maintain maximum control over expenditure of funds.

The final point is perhaps the most salient from the cost planning perspective. It is also inevitable that problems may arise when a naive client attempts to control a process about which it has very little knowledge and/or understanding.

5.4 Types of clients – user clients and paying clients and the stakeholder perspective

A thorough understanding of the client organisation (system) provides the foundations for producing the most effective brief. We have briefly explored the contention that clients can be seen as somewhere between naïve and sophisticated; however, there is a more detailed consideration of the client organisation which should be carried out in order to help develop the brief effectively.

Often, the client is considered to be the person or organisation that will ultimately foot the bill for the project. However, the design team must look beyond this and consider how their client's organisation is made up, its business operations and interaction with the wider economy.

Kamara *et al.* (2002) recognises the complexity that differentiation between user clients and paying clients brings to the briefing process. The emergence of construction projects from a desideratum to satisfy a commercial or societal need of the client is thus an important contention. The hypothesis is that satisfaction of the client need requires an understanding and resolution of the 'different perspectives represented by the client (definition and analysis of requirements), and restating his or her business need in construction terms (translation of requirements into solution-neutral specifications)'.

The term stakeholder is often used to describe various individuals or organisations involved in the procurement life-cycle; however there is often confusion surrounding the difference between a stakeholder and a client. Typically in construction, clients are those who pay for the building, but are they stakeholders as well? How can we differentiate between the two, or is a client a subset of the stakeholder group? The Oxford English Dictionary defines a stakeholder as 'a person, company, etc., with a concern or (esp. financial) interest in ensuring the success of an organisation, business, system, etc.'.

This helps little in addressing the ambiguity of terms but Adizes (2004) offers an interesting take on the two concepts:

Stakeholders are: 'those who have a stake in the organisation, i.e. who have a certain interest in the existence of the organisation, but the organisation does not *exist* for the stakeholder. It tries to satisfy the needs of stakeholders by satisfying the needs of its clients. Stakeholders are the driven force. Not the driving force.'

Clients are: 'the purpose for which the organisation exists' and thus 'stakeholders are all those interests, internal and external, that [come] together for the purpose of satisfying [the] client needs and in doing so expect some return for their effort'.

Whatever view is taken here, the message is clear – in order to provide the best possible product to the client, a comprehensive understanding of their needs is critical.

5.4.1   Case study: new railway station

This case study exemplifies 'systems thinking'. Consider a new railway station that is being constructed in a commuter suburb on the outskirts of a large regional city. The outline brief requires the construction of four new platforms, station buildings including offices for staff and waiting areas for passengers, entrance concourse with retail facilities, car parks, public address and information systems and the construction of new permanent way, signalling and overhead line equipment (OHLE). The client system may on first inspection appear quite simple – Network Rail or one of the Train Operating Companies (TOCs) that is the 'paying client'. This is not the full story, however, and the client map will have to reflect the engagement of the user clients and the other project stakeholders, as mentioned above. An exemplar client map in Figure 5.1 shows the potential complexity of the client system and thus the importance of reconciling all of these within the brief. Within the map, the three groups of paying clients, user clients and other project stakeholders are included and their relationships with the key elements of the brief are shown. This example is rather simplistic and a proper investigation would perhaps reveal more complexity. However, the figure does serve as a useful example of how client systems need to be explored and thoroughly understood.

5.5   The brief and the process of briefing

Briefing is a vital aspect of every construction project. It is the process by which the client's requirements are investigated, developed and communicated

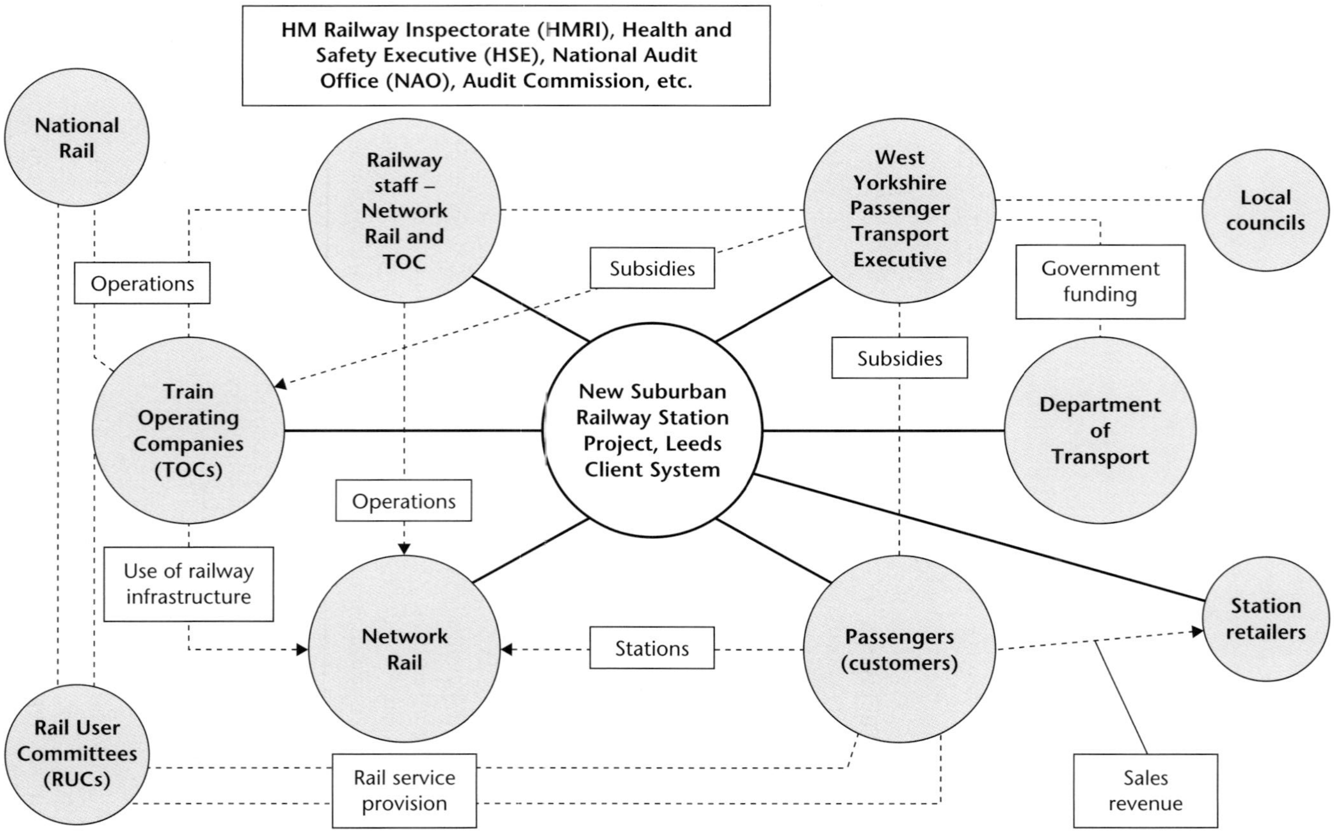

Phase 1

Figure 5.1 Simplified client system mapping.

Diagram labels:

- Local councils
- Station retailers
- Government funding
- Department of Transport
- Sales revenue
- West Yorkshire Passenger Transport Executive
- Subsidies
- Passengers (customers)
- HM Railway Inspectorate (HMRI), Health and Safety Executive (HSE), National Audit Office (NAO), Audit Commission, etc.
- New Suburban Railway Station Project, Leeds Client System
- Subsidies
- Stations
- Railway staff – Network Rail and TOC
- Subsidies
- Operations
- Network Rail
- Operations
- Rail service provision
- National Rail
- Train Operating Companies (TOCs)
- Use of railway infrastructure
- Rail User Committees (RUCs)
- Operations

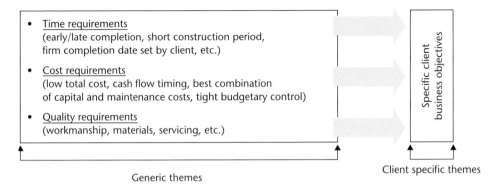

- Time requirements
  (early/late completion, short construction period,
  firm completion date set by client, etc.)

- Cost requirements
  (low total cost, cash flow timing, best combination
  of capital and maintenance costs, tight budgetary control)

- Quality requirements
  (workmanship, materials, servicing, etc.)

Specific client
business objectives

Client specific themes

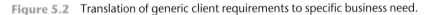

Generic themes

Figure 5.2   Translation of generic client requirements to specific business need.

to the 'supply side' in the construction industry (Figure 5.2). Briefing of some kind always occurs in a construction project. However, the quality of briefing varies considerably; good briefing is not always easy to achieve, but recent studies have suggested that improvements to the briefing process almost always lead to clients being more satisfied with the final building that is handed over.

Buildings are expensive acquisitions and inevitably have an important effect on the operations of the organisations that occupy them. Poorly performing buildings may result in low productivity, an unsatisfactory working environment and low staff morale. In a competitive world, such shortcomings can be very serious. Furthermore the costs of disposing of a poorly performing building and obtaining a satisfactory replacement can be high and sometimes prohibitive. A client with such a building may be stuck with it for the foreseeable future. Good designers will do their best to give clients the buildings they want. However, if they are unable to determine what clients really need, their task is difficult, if not impossible.

Good briefing seeks to minimise the likelihood of a client receiving an unsatisfactory building by ensuring that project requirements are fully explored and as clearly communicated as possible. Whilst good briefing cannot guarantee that a building will be perfectly adapted to its occupants, it can help avoid serious mistakes.

Briefing is often regarded as an early stage activity during which the client's requirements are written down in a formal document called the brief. This document should also provide the benchmark to guide the development of the designs. Early stage briefing is also vital since theoretically, potentially expensive design variations can be avoided by identifying all options at the outset. Figure 5.3 shows the relationship between cost and opportunity of change through the project life-cycle. The message here is that if changes occur later in the life-cycle, the ability to make the changes is constrained and

Phase 1

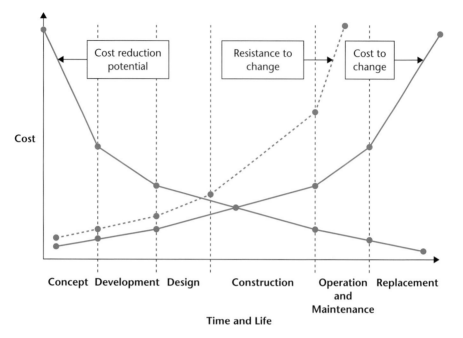

Figure 5.3 Relationship between cost and opportunity of change through the project life-cycle (based on BRE *Value from Construction – Getting Started in Value Management* guide and BS 3843:1992).

the cost of doing so is much higher. This is not to say that the client, particularly the naive or inexperienced client, cannot be expected to identify all the functional and performance requirements of the building at the outset. However, the client should not view the initial brief as the end of the process; it is simply folly to expect that a quality design will emerge from it without further effort. Full participation throughout the project is thus vitally important. Whilst a clear initial brief can be a great asset, it is not the end of the story. The important thing is to make decisions appropriate to the particular stage of project development. Strategic decisions will need to be made early on and the detail left until a later stage, and the client should not withdraw from the process once an initial brief has been drawn up.

Briefing looks beyond the micro level; it requires 'blue skies thinking' that encapsulates the business objectives of the client as a function of the need for a building. The briefing process is also about creating or adding value; to argue the case for a new building in terms of business process improvement represents a significant step forward in conventional thinking. The briefing process should not be underestimated, however; developing an understanding of the client organisation complexities can be time consuming and costly.

5.6   Format and content of the brief

Ultimately, there is no verbatim method of producing a brief. There has been significant research in the general area of briefing and many proposals to improve the process from a qualitative and quantitative perspective, including the use of IT and decision support systems. Practically, the brief can come in a variety of guises. In the most basic form it is purely verbal – a set of instructions between the client and architect, for example. This is probably just about sufficient for a basic house extension, but for more complex projects there should be something more substantial. Working drawings and standard forms of contract could also technically be viewed as briefing documents since these facilitate the transmission of information between the client and the design/construction teams. It is advocated though that a bespoke, formal document is constituted and this should be designed in such a way that it is beneficial to all parties in the project. The Constructing Excellence guide to briefing (2004)[5] identifies the following facets of good briefing.

(1) *Establish the need to build:* Is a new building required? It may be that refurbishment could be better, or indeed a change in business practice may negate the need for a building! The designers should understand the business case.

(2) *Adequate resources:* In terms of both capital investment and client involvement. The client should resource the briefing stage fully to ensure delivery of the best product.

(3) *Careful management:* The client should ensure that the information provided to the design team is clear and unambiguous. The design team should ensure that the information is translated into correct design.

(4) *Good teamwork:* Interpersonal relationships are vital, including the element of trust. Therefore good communication channels must be established.

(5) *Clear communication:* The construction professional should avoid using confusing jargon and should not expect clients to interpret complex engineering design documents. Conversely, the client should ensure that ambiguous and contradictory information is not provided to the design team.

(6) *An approach appropriate to the project:* Briefing techniques should be adapted to the particular project. Take stock of the particular characteristics of a project and design an appropriate approach. Complex projects pose more significant challenges than smaller projects.

(7) *Involve the end users:* The client is often not the end user, so differentiate between paying clients and user clients. The designer may need to reconcile between the two. Don't assume that the paying client knows everything about their business.

(8) *Formal information gathering techniques:* More structured techniques may be required for complex projects such as hospitals: schedule of accommodation, environmental conditions, etc.

Phase 1

5.7 The client's budget

Having examined the briefing process, the next stage is to consider how the information from that process can be translated into the budget planning phases. This section looks at the principal components of the budget and some simple techniques that may be used to build it up.

5.7.1 Total cost

When any type of building development takes place the total expenditure on the project will be much higher than the net costs of the building fabric, comprising also: cost of land; legal and other costs of acquiring and preparing the site and obtaining all necessary approvals; demolition or other physical preparation of the site such as decontamination, archaeology; building cost; professional fees in connection with the above; furnishings, fittings, machinery, etc.; costs in connection with disposal (sale, letting, etc.) where the building is to be disposed of at completion; value added tax on above items where chargeable; cost of financing the project (this principally represents interest etc. on the money which has to be spent before any return is obtained either by way of income or of use); cost of management, operation and maintenance where the building is to be retained by the client for use or only partially disposed of, or where the building is let to a tenant but the owner has accepted responsibility for some or all of these costs.

In commercial development there is a very close relationship between how much the project will produce in revenue and its commercial value, the latter being largely the capitalised value of the former, and the total costs of the scheme must not exceed this commercial value. In social development this relationship may be expressed by quantifying what are considered to be social benefits, but the issue is never clear-cut and other criteria may be more important in assessing the amount that can be spent on the scheme.

5.7.2 Building cost

It is important to realise that building cost is only one part of the client's total cost, and it may be thought that many cost planning exercises are unduly focused on this one item rather than on the whole package. There is some truth in this, probably because many of the techniques were developed in the public service in the days when building costs were closely examined but total costs were rarely considered. Today this is rarely the case. In extreme circumstances, indeed, keeping the building cost down may not be particularly important. For example, the cost of the land may be such an important factor in the costing exercise that everything else has to be fitted round it and the problem then is to get the maximum amount of accommodation that the planning authorities will allow, almost regardless of cost efficiency. Alternatively, the building may be only part of the total project, as in the setting up of a television transmitter station where the cost of the mast, electronic equipment and cable laying dominate the project and the transmitter building is of minor importance.

In these circumstances it would be inappropriate to consider the cost of the building in isolation; it is much more important that the building should phase in with the general programme and that its completion should not be a source of worry to the engineer in charge. It may not be necessary to cost plan the building at all. In the later stages of the cost planning process building costs will dominate the calculations. At the early budget stage, however, they can only be estimated very generally because so little is known about the project.

5.7.3   Land values and development

One of the most important items on any developer's balance sheet is the cost of the land. Building development of any kind requires a plot of land on which to take place and which, once used, will no longer be available for any other development unless the first one is demolished or converted. This makes building development quite different from other enterprises, because no developer can start work until the firm or organisation has acquired an interest in the land on which the development is to stand. This has all sorts of economic consequences because the supply of land is largely fixed. As a now famous estate agent's advertisement put it: 'Buy now! They have stopped making it!' Like the prices of other commodities, land values are affected by the laws of supply and demand. However, unlike other commodities the amount of land available in any one area is finite and cannot be increased at times of high demand by manufacturing at a higher rate or importing from outside. This means that if there is high demand in a particular area the prices of land in the locality will rise very steeply, even though similar land may still be cheap 50 miles away.

5.7.4   Role of the valuer

Valuation surveyors are the experts on the value of land or buildings for investment and the income that various types of development can be expected to produce, but anyone who is involved in cost planning of buildings needs to have some knowledge of the factors affecting the costs of land for development.

Phase 1

5.7.5   Development value

The development value of a piece of land is the difference between the cost of erecting (or converting) buildings on it and the market price of the finished development including the land. Nobody can commission building operations on a piece of land unless they have a financial interest in it; if possible they would wish to own it so as to obtain the full benefit of the development value, although in some areas where land is in very short supply they may have to lease it.

The development value of the land will be determined by:

- Location, both the geographical region and the local position with regard to amenities, communications, etc. Real estate agents often emphasise the importance of this by saying that the three most important factors in land value are 'location, location and location'.
- Restrictions on use of the land, imposed either by the vendors in the form of covenants or by the community in the form of planning restrictions.
- Any easements[6] that are associated with the land, such as a right of way across it.
- The physical state of the land: whether it is level or very hilly and whether there are buildings on it which need to be demolished.
- The current state of the economy.

5.7.6   Effect of building cost on land value

We can express the value of land for development as an equation:

Value of development = price of undeveloped land + building costs + profit

As the value of the development in a free market is its worth to consumers (as compared to other things they can spend their money on), this in turn largely fixes the price of the undeveloped land. The equation can therefore be better expressed as:

Price of undeveloped land = value of development − (building costs + profit)

The people selling the undeveloped land should be as capable as the developer of doing this calculation, and will fix their selling price accordingly. It is therefore not really correct to say that the cost of land pushes up the price of housing or other accommodation; in fact it is the other way around – the market price of development pulls up the cost of land.

In conditions where there is a shortage of suitable building land and a constant rise in development values, as in the south-east of England, fluctuations in building costs will have little effect on land prices. However, if the situation were reversed and development values remained constant or even fell, then an increase in building costs would have the effect of reducing undeveloped land values. It is normally the price of undeveloped land that is the result of the equation, when the other figures have been filled in.

5.7.7   Social considerations of land use

Most people agree that a completely free market in land for development is socially undesirable, and the setting of limits on the scale and nature of development under town and country planning legislation brings a measure of public control to the process. Even so, there is a strong body of opinion that wants to see the public obtain much of the benefit of any increased development value of private land, and this would certainly make it much easier to undertake social development in urban areas. The more radical members of this group dislike the whole idea of anybody making money which they have not actually earned, and it would be difficult to satisfy them, but the more moderate members base their arguments on the following: the loss of green land or other pleasant environment to the community when development is undertaken for profit; the fact that public expenditure on infrastructure (roads, services, transport systems, etc.) has often contributed to the rise in value; the way in which decisions taken by the planning authorities are handing windfalls to some lucky landowners and withholding them from others, thus providing strong incentives for undesirable (or even improper) influence on those decisions.

Several unsuccessful attempts have been made by governments to deal with the problem, of which the most important were the Town and Country Planning Act 1947 and the Community Land Act 1975. However, these Acts tended to inhibit development of any kind, as landowners had little incentive to allow their land to be built on and preferred to wait until a more right-wing government got into power and repealed the legislation. More recent developments include environmental legislation and new planning guidelines favouring brownfield sites for redevelopment rather than building on undeveloped land. However, so that the community does not actually lose money over private development, the developer may often be asked to contribute towards infrastructure costs as a condition of the granting of planning approval, and sometimes a developer may make such an offer as part of the planning application. Cost advisers on both sides should obviously be involved in the agreement of these costs. An interesting technique was used in Hong Kong in the 1980s, when the then colonial government was able to act unilaterally and the new underground railway system was largely financed from a tax on the increase in development values in the areas around its suburban stations.

5.7.8   Effect of land values on cost planning techniques

It is possible to try and optimise building costs by looking at such factors as low wall/floor ratios, the avoidance of multi-storey construction, the centralisation of services, and so on, in isolation from land costs. Where land prices

Phase 1

are relatively low, or in a public sector organisation where land costs and building costs are not considered together, this may seem a valid way of looking at the problem. However, where development values and land prices are high, the picture alters.

Example

Suppose that the current price of flats in a fashionable urban area is 20% building cost and 80% land cost (and profit) and a piece of land has been bought at a price which assumes that 20 flats can be built on it. If it proves to be possible to build more than 20 flats on the plot, the building cost of these extra flats would be only 20% of their market value, leaving 80% profit.

It would therefore be worth accepting a less efficient building configuration to provide these extra flats. Even if this increased the building cost slightly there would still be a handsome profit. In these circumstances cost planning may develop into an exercise to secure the maximum number of accommodation units on a given site, at the same time optimising the design with regard to any available grants and subsidies. Unfortunately for the developer, a scheme which squeezes the utmost money out of the site is rarely acceptable to the planning authorities, who have more regard than the developer for the amenities and appearance of the neighbourhood. Their veto on a scheme is final, apart from the possibility of a lengthy and time-consuming appeal to central government, which has the same general motivation. On such projects the costing and income appraisal of many alternative schemes will be required, and this will be the main cost planning contribution. Once a scheme has been approved by the authorities speed may be vital (because of the need to recover a massive investment as soon as possible), and the cost planner will be required to advise on a suitable contractual system to achieve speed with an appropriate measure of cost control.

However, not all private development is of this speculative kind. The client organisation may wish to erect a building for their own use, such as offices, warehouse, store, etc. In this case the effect of high land costs on the cost planning exercise should lead to an examination of the economics of possible alternatives. These might include carrying out the development in a less expensive area, or possibly changing the client's requirements for a building by solving the prob-lem in another way. An example of this might be to avoid the cost of branch warehouses by a system of daily distribution from the factory. There might be an advantage to a user–developer in building in a high-cost area because the firm would always have the value of the site on its books, but there are many disadvantages in having too many of a firm's assets tied up in site values.

5.7.9   Grants, subsidies and taxation concessions

Action under town and country planning legislation is essentially passive, and while it can prevent undesirable development it cannot cause socially desirable development to be undertaken. It can certainly earmark a certain area for a particular type of development, but unless the development itself is to be carried out by a public agency, or unless it is obviously going to be highly profitable, the matter will rest there. In order to make such development attractive to a private investor, central or local government may sometimes offer special financial inducements. These are usually offered on a sector basis (e.g. for a hotel or other tourist building), on a regional basis or both (e.g. for factories in an area of high unemployment). These incentives will need to be taken into account in preparing budgets for the type of development concerned, and may be of crucial importance in deciding between alternative sites. The incentives may be of several different kinds.

Low rent or rates. Land (or even completed buildings) may be made available by a public body at a very low rent or with substantial relief from local authority charges, or both. This concession usually runs for a limited number of years, after which more normal conditions will apply, so some kind of discounted cash flow analysis will probably be necessary.
A grant towards capital costs. The Department of Trade and Industry used to provide grants towards the cost of factory building and plant located in 'areas of expansion', and an increasing number of authorities provide partial funding for development in the fields of tourism, leisure, conservation, etc.
Taxation relief. Taxation relief may be given under some circumstances on development and plant costs. The main disadvantage of taxation relief compared to a grant is that there must be a tax liability against which the relief can be set. It is therefore necessary to make a profit before the benefit can be obtained.

The situation is always changing over the years as different political and economic priorities come into play. A most important factor in assessing the value of a grant is its timing in relation to the developer's expenditure, particularly whether it is a reimbursement or an advance. It should also be noted that other forms of financial encouragement, such as a subsidy paid for each person employed, may need to be taken into account in a development budget, even though strictly speaking it is not a grant towards the development itself. The extent to which government incentives may be nothing more than compensation for straightforward commercial disadvantages, such as high transport costs, will certainly emerge from the cost planner's calculations. It is also worth pointing out that it is possible for government to impose financial disincentives for types of development of which it does not approve. The lack of grants for industrial building in some regions can be seen as such. A more extreme example of disincentive in post-World War II years was the long-term discouragement of private rental housing development by the control of rents at uneconomically low levels. Although the UK government now wishes to reverse this trend, prospective developers need to be able to look ahead for more than the five-year term of a government, and the treatment of housing (and other development matters) as a political football by the two

Phase 1

main British parties has discouraged a stable property rental market such as exists in other European countries.

5.7.10   Land costs and social development

In the past the costs of site acquisition have usually been ignored in cost planning social development, at any rate as far as the cost planner is concerned. Without any profit a profit–investment appraisal cannot be prepared for such projects, and the cost of the land is just one of the many items on the expenditure side with which no comparison with income can be made. Also, there are still many projects where the budgeting stage is bypassed, and the cost planner is concerned with nothing more than meeting a target for building costs which may have been set using some formula or other. However, there is an increasing tendency to use the profit–development type of calculation on any social projects, such as housing, where there is a real or hypothetical income.

5.8   Calculating building costs

Although building costs are only one of the ten items making up the total budget, they are still likely in most cases to form the largest single item. However, it must be remembered that at this stage almost nothing is likely to be known about the building except its general size, and therefore it is pointless to go into detail about cost before any designing has been done. This is the place to use one of the traditional 'single price rate' methods of estimating cost introduced in Chapter 1; the size of the building is measured in one form or other and the resulting quantity is multiplied by a single price rate to give the estimated cost.

5.8.1   The cube method

The cube method was the traditional method. All QS's offices used to keep a 'cube book'. Whenever a contract was signed the amount of the accepted tender was divided by the cubic content of the building and the resultant cost per cubic foot entered in the book. When an estimate was required for a new project its volume in cubic feet would be calculated and an appropriate rate per cubic foot taken from a previous job in the cube book. The method has largely died out as a result of a number of factors, the main one being that the cost of a building is usually more closely related to its gross floor area than its cubic capacity.

Even with a blunt instrument such as a cube, it was necessary to have common standards of measurement so that the rate obtained from one project might be fairly compared with another. The Royal Institute of British Architects' (RIBA) pre-metric set of rules was the one commonly accepted; these involved multiplying the plan area measured over the external walls by the height from the top of the concrete foundation to halfway up the roof if pitched, or to 2 ft (0.60 m) above the roof if flat. Extra allowances were made for projections such as porches, bays, oriels, turrets and fleches, dormers, chimney stacks and lantern lights. This formula had little to recommend it except the vital matter of uniformity. The additional allowance for a pitched roof (which may well have been cheaper than a flat roof) and the allowance for fluctuations in foundation depth were both very arbitrary. The Standard Form of Cost Analysis (published by the Building Cost Information Service of the RICS) still makes provision for cubing but uses a different set of rules.

5.8.2   The superficial area method

This is an alternative single price rate method in which the total floor area is calculated. The convention usually accepted is that the area of the building inside the external walls is taken at each floor level and is measured over partitions, stairwells, etc. The resulting figure is known as the gross internal floor area. In cases where columns to frames are situated inside or beyond the external walls, these columns are ignored and the dimensions are still taken from the internal face of perimeter walls. Where there is no external wall at floor level (as in the case of freight sheds) the dimensions are taken from the external faces of columns.

More recent than the cube method, the superficial area method first came into use in the 1940s and 1950s for such projects as schools and local authority housing where the storey heights were reasonably constant. It has the virtue of being closer to the terms in which the client's requirements are expressed, as the accommodation is more likely to be related to floor area than to cubic displacement. Sometimes when a very early estimate is required, the only data available may be the approximate floor area, so this method is still widely used for budgeting as the first stage in a cost planning process.

5.8.3   The unit method

This technique is based on the fact that there is usually a close relationship between the cost of a building and the number of functional units it accommodates. By functional units we mean those factors which express the intended use of the building better than any other. For example, it may be number of pupils in a school, number of vehicle spaces in a car park, number of seats in a theatre, number of people housed in a housing development, or number of beds in a hospital ward. Several cost yardsticks operated by

Phase 1

government departments were based on this relationship and provided a benchmark of what was reasonable for a given scheme. After all, by costing a single functional unit instead of, say, floor area, the designer has greater flexibility in choosing between the quantity of building and its quality, and the funding organisation has the assurance that it has not paid beyond what is reasonable. This example shows the more general use of the technique.

Example

Suppose a multi-storey car park for 500 cars had recently been constructed for the sum of £3,000,000. The cost per car space would obviously be: {£3,000,000/500} = £6,000 per car space. If a similar car park is to be built for 400 cars then it might reasonably be assumed that the cost would be: $400 \times £6,000 = £2,400,000$, *ceteris paribus*[7].

5.8.4   Single price rate methods generally

Unfortunately all things are not usually equal and consequently a number of judgements must be made about:

- prevailing price levels (due to inflation and market conditions) at the proposed date of tender;
- site difficulties;
- specification changes;
- different circulation and access arrangements, etc.

Once again, as with all single price estimates, these adjustments to the analysed figure would be based on the cost adviser's judgement and are critical to success.

5.8.5   Cost of financing the project

Having looked at the role of land costs and building costs in relation to the project, it is now time to look at financing costs. This is the intangible component of total building cost; all the other things like buying land, paying a building contractor, paying professional fees, etc. involve paying money to somebody else in exchange for property or services. As soon as this intangible expenditure starts, however, the developer will be laying out money, and there will be nothing to show for it in terms of income (or use) until the building is ready for occupation or can be disposed of. Hence the cost of lending this money to the development for a period of possibly several years becomes part of the cost of the development itself. When interest rates are high this financing cost can be considerable, as already pointed out, so the project should be planned to avoid unnecessarily early expenditure on any part of it.

5.8.6   Where the money comes from

The capital for the development may come from a number of alternative sources:

- *Sources of finance for private development.*
- *Bank overdraft.* This is rarely available as a means of financing a whole development, but may be useful for short-term bridging purposes; the money would probably be lent on the security of the development.
- *Loan from merchant bank or insurance company.* This is similar to the last, except that it is possible to obtain longer-term finance from these sources. Interest rates are usually somewhat higher.
- *Capital account.* Where the development is being undertaken for a firm's own use, such as factory, warehouse, etc., the capital for the development can be regarded as part of the general capital expenditure of the firm.
- *Trading funds.* Property companies will have funds for carrying on their business.
- *Shares in the development.* Speculative development is sometimes financed by paying the builder with shares instead of cash, or by allowing the building firm to develop part of the site for itself. Alternatively a joint venture may be undertaken with another developer.
- *Finance by the intended occupier.* It may be possible to persuade the intended occupier to pay part of the sale price during the erection of the building. This is sometimes the case with houses built for sale.

5.8.7   Sources of finance for public (social) development

- *Government funds.* These will be available either as direct finance for government building (e.g. defence works) or as a grant or subsidy for other public sector or social building.
- *Loans from government funds.* These are usually available at a lower rate of interest than money raised on the open market, but are very restricted as to eligibility and amount.
- *Issue of stock.* Local authorities and public boards are empowered to issue fixed-interest loan stock on the Stock Exchange. Full market rates of interest have to be paid and there is a long-term commitment to these, which can prove very expensive if interest rates subsequently fall.
- *Local authority mortgages.* These are a useful way of raising money as there is a firm commitment by both sides, but for a short period of three to five years only. Interest rates are usually a little higher than the current yield on gilt-edged securities.
- *Loans from other funds.* A public authority can borrow from its other funds (if it has any) for development purposes.

Phase 1

- *General income.* It is common for a local authority to raise money for development from its charges or council tax; this has the advantage that the money does not have to be paid back!
- *The National Lottery and the National Lottery Charities Board (now called the Big Lottery Fund).* Grants are also available for suitable projects by application to the Arts Council or the Sports Council for England. There is strong competition for money from these sources.
- *The European Union.* The so-called structural funds such as Objective One provide capital funding for development schemes in areas of the EU that are considered to be at risk from social and economic deprivation. Liverpool in the UK is one example of a city that has been transformed by the impact of this funding.
- *Private developer.* In large-scale urban redevelopment a private developer can be required to undertake some social development in exchange for being allowed to pursue a profit-related scheme. Alternatively, a developer can be offered some profitable role within a social development if it contributes to the cost of the public part.

5.8.8   Budgeting for refurbishment work

Before drawing up a budget for refurbishment work, or work involving major alterations to existing premises, a number of difficult decisions have to be made and it is necessary to consider very carefully how the work will be carried out before establishing even an initial budget.

5.8.9   Budgeting for new-build projects on very restricted urban sites

Such projects have many problems in common with refurbishment work, and reference to Chapter 19 would be worthwhile.

5.9   Budgetary examples

This section looks at some simple budget examples. These are in no sense typical in that every project has its own particular problems and priorities, but they give an idea of the type of basic calculations involved. The method of arriving at the various estimated costs has not been shown, as the concern here is how to manipulate the answers. The interest rates used for most of the calculations are based on a criterion rate of return – the return that the developer requires in order to make a reasonable profit.

5.9.1 A budget for minimum economic selling price (24 town houses in a provincial city)

Costs	£	
Cost of land	1,000,000	
Legal and professional cost of acquiring land, obtaining planning permission, etc.	100,000	
Demolition of existing property	60,000	
Building cost and associated professional fees	960,000	
Site layout, ditto	40,000	
Agent's charges for selling	20,000	
	2,180,000	2,180,000

Cost of finance (2% per month compound interest – criterion rate of return)

Land, legal costs, demolition – finance on £1,160,000 for 12 months	313,200	
Building, professional fees, and site layout – finance on £1,000,000 for 4 months (av)	80,000	
Agent's fees are paid after sale, no finance charge	–	
	393,200	393,200
		2,573,200

Income

Minimum economic selling price of each house would therefore be £2,573,200/24 = £107,210 (say £108,000)

Anything less than this will not produce the criterion rate of return, and if there is any doubt about obtaining such a figure then the development should not go ahead based on the information provided here. It will be seen that instead of discounting all expenditure and receipts to the commencing date of the scheme, they have been carried forward to the end of the scheme and compound interest added. This has the same effect; it is easier to do it this way in this instance as it is the future selling price which we are interested in and not PV.

5.9.2 A budget for a profit development (for rental offices)

Costs	£	
Cost of land	500,000	
Legal and professional cost of acquiring land, obtaining planning permission, compensation, etc.	170,000	
Demolition	60,000	
Building cost and associated professional fees	700,000	
	1,430,000	1,430,000

Cost of finance (2% per month compound interest – criterion rate of return)

Land, legal costs – finance on £670,000 for 36 months	696,800	
Demolition – finance on £60,000 for 18 months	25,800	
Building, professional fees, and site layout – finance on £700,000 for 10 months (av)	154,000	
	876,600	876,600
Total capital cost at completion		2,306,600
	(say)	2,310,000

Income

Annual income from rents (figure given by valuer)	600,000	600,000

Less

Allowance for vacant tenancies (voids)	50,000	
Maintenance, repairs and redecorations (excluding tenants' responsibilities)	30,000	
Cleaning, heating, lighting and council tax on public part of building (staircases, entrance halls)	50,000	
Management expenses, including caretaker, arranging lettings, and collecting rents	18,000	
Sinking fund to replace capital at end of 40 years at 2.5% pa on £2,310,000 less cost of land (which will still be there when the building is demolished) £1,810,000 at 1.5p	27,150	
	175,150	175,150
Net annual income		424,850

This gives a return of 18.4% on the capital of £2,310,000.

An alternative way of setting out this last budget would be to start with the estimated income as the criterion. This is the approach a developer would be likely to favour.

5.9.3 A budget for a profit development (alternative approach)

Income	£
Estimated annual income from rents	600,000
Less expenses (excluding sinking fund)	148,000
	452,000

Note: The cost of the sinking fund cannot be accurately assessed at this stage, as the capital value of the buildings, etc., is not known. It could be approximated thus: Capitalised value of income at 18.5% is approximately

£2,443,000. Say 25% of this represents residual land value, which will not need to be replaced.

Sinking fund to replace £1,832,000
 at end of 40 years at 2.5% (at 1.5p)   27,483
Estimated net annual income   424,517
Net value of 40 years' annual rent £424,517 at 18%
 (criterion rate of return) at £5.40   2,292,400

This represents the capitalised value of income at the date of completion in 36 months' time. Discounted to PV at 2% per month at 49.0p, PV of income = £1,123,270

Costs
Estimated legal and other costs of acquiring land   170,000
Estimated cost of demolition of existing
 buildings (PV of £60,000 in 18 months' time at
 2% per month = 70.0p per £)   42,000
Estimated cost of building and professional fees
(PV of £700,000 in (average) 26 months' time at 2%
 per month = 59.8p per £)   418,600
                                                             630,600

The maximum amount which can be paid for the land at the present time to give the required rate of return would therefore be £1,123,270 (PV of income) minus £630,600 (PV of expenses) = £492,670 (say £500,000). At this stage the calculations would have to be based on the minimum amount of accommodation for which the client could expect to get planning permission, unless an outline scheme had already been approved. It should be noted that taxation provisions, which are constantly changing, have been ignored in these calculations but cannot be ignored in practice.

5.9.4   Sensitivity analysis

It is possible to examine the effect of changing the values of the variables within this model either singly or in combination using sensitivity analysis. This will enable those variables or assumptions which are significant to the calculations to be identified and quantified. Very often in profit development the overall time for designing and building the project is of supreme import-ance, and simulations (ordinarily using Monte Carlo techniques) might well be carried out to assess the effect on construction costs and financing charges of different timescales and fast-track methods of procurement. Certainly it will be seen in the above examples that a large proportion of the total cost is represented by financing charges. It would be easy to reduce the length of the programme on paper; the more important matter is whether this would be achievable in practice, allowing for all the things that can go wrong.

5.9.5   Need for caution

Schemes founded on over-optimism tend not to have a happy outcome. In fact, some developers require that budgets be prepared on a mid-to-worst-case scenario, knowing that construction costs tend to escalate but funding has to be obtained up-front. In one case the developer actually told the cost planner to presume a figure for construction costs which was likely to reduce during the design development and construction stages, and which would not be exceeded without very good reason.

5.10   Key points

- The total expenditure on the development will be much higher than the net cost of the building fabric, and will include cost of land, legal costs, demolition or other physical preparation, professional fees, furnishings and fitting out, VAT, cost of leasing or selling, cost of finance and costs of maintenance or disposal.
- Building development of any kind requires a plot of land on which it can take place and which, once used, will no longer be available for any other development, unless the first one is either demolished or converted.
- The development value of a piece of land is the difference between the market price of the finished development including the land, and the cost of erecting (or converting) buildings on it.
- Where land costs are very high, cost planning may develop into an exercise to secure the maximum number of accommodation units on a given site, at the same time optimising the design from the point of view of grants and subsidies (if available).
- As soon as tangible expenditure on land purchase, building work, etc., starts, the developer will be laying out money, and there will be nothing to show for it in terms of income (or use) until the building is ready for occupation or can be disposed of. Hence the cost of lending this money to the development for a period of possibly several years becomes part of the cost of the development itself.

Notes

(1)   Tavistock Institute (1966) *Interdependence and Uncertainty: A Study of the Building Industry*. Tavistock Publications, London.

(2)   The theory of complexity and complex systems has become an important part of management science. For information on the application of complexity theory within the built environment, refer to CIB TG-62 Complexity and the Built Environment at www.cibworld.nl and http://pcwww.liv.ac.uk/~kirkham/CIBTG62.htm

(3)   Members of the Scottish Parliament.

(4)   BPF (1983) The British Property Federation System for the Design of Buildings. British Property Federation, London.

(5)   Constructing Excellence (2004) Briefing Factsheet http://www.constructingexcellence.org.uk/pdf/fact_sheet/Briefing_Team.pdf (accessed 14 February 2007)

(6)   An easement is the right to do something or the right to prevent something over the real property of another.

(7)   *Ceteris paribus* is the classical economic term meaning other things being equal.

References

Adizes, I. (2004) Difference between clients versus stakeholders. *Adizes Insights*, May 2004. http://www.adizes.com/Insights9/01.htm (accessed 22 February 2006).

Kamara, J.M., Anumba, C.J., and Evbuomwan, N.F.O. (2002) *Capturing client requirements in construction projects*. Thomas Telford Ltd, London.

Masterman, J.W.E. and Gameson, R.N. (1994) Client characteristics and needs in relation to their selection of building procurement systems. In: Rowlinson, S. (ed.) *CIB 92: East Meets West*, Symposium, Department of Surveying, University of Hong Kong, December. CIB Publication No. 175, 79–87.

Phase 1

Chapter 6
The Economics of Cost Planning

6.1 The time value of money

Anything that involves the forecasting, estimating or prediction of costs into the future is governed by the concept of what is known in economics as the time value of money. Strangely, many people find this confusing but in reality it is simple and is usually exemplified in the following example:

You are offered the option of accepting £100 today, or £100 in a year's time. Which option would you accept *ceteris paribus*[1]?

Naturally, most rational people would accept the £100 today rather than in a year's time, for many reasons such as risk, uncertainty and opportunity cost[2]. However, it is clear that a sum of money in the future will always be worth less than the same sum today, and this will depend on the length of time between options, future risk and uncertainty and future interest rates. The determination of this future sum is obtained by the use of standard discounting calculation methods. The standard formulae will be covered later in this chapter.

The concept of discounting is a fundamental aspect of WLCC since this technique involves the forecasting of future sums over longer periods of time. WLCC is covered in greater depth in Chapter 7. However, in the shorter term, cost planners are often required to perform calculations on a monthly or even weekly basis, and consequently annual interest rates are difficult to deal with. Cost planners will generally use an equivalent rate for the period, dividing the annual rate by 12 (months) or 52 (weeks). Naturally, this is interpolating to some extent since it ignores the effect of compounding[3]. The exact annual equivalent of a monthly interest rate i is not $12i$ but $(1 + i)^{12} - 1$. The yearly equivalent of 1% per month is therefore 12.68%, and conversely the monthly equivalent of 12% per annum is 0.94888%. Where comparisons between two or more alternatives are being considered, this difference is rarely of much significance, but it should be allowed for if specific annual interest rates are an important factor.

6.2   Interest – the cost of finance

The issue of interest and financing charges will normally apply to all construction projects. Large, complex building procurements in particular will require some form of financing in order to allow the developer to realise the project. This is due simply to the fact that significant funds are normally necessary to resource a project in terms of materials, labour, plant, land fees, professional fees, insurances, taxation, etc.

The cost of borrowing money is quantified by the interest rate. Some economists will describe the interest rate as a measure of the 'rent' paid to borrow a sum of money. Why does borrowing money always involve interest? In the classic economic scenario, the lender must be incentivised to release funds for others to borrow; this incentive comes in the form of a 'compensation' to take into account the lender's decision to defer consumption of the principal. The original amount lent is normally referred to as the principal, and the percentage of the principal which is payable over a period of time (usually one year) is the interest rate.

In the UK, the Bank of England is responsible for setting the interest rate and it is often referred to as the Bank of England base rate. Since 1997, the Bank of England Monetary Policy Committee (MPC) has held sole responsibility for setting interest rates in an attempt to depoliticise economic decision-making. On 14 February 2007 the rate stood at 4.75%. However, the cost of borrowing money can depend on a variety of circumstances and often this cost will be encapsulated in a rate well above the Bank of England base rate. This is loosely referred to as a credit-worthiness premium, which is based on an assessment of the borrower's ability to repay the principal plus the estimated accrued interest. Other factors such as liquid assets, credit history, guarantors and so on will affect the interest rate; this can sometimes be a major problem for small contractors who experience regular problems with cash flow and timing of payments. In other words it represents a risk premium set by the lender.

Risk is an important concept in construction (it is distinct from uncertainty) and represents (in this context) a quantification of the probability of default. The higher the probability of default, the higher the risk premium and vice versa. The risk premium can be found by comparing the project with the rate of return required from other projects with similar risks.

On a macro scale, interest rates are used as a method of controlling inflation within the economy. The higher the rate of interest, the more attractive investing and saving becomes over spending and borrowing. This reduces

Phase 1

demand in the economy and should in turn lead to a steady state or fall in prices. In some countries, this can be a real headache for cost planners, who must account for potential cost increases in raw materials such as steel. The Far East, for example, has experienced highly volatile raw material prices since the mid-1990s.

Cost planners should, where possible, understand the national/international supply/demand for key materials. In some specific cases, cost planners have advocated the purchase of raw materials well in advance of the start of a project in order to mitigate the effects of a potentially significant price rise further down the project life-cycle. This will of course affect cash flow and increase costs for storage and security and such a decision would need to be assessed on its individual merits.

Cost planning is not so much a science as an art, one which is perhaps impossible to truly master given the externalities impacting on construction projects.

6.3   Is interest always taken into account?

Yes. Interest and finance charges apply to all projects and developments, simply because the capital used to fund the project is either being borrowed and is thus attracting interest charges, or is the developer's own money, which because it has been spent on the development is not available for investing elsewhere and is therefore not accumulating interest (opportunity cost). Strangely enough, although most cost planners are aware of this factor, some clients' accounting systems do not recognise it and so in practice cost planners are not always able to demonstrate any benefit from optimising payment in relation to time. However, private developers in particular are certainly aware of the importance of the timing of income and expenditure; and building contractors have always recognised its significance, even if they have not always been able to formalise their methods of dealing with it.

6.4   The importance of understanding the cost–time relationship

The construction industry delivers some of the most prestigious public projects from roads and bridges to schools, hospitals and corporate developments. Many of these projects will take years to complete, and so will be susceptible to market forces such as inflation and interest rates. With the increasing complexity of large building projects it is not uncommon for some years to elapse from the start of expenditure on the project (outline design) through to the time when practical completion/handover is achieved or the building becomes a revenue-earning asset.

The interest rate is a key variable in the cost/time relationship as this has a direct impact on the construction industry generally. The demand for construction work is usually cyclical and regionally volatile. During periods of high demand for products and services in the economy, interest rates will more likely rise at some stage to check the demand. This of course will have a knock-on effect for projects that are heavily financed. Therefore cost planners engaged on long duration projects will require extra diligence when assessing likely changes in the economy, both on a local, national and international basis.

Many larger contracting organisations now employ in-house economists who monitor market trends in order to help cost planners deal with inflationary issues. There has also been a large increase in the number of companies (many part of a PQS organisation) which concentrate on providing specialist cost advice to clients.

6.5 The application of simple development economics in cost planning: terminology and nomenclature

Conventionally, the cost planner will be required to perform calculations to account for future project income and expenditure. On shorter project durations, the selection of the appropriate methods of evaluation is not as critical as it is on long duration schemes. Public Private Partnerships (PPPs) and in particular, PFI schemes are one example of the latter. In this instance, cost planners are typically involved in 25-year concession period cost plans which involve not only design and construction costs, but also operational and maintenance expenditure such as energy and facilities management services. In this context, the sums become slightly more complex. This subject is covered in greater depth in Boussabaine (2006). Notwithstanding, cost planners should be competent in the use of standard tools and methods that are appropriate for any type of scheme, and these are explained in the remainder of this chapter.

6.5.1 Simple interest and compound interest

Simple interest (equation 1) is the amount of interest paid on the principal only. If a sum of money is invested for a number of years it will have earned some interest by the end of the first year. Compound interest (equation 2) assumes that this earned money is immediately added to the principal and re-invested on the same terms, this process being repeated annually. Many forms of investment provide for this automatically, but it is in any case the correct method to use when making investment calculations since it should not be assumed that money would be allowed to lie idle. Note that this formula is also useful for extrapolating inflation rates over a period of years.

Simple interest $= P \times i \times n$ $\qquad$ (1)

Compound interest $= P(1 + i)^n$ $\qquad$ (2)

where P = principal, i = interest rate, and n = time (number of years).

Example

A property developer decides to invest £128,000 in a guaranteed bond at 6.76% compound interest to offset against future maintenance costs for his estate portfolio. Calculate the value of this investment after 12 years.

Using (2): £128,000 $\times (1 + 0.0676)^{12} =$ £280,616.21

6.5.2   Future value and present value

In cost planning there will invariably be a need to calculate the value of sums of money in the future and also to be able to convert future sums to today's value. In order to do this the future value (FV) and present value (PV) formulae, respectively, are used.

6.5.2.1   Future value

FV measures what money is worth at a specified time in the future assuming a certain interest rate. As in the previous example, equations for FV are available for simple and compound interest:

FV (without compounding) $= P(1 + in)$ $\qquad$ (3)

FV (with compounding) $= P(1 + i)^n$ $\qquad$ (4)

where P = principal, i = interest rate, and n = time (number of years).

Furthermore, it is possible to calculate the FV of a sum of money invested at regular intervals over time; this is known in economics as an annuity. An annuity is a flow of fixed payments over a specified time period. So instead of the worked example given before where a single sum is invested, an annuity involves a sum of money regularly invested at a specified time.

Annuities fall into two groups, annuity-immediate (ordinary annuity) and annuity-due. The former is where payments are made at the end of each period and the latter where payments are made at the start of each period. Thus, the FV of an annuity-immediate and an annuity-due are given in equations (5) and (6) respectively:

$$FV = A \times \frac{(1 + i)^n - 1}{i}$$ $\qquad$ (5)

$$FV = A \times \frac{(1 + i)^n - 1}{i} \times (1 + i)$$ $\qquad$ (6)

where A = the annuity payment, i = periodic interest rate, and n = time (number of years).

Example

The sum of £2,850 is invested annually into an account, again at a compound interest rate of 6.76%. The period of the investment is 9 years. Calculate the value of the investment at the end of Year 9, assuming the investment is made at the end of each year.

Using (5) $FV = £2,850 \times \dfrac{(1 + 0.0676)^9 - 1}{0.0676} = £33,798.54$

If the investment is made at the start of each year, since each annuity payment compounds one extra period, the value of an annuity-due is equal to the value of the corresponding annuity-immediate multiplied by $(1 + i)$.

Using (6) $FV = £2,850 \times \dfrac{(1 + 0.0676)^9 - 1}{0.0676} \times (1 + 0.0676) = £34,027.02$

6.5.2.2 Present value

PV of a future cash flow is defined as a sum of money at some time in the future, discounted[4] to today's value (accounting for the time value of money). The Treasury (HM Treasury 2003) defines PV as the FV expressed in present terms by means of discounting. The importance of the discount rate calculation methods is explored in greater depth later in this chapter. The calculation of PV is important in cost planning as it allows appraisal options on a like-for-like basis.

In the compound interest example (2) the proof shows that £128,000 invested for 12 years at 6.76% compound interest grows to £280,616.21. The inverse of this is simply that the PV of £280,616.21 in 12 years' time at 6.76% interest is £128,000. Therefore, PV of £1 is expressed mathematically as the reciprocal:

$$PV = \frac{1}{(1 + i)^n} \tag{7}$$

Example

Calculate the PV of £1,200 in 35 years' time, discounted at 15% per annum.

From (7) $PV(£1) = \dfrac{1}{(1 + 0.15)^{35}} = £0.008$

Thus, $£1,200 \times £0.008 = £9.60$

In this example you can see that over a very long period, the value of the principal diminishes significantly. With high interest rates this is to be expected. In reality, the Treasury *Green Book* gives guidance on the use of discount rates and time. It recommends that the test discount rate becomes 3.5% for years 0–30, 3.0% for years 31–75, and 2.5% thereafter (in the public sector). This is basically to mitigate the effects shown in the example above. These kinds of long-term predictions would, however, be more common in WLCC appraisals rather than conventional short duration project cost planning *per se*.

6.5.2.3   Present value of £1 payable at regular intervals

In (7) we calculated the PV of a sum of money at a point in the future. However, as in (5) and (6), the cost planner may be required to calculate the PV of regular future payments or receipts. Therefore, the PV of £1 payable at regular intervals is given:

$$\text{PV of £1 payable at regular intervals} = \frac{[(1 + i)^n - 1]}{[i(1 + i)^n]} \tag{8}$$

Example

Calculate the PV of £1,200 payable annually for 10 years assuming an interest rate of 8% per annum.

From (8) the PV of £1 paid annually is £6.71. The PV of £1,200 annually is therefore £1,200 × £6.71 = £8,052. This is the sum which would have to be invested today at 8% compound interest in order to discharge an obligation.

It is interesting to see what difference would result if the money were to be paid in quarterly instalments of £300 instead of at the end of each year. 8% per cent per annum is equivalent to 2% per quarter so from (7) the PV of £1 over $10 \times 4 = 40$ periods at 2% is £27.36. The PV of £300 paid quarterly for 10 years is therefore:

$300 \times £27.36 = £8,208$, compared to £8,052 for the yearly payments of £1,200

6.5.3   Annuity purchased by £1

This is the reciprocal of (8) and gives the annuity (or regular annual payment) purchased by a lump sum payment of £1. At the end of the given number of years the money will be exhausted. It is therefore useful for calculating the annual equivalent of a given present-day lump sum (whereas (8) calculated the present-day lump sum equivalent of a given annual amount):

$$\text{PV of £1 payable at regular intervals} = \frac{[i(1 + i)^n]}{[(1 + i)^n - 1]} \tag{9}$$

Example

What annual saving in maintenance costs over a period of 10 years would justify an increase in capital costs of £90,000, assuming an interest rate of 8%?

The annual equivalent of £1 from (9) is 14.9p. The annual equivalent of £90,000 is therefore £90,000 × 14.9p = £13,410. If the saving in maintenance costs exceeds this amount then the additional capital investment of £90,000 would be justified.

6.5.4   Sinking fund

A sinking fund is a method by which an organisation sets aside money over time. More specifically, it is a fund into which money can be deposited, so that over time its preferred stock, debentures or stocks can be retired. In general finance, sinking funds have the benefit of the principal of the debt (or at least part of it) being available when due. Naturally, sinking funds can be used as a risk premium reduction tool since the fund reduces the risk of default on maturity of the debt. The equation for sinking fund (10) is the reciprocal of (5) since it returns the value of how much must be invested each year to accumulate a certain sum over a certain number of years. Sinking funds are useful when planning life-cycle replacements for equipment such as mechanical and electrical (M&E) services.

$$\text{Sinking fund £1} = \frac{1}{[(1 + i)^n - 1]} \tag{10}$$

Example

How much must be invested at the end of each year at 7% per annum to amount to £20,000 at the end of 12 years?

From (10), 5.6p has to be invested each year to realise £1 after 12 years. For £20,000 to be realised therefore, again from (10):

£20,000 × 5.6p = £1,120 per annum sinking fund

6.6   Project cash flow

The basic economic concepts described in the earlier part of this chapter must be conceptualised within the wider context of how cost planners need to be

Table 6.1   Example cash flow statement for period 1.

Description	Cash in (£)	Cash out (£)	Net cash flow (£)
Loan from finance institution	£1,456,000		£1,456,000
Sale of previous assets to finance project	£312,500		£1,768,500
Prime cost items		£789,000	£979,500
Design and professional fees		£80,000	£899,500
Land charges		£40,000	£859,500

aware of the importance of timing various payments and receipts into a construction project (this is critical on long duration schemes such as PFI). These techniques are the foundation of sound project management and planning. The integrated process of planning these payments and receipts is conventionally referred to as cash flow management.

Simply, cash flow is the difference between income coming into a project and expenditure going out. A project is in a position of negative cash flow when more money is going out of the project than coming in, and vice versa for positive cash flow. Table 6.1 gives a basic example for one period. For *n* number of periods you would be able to produce an X–Y line graph showing cash flow visually over the project duration. For large international contracting organisations, cash flow is not so much a problem (assuming normal economic conditions) since these companies are involved in many other ventures which provide for a relatively stable overall company balance sheet. However, for smaller contractors cash flow can be critical and can be the difference between being in business or not. Poor cash flow is the antithesis of good business performance for most small contracting organisations, so accurate and effective cost planning is all the more critical.

The principle of cash flow is basic but the complexities of the process within are not. Consequently, work on cash flow modelling has been the focus of a good deal of research in the UK industry, with notable contributions from Kaka and Lewis (2003) – 'Development of a company-level dynamic cash flow forecasting model (DYCAFF)'; Boussabaine and Elhag (1999) – 'Applying fuzzy techniques to cash flow analysis'; Lam *et al.* (2001) – 'Using an adaptive genetic algorithm to improve construction finance decisions'; and Blyth and Kaka (2006) – 'A novel multiple linear regression model for forecasting S-curves'. There are a wealth of other factors that will affect cash flow, including materials prices, labour rates and availability, the general economy, energy prices, and interestingly the procurement route selected.

Whilst the examples of research referenced above deal with the mathematical and scientific nuances of cash flow forecasting and modelling, there are a good number of techniques that cost planners can adopt to help control project costs on a practical level including the following.

Adequate cash forecasting. One of the prevalent issues on construction projects is failing to adequately project cash requirements in the future. In many normal retail scenarios cash transaction can be swift (i.e. within minutes), but in construction there can be delays in receiving payments for such reasons as credit holidays or debtor problems.

Plan for infrequent expenses and variations. Cost planners should forecast expenses that are not due every month, such as annual insurance premiums. Variations in the project can also be a key problem in terms of cash flow; steps should be taken at the briefing stage in order to minimise the impact of variations.

Cost control. Forecasting is only half the story of the cost plan; of equal importance is the control, so the cost planner must monitor project income and expenditure efficiently.

Reduce debtor days (credit holidays). Longer credit periods for payments will inevitably lead to cash flow problems. The planner must strike the right balance between creditor days and debtor days; ideally these should be equal in order to balance out cash flow but if the payments are being made quicker than received this can create difficulties.

Just-in-time. Cash flow can be affected by materials being purchased but not used on time. Delays in valuation certificates being issued will invariably affect the cash flow profile for the project. Just-in-time ordering is a system where materials are ordered more or less on the project's critical path. Sometimes however, this is not cost effective so caution must be exercised.

6.7   Discounted cash flow techniques

Earlier in this chapter the concept of discounting as part of PV calculations was introduced. These techniques form the basis of what is known commonly in finance and economics as the discounted cash flow (DCF) technique. The Treasury define DCF as 'a technique for appraising investments. It reflects the principle that the value to an investor (whether an individual or a firm) of a sum of money depends on when it is received'. DCF methods determine the PV of future cash flows by discounting (as shown in the example for equation 7). This is necessary because cash flows in different time periods cannot be directly compared due to the time value of money notion. A typical discounted cash flow appraisal is shown in Table 6.2.

In Table 6.2, project 1 is seen to have an NPV of £2,457. This means that if we were to finance it at a rate of interest of 15%, the income that it generated would be sufficient to pay back all the borrowed money, interest on it at 15%, and there would still be a sum of money, whose PV is £2,457, left over at the end.

Table 6.2   Example DCF calculations for two investment options (at 15%).

Project 1

Year	Costs	Net Income	Net Cash Flow	PV of 1 @ 15%	DCF @ 15%
0	−£10,000		−£10,000	£1	−£10,000
1	£0	£1,000	£1,000	£0.8696	£870
2	£0	£2,000	£2,000	£0.7561	£1,512
3	£0	£3,000	£3,000	£0.6575	£1,973
4	£0	£3,500	£3,500	£0.5718	£2,001
5	£0	£3,500	£3,500	£0.4972	£1,740
6	£0	£4,000	£4,000	£0.4323	£1,729
7	£2,000	£5,000	£7,000	£0.3759	£2,632
NPV					**£2,457**

Project 2

Year	Costs	Net Income	Net Cash Flow	PV of 1 @ 15%	DCF @ 15%
0	−£12,000		−£12,000	£1	−£12,000
1	£0	£2,500	£2,500	£0.8696	£2,174
2	£0	£4,000	£4,000	£0.7561	£3,025
3	£0	£4,000	£4,000	£0.6575	£2,630
4	£0	£5,000	£5,000	£0.5718	£2,859
5	£0	£3,000	£3,000	£0.4972	£1,492
6	£0	£2,500	£2,500	£0.4323	£1,081
7	£3,000	£2,000	£5,000	£0.3759	£1,880
NPV					**£3,139**

The decision rule for this kind of analysis is that the project with the highest NPV should be chosen provided that the rate of discount used is greater than the firm's cost of borrowing.

6.7.1   Discounting

Discounting is a method used to convert future costs or benefits to PVs using an applied discount rate, and is the cornerstone of the time value of money concept. Discounting is often confused with inflation but they are separate concepts. The discount rate is therefore not the inflation rate but is the investment premium over and above the rate of inflation. In techniques such as WLCC, however, it is standard practice to exclude inflation effects. Inflation or cost escalation would only be considered where there is evidence to suggest that inflation in the prices of one element within a model would be significantly higher or lower than the other elements. In other words, the assumption is that inflation will affect all prices equally, so all values are expressed in constant prices at a given date.

For the cost planner, setting the correct discount rate is one of the key decisions, since the value is common through the whole DCF calculation. All PVs are based on the discount rate, as is the calculation of indicators such as net present value (NPV) (see section 6.7.3.1).

The discount rate is set by the client and includes the degree of risk on return required in a commercial context, or the rate of interest payable where loans are required to finance the construction work. If it is set too high, future costs will appear insignificant and will be favoured by the calculation. If it is set too low, higher capital costs will be discouraged but high operational costs may result. If inflation is taken into account in the discount rate and if rates are substantially different in practice, the calculation may lead to inappropriate choices.

6.7.2   Dealing with inflationary issues

One can use either real or nominal discount rates in the cost plan. If one uses real discount rates, each expected future cash flow is forecasted at today's prices. The discount rate used to convert these 'constant price' cash flows to their PV is based on the real rate of interest (this is the nominal rate of interest minus the expected rate of inflation) plus a risk premium. Nominal discount rates, on the other hand, see future cash flow forecasted in terms of the expected quantity of goods or services multiplied by the unit price expected to prevail at the time of the expenditure (future prices). The discount rate used to convert these cash flows to their PV is based on the nominal rate of interest (required real rate of interest plus the expected rate of inflation) plus the risk premium.

To calculate the nominal discount rate[5], if the real discount rate and inflation rate are known, one uses the following formula (11):

$$R = [(1 + r) \times (1 + i)] - 1 \tag{11}$$

where R = nominal rate of interest or discount rate, r = real rate of interest or discount rate and i = annual implied inflation rate.

In public sector projects, the Treasury *Green Book* advocates the use of a 3.5% discount rate. This calculation is based on the 'social time preference' (STPR) and is defined as the value society attaches to present, as opposed to future, consumption. STPR is a rate used for discounting future benefits and costs and is based on comparisons of utility across different points in time or different generations. The STPR has two components (HM Treasury 2003):

■ The rate at which individuals discount future consumption over present consumption, on the assumption that no change in per capita consumption is expected.

■ An additional element, if per capita consumption is expected to grow over time, reflecting the fact that these circumstances imply future consumption will be plentiful relative to the current position and thus have lower marginal utility. This effect is represented by the product of the annual growth in per capita consumption (g) and the elasticity of marginal utility of consumption (μ) with respect to utility.

The equation for calculating the STPR is therefore:

$$r = \rho + \mu \cdot g \tag{12}$$

where r = real discount rate, ρ = catastrophe risk and pure time preference[6], μ = elasticity of the marginal utility of consumption and g = output growth.

Thus the real discount rate in the *Green Book*:

$$r = 0.015 + 1.0 \times 0.02 = 3.5\%$$

6.7.3   DCF methods of appraisal

There are many methods of appraisal that lie at the disposal of the cost planner; many are esoteric for the purposes of this book, so we shall concentrate on some of the most common methods:

■ net present value (NPV);
■ internal rate of return (IRR) and the adjusted internal rate of return;
■ net savings (NS);
■ equivalent annual cost (EAC);
■ discounted payback (simple payback also covered).

6.7.3.1   Net present value

The NPV method is one of the most well-known methods of financial appraisal and is defined in the *Green Book* as follows: 'The discounted value of a stream of either future costs or benefits . . . (NPV) is used to describe the difference between the PV of a stream of costs and a stream of benefits'.

The NPV method is useful for evaluating various competing long-term projects and is used widely in WLCC in particular. It measures the excess or shortfall of cash flows, in PV terms, once financing charges are met. Strictly speaking, all projects with a positive NPV should be undertaken. However, in reality NPV will not be the only decision-making criterion; there are always other intangibles that would be considered as well (multi-criteria decision-making (MCDM) techniques are used to deal with this).

NPV is given by the formula:

$$\text{NPV} = \sum_{t=1}^{T} \frac{C_t}{(1+r)^t} \text{ or } C_0 + \frac{C_1}{(1+r)^1} + \frac{C_2}{(1+r)^2} + \frac{C_3}{(1+r)^3} + \ldots \frac{C_t}{(1+r)^t} \quad (13)$$

Example

This example will require the use of equations (7) and (13). In this example, a developer wishes to consider three short-term investment options. Each investment option has the same capital cost but different anticipated future revenues. Thus, the capital cost of each investment is £175,000[7] and the yearly cash flows are shown in Years 1 to 5 in the table below (£000):

Year	0	1	2	3	4	5
Investment Option A	(175)	10	20	56	76	37
Investment Option B	(175)	35	25	30	56	12
Investment Option C	(175)	50	25	30	50	34

When carrying out NPV calculations, the first year in which capital is invested is always referred to as Year 0. Then each year's cash flows from 1 . . . n. When carrying out the NPV calculation, the initial capital invested is treated as a negative number (shown in parentheses, as in the table).

So let us assume that we are required to calculate the NPV of each investment option at a discount rate of 7%. The following method would be adopted.

1. Using (7) or the PV (£1) table in Appendix B, calculate the PV of £1 for each year (1–5). This will give:

Year	0	1	2	3	4	5
Discount factor at 7%	1.000	0.935	0.873	0.816	0.763	0.713

2. Next, multiply each yearly cash flow by the yearly discount factor calculated in the table above.

For example:

For Option A, in Year 1 PV = £10,000 * 0.935 = £9,350

For Option B, in Year 1 PV = £35,000 * 0.935 = £32,725

3. Once you have calculated all of the PVs for each year, sum these up beginning with the initial capital outlay, so this would be (Option A):

−£175,000 + £9,350 + . . . = NPV

Phase 1

4. Complete the remaining calculations and make sure you are able to calculate NPV correctly. The preferred project is of course the one that returns the highest NPV because we are dealing with the net difference between income and outlay. Also try working out the simple payback for each project (solution at the end of the chapter).

6.7.3.2   Internal rate of return and adjusted internal rate of return

The internal rate of return (IRR) is disputed by many as simply a theoretical arithmetic result as opposed to an economic measure of performance (see AIRR). The IRR of a project can be defined as the rate of discount (r) which, when applied to the project's cash flows, produces a zero NPV, so in general terms the IRR is the value for r which satisfies the expression:

$$\sum_{t=0}^{N} \frac{A}{(1+r)^t} = 0 \qquad (13)$$

The decision rule for IRR is that only projects that have an IRR greater than or equal to a predefined cut-off point should be accepted. This cut-off rate is usually the market rate of interest (i.e. the discount rate that would have been used if an NPV analysis were undertaken instead). All other investment project opportunities should be rejected. The logic behind IRR is similar to that of NPV. The market interest rate reflects the opportunity cost of the capital involved. Thus, to be acceptable, a project must generate a return at least equal to the return available elsewhere on the capital market.

Referring back to project 1 in Table 6.2, at 15% this project has a positive NPV of £2,457. The discount rate necessary to bring this down to zero is therefore going to be greater than 15%. If a rate was applied to the net cash flow which resulted in a negative NPV, then clearly the project would be yielding a rate of return of less than that rate. By trial and error we can establish the rate which will return an NPV of zero (Table 6.3).

At 20% the NPV is barely positive; at 22% it is marginally negative. The precise rate of discount which will bring the NPV to zero can be found graphically or by interpolating between these two rates – i.e. it will lie $347/(347+356)$ of the way between 20% and 22%:

Thus, by interpolation: $20\% + [(347/703) \times (2/100)] = 20.98\%$

In this case the IRR works out at $22\% + [(308/57) \times (2/100)] = 22.94\%$.

As the decision rule in cases where the IRR is used is that the project showing the *highest* rate should be chosen (provided that the rate is greater than the investor's cost of finance), then project 2 would be the preferred option.

The adjusted internal rate of return (AIRR) is a measure of the annual percentage yield from a project investment over the project life. Like the net savings method (see next section), it is a relative calculation that needs a

Table 6.3   IRR calculations for project 1.

Project 1

Year	Net Cash Flow	DCF @ 15%	DCF @ 20%	DCF @ 22%	DCF @ 25%
0	−£10,000	−£10,000	−£10,000	−£10,000	−£10,000
1	£1,000	£870	£833	£820	£800
2	£2,000	£1,512	£1,389	£1,344	£1,280
3	£3,000	£1,973	£1,736	£1,652	£1,536
4	£3,500	£2,001	£1,688	£1,580	£1,434
5	£3,500	£1,740	£1,407	£1,295	£1,147
6	£4,000	£1,729	£1,340	£1,213	£1,049
7	£7,000	£2,632	£1,954	£1,740	£1,468
NPV		**£2,457**	**£347**	**−£356**	**−£1,286**

base case for comparison. AIRR is used to compare against the minimum acceptable rate of return (MARR), which is the smallest amount of revenue considered acceptable for an organisation to undertake a project. Typically, MARR is equal to the cost of capital plus a return, sometimes referred to as the hurdle rate.

This is generally equal to the discount rate used in the calculations. If the AIRR is greater than the MARR, then the project can be defined as economic; if less it is deemed unworthy of investment. If AIRR is equal to the discount rate, this is break-even and hence economically neutral. AIRR can be used in the same fashion as Savings to Investment Ratio (SIR). SIR is derived by the division of present value of savings and present value of investment.

The AIRR, in contrast to the IRR measure, explicitly assumes that the savings generated by the investment decisions can be reinvested at the discount rate for the remainder of the service life. If these savings could be reinvested at a higher rate than the discount rate, then the discount rate would not represent the opportunity cost of capital. IRR implicitly assumes that interim savings can be reinvested at the calculated rate of return on the project, an assumption that leads to overestimation of the project's yield if the calculated rate of return is higher than the reinvestment rate. AIRR and IRR are only the same if the investment yields a single, lump sum payment at the end of the service life, or in the unlikely case that the reinvestment rate is the same as the IRR.

As discussed earlier, some people dispute IRR as a performance measure in that more than one rate of return may make the value of the savings and investment streams equal, as required by the definition of the internal rate of return. This may be the case when capital investment costs are incurred during later years, giving rise to negative cash flows in some years. The formula for calculating AIRR is:

$$\frac{\sum_{t=0}^{N} S_t (1 + r)^{N-t}}{(1 + i)^{N}} - \sum_{t=0}^{N} \frac{\Delta I_t}{(1 + r)^t} = 0 \qquad (14)$$

where S_t = annual savings generated by the project, reinvested at the reinvestment rate, r = discount rate and $\Delta I_t \div (1 + r)^t$ = PV investment costs on which return is to be maximised.

6.7.3.3 Net savings

The net savings (NS) method calculates the net amount in PV terms that an investment decision is expected to save over the specified time period. As NS is expressed in PV terms, it represents savings over and above the amount that would be returned from investing the funds at the minimum expected rate of return (i.e. the discount rate). The NS for a project, relative to a designated base case, is calculated by simply subtracting the cost of the alternative project under consideration (X_2), from that of the base case (X_1).

$$NS = X_1 - X_2 \qquad (15)$$

Generally, if the NS value returned is greater than zero, then the project under consideration is economically cost effective relative to the base case. The use of NS can also be extended to individual cost differences between the base case and the alternative. Examples such as capital costs, maintenance cost, etc. can benefit from this. However, this does require additional calculations compared with the simple method above, but it is useful as this value is needed in the calculation of other methods such as savings to investment ratio (SIR) and AIRR. Calculating NS using individual cost differences is useful as a check to ensure that SIR and AIRR calculations are based on correct intermediate calculations. That is, the NS should be exactly the same whether computed by the comparison of projects or by using individual cost differences. For the latter, the following equation can be employed in the calculation of NS:

$$NS_{X_1:X_2} = \sum_{t=0}^{N} \frac{S_t}{(1 + r)^t} - \sum \frac{\Delta I_t}{(1 + r)^t} \qquad (16)$$

where $NS_{X_1:X_2}$ = NS, in PV, of the alternative (X_2), relative to the base case (X_1), S_t = savings in year t in operational costs associated with the alternative, ΔI_t = additional investment related costs in year t associated with the alternative, t = year where base date $t = 0$, r = discount rate and N = number of years in study period.

6.7.3.4 Equivalent annual cost

At the early stages of a briefing exercise, the possible investment projects under consideration may be of unequal time periods. Where this is the case, simply comparing the NPV of each project is incorrect because there is a possibility that the net cash inflow generated from the shorter duration project could be rein-vested elsewhere for the remaining time difference, thus generating additional NPV which may total more than other project NPV. This fact is often ignored as only the NPV of projects at the end of their respective lives is compared.

Where this is the case (and assuming that each project carries the same level of risk), then the equivalent annual cost (EAC) approach can be used (Idowu 2000), as follows.

EAC calculates the PV of costs for each project over time *t*, expressing the PV in an annual equivalent cost using the appropriate annuity factors for each cycle. The annual equivalent of NPVs of the two or more projects can then be compared. Having calculated the EAC for each cycle and each pro-ject, then compare the EACs. The project that has the lowest EAC over the cycles is the better one if lowest outlay is the objective, or the higher EAC would be preferred if the highest revenue were the objective. The use of EAC is generally in very early stage project feasibility assessments rather than in conventional cost planning, but is closely related to the concepts of NPV so cost planners should be aware of the power of EAC when dealing with unequal project life spans for the purposes of comparison.

6.7.3.5   Discounted and simple payback

Both simple payback and discounted payback are the most basic measures of the amount of time it takes to recover the initial investment in capital. Both are expressed as the period of time that has elapsed between the beginning of the study period and the time at which cumulative savings are just sufficient to cover the initial capital cost of the investment decision. An exemplar devel-opment appraisal using simple payback is shown in Figure 6.1. Payback is rarely used in contemporary cost planning since the problem with both mea-sures is that they are not valid for comparing multiple, mutually exclusive project alternatives, nor for ranking alternatives. The irony of this is that cognitively, many of us consider payback when we purchase something (i.e. a new domestic wind turbine or domestic photovoltaic system). As a rule of thumb, it is best employed as a screening method for projects that are so clearly economical that a fuller cost planning exercise of the project is uneco-nomical and unwarranted. The general formula for calculating discounted payback on an option appraisal basis is given by:

$$\sum_{t=1}^{y} \frac{(S_t - \Delta I_t)}{(1 + r)^t} \geq \Delta I_0 \tag{17}$$

where y = minimum length of time over which future net cash flows have to be accumulated in order to offset initial capital cost investment, S_t = savings in operational costs in year t associated with an alternative project, ΔI_0 = initial investment costs associated with the alternative, ΔI_t = additional investment related costs in year t, other than investment costs, r = discount rate and t = time.

PROJECT: Office Development, Sometown
Estimated income
Gross area (m²)   4200
Lettable %   80
Annual rental (£/m²)   210.00

Annual income = £705,600
Area × Rental × Lettable % /100

Estimated costs
Land cost   1,600,000   £1,600,000
Fees, duties, etc. %   3   £48,000 *Land cost × Fees etc. /100*

Total land cost = £1,648,000 *Land cost ×*
(1 + Fees etc. /100)

Building cost (£ per sq m)   1,125.00   £4,725,000 *Building cost × Gross floor area*
External works   70,000   £70,000
Demolition   30,000   £30,000
Prof. fees etc. %   12   £579,000 *(Building + Externals + Demolition) ×*
                                                         Prof. fees /100

Total construction cost = £5,404,000
(Building + Externals + Demolition + Prof fees)

Pre-constr'n period (yrs)   1.00
Construction period (yrs)   1.83
Interest rate % (p.a.)   7.50

Land interest   £386,184   *Interest on total land cost over the entire*
                                                     development period compounded quarterly
Construction interest   £380,910   *Interest on total construction cost assumed over half*
                                                     construction period compounded quarterly
Charges, legal, etc. %   1   £70,520 *Percentage charge on total loan,*
                                                         i.e. Constr'n cost + Land cost

Total finance cost = £837,614

Marketing & agents'   20   £141,120 *One-off costs based on percentage of*
fees etc. %   *annual rental income*

Total letting cost = £141,120

Grand total cost = £8,030,734 *Land + Building*
+ Finance + Letting

Notional payback period

Period (years) = 11.38 *Grand total cost/Annual*
income

Figure 6.1   Example of development appraisal calculation.

(margin, rotated) **Phase 1**

6.8   Key points

- Due to our preference for money today rather than the same sum in the future, cost planning is governed by the time value of money concept.
- Discounting is the process of recognising this phenomenon.
- Most standard methods of appraising and planning the costs of projects (particularly long duration schemes) use discounting methods, such as the discounted cash flow techniques.
- The cost of borrowing money is quantified by the interest rate.

Notes

(1)   *Ceteris paribus* is the classical economic term meaning other things being equal.

(2)   The opportunity cost is the cost of something in terms of an opportunity forgone (and the benefits that could be received from that opportunity). As an example, consider a local authority (LA) that wishes to procure a new social housing scheme on some land owned by the LA. The opportunity cost is the value that is forgone in selecting this option. So, in procuring the new housing scheme, the local authority has forgone the opportunity to procure something else.

(3)   Compounding describes the situation where a sum of money (principal) is invested for a number of years and it is assumed that interest earned each year is immediately added to the principal and re-invested on the same terms, this process being repeated annually.

(4)   Discounting is a technique used to compare costs and benefits that occur in different time periods. It is a separate concept from inflation, and is based on the principle that generally people prefer to receive goods and services now rather than later. This is known as time preference (HM Treasury 2003). Detailed consideration of discounting appears later in this chapter.

(5)   If a nominal discount rate is used, then future cash flows are escalated by a factor (known as the relative price of escalation). The FVs are then said to be presented as prices expected to prevail at the time of the expenditure. For example, if one assumes a real discount rate of 3.5% and an annual inflation rate of 2.5%, then the nominal discount rate is 6.09%. If one assumes that future cash flows are going to escalate at the same price as inflation, then one uses a relative escalation rate of 2.5%. If one expects the cash flows to escalate faster than the inflation rate, then one uses a higher relative price of escalation (e.g. 5%).

(6)   From the Treasury *Green Book* (2003): 'The first component, catastrophe risk (L), is the likelihood that there will be some event so devastating that all returns from policies, programmes or projects are eliminated, or at least radically and unpredictably altered. Examples are technological advancements that lead to premature obsolescence, or natural disasters, major wars, etc. The scale of this risk is, by its nature, hard to quantify. The second component, pure time preference (δ), reflects individuals' preference for consumption now, rather than later, with an unchanging level of consumption per capita over time. The evidence suggests that these two components indicate a value of around 1.5 per cent a year for the near future.'

Phase 1

(7)   It is conventional to show the initial capital outlay in Year 0 in parentheses or as a negative number.

Further reading

Arnold, J., Hope, T., Southworth, A. and Kirkham, L. (1994) *Financial Accounting*. Prentice Hall, Harlow.

Myers, D. (2004) *Construction economics: A new approach*. Spon Press, London.

References

HM Treasury (2003) *The Green Book*. http://greenbook.treasury.gov.uk/index.htm

Blyth, K. and Kaka, A. (2006) A novel multiple linear regression model for forecasting S-curves. *Engineering, Construction and Architectural Management*, 13(1), 82–95.

Boussabaine, A.H. (2006) *Cost Planning of PFI and PPP Building Projects*. Taylor & Francis, London.

Boussabaine, A.H. and Elhag, T.M.S. (1999) Applying fuzzy techniques to cash flow analysis. *Construction Management and Economics*, 17(6), 745–55.

Idowu, S.O. (2000) *Capital investment appraisal – Part 2*. Association of Chartered Certified Accountants, London.

Kaka, A.P. and Lewis, J. (2003) Development of a company-level dynamic cash flow forecasting model (DYCAFF). *Construction Management and Economics*, 21(7), 693–705.

Lam, K.C., Hu, T., Ng, S.T., Yuen, R.K.K., Lo, S.M. and Wong, C.T.C. (2001) Using an adaptive genetic algorithm to improve construction finance decisions. *Engineering, Construction and Architectural Management*, 8(1), 31–45.

Solution to example in section 6.7.3.1.

Year	0	1	2	3	4	5
Cash flow	(100)	30	30	30	50	10
Discount factor at 12%	1.000	0.893	0.797	0.712	0.636	0.567
Present value	(100)	26.79	23.91	21.36	31.80	5.67

Thus, NPV = £9,530 and Payback = 3.2 years (−100 + 30 + 30 + 30 + 10) so 3 years and 10/50 which is 0.2 of a year. Also can be given as 3 years and 2.4 months.

The preferred project on financial grounds is Project C as the NPV is the highest. This means it generates the required return of 12% plus £9,530 more. Project B pays back most quickly, but would actually destroy value as the NPV is negative, i.e. it does not even give the required 12% return. So it would be financially better to invest the money in Project C as it gives the highest return, measured as NPV, of the three options. It has a slightly longer payback than Project B, but shorter than Project A.

Chapter 7
Whole Life-cycle Costing and Design Sustainability

7.1 Introduction

Whole life-cycle costing (WLCC) is a methodology that involves the systematic analysis of the long-term cost implications of procurement decisions, whether for new capital projects or in the development of asset maintenance strategies for existing buildings. WLCC studies take into account not only the upfront capital costs of a project, but also the costs that will accrue throughout the life of the building, such as maintenance, energy and finance costs. In summary, WLCC is concerned with assessing the cost of an asset 'from cradle to grave'.

This chapter will explore the concept from first principles and explain the methodology and application of appropriate mathematical and financial tools. The issues relating to risk and uncertainty will also be developed, with some practical examples of WLCC modelling. The use of the formulae in Chapter 6 will be demonstrated and the impact of government procurement regulations on WLCC decision-making will be considered. It is perhaps pertinent to begin this chapter with some disambiguation, since many believe that WLCC is basically the same as the more well-known life-cycle costing (LCC); recent research suggests that this is not the case.

7.2 History of WLCC

The underlying thinking behind the 'whole life' movement as an appraisal methodology has enjoyed a relative renaissance within the construction and civil engineering sectors since the late 1980s. Renaissance is possibly an apposite depiction given that the fundamentals of WLCC within the UK are entrenched in the early 1970s 'terotechnology' initiative carried out by the then Department of Industry. This report became the focus of significant interest within both academia and industry. Major manufacturing companies assumed a lead role in the development of terotechnology processes, with organisations such as Rolls Royce and Royal Ordinance being prominent. The 1990s witnessed the crystallisation of this work by defining LCC in BS 3843 (1992), as well as a raft of publications dealing with specific industry themes. Whilst BS 3843 brought an element of standardisation to the process in the UK, it did not act as the catalyst that many observers expected. Concurrently, the use of LCC techniques in the USA developed rapidly through the US Federal Energy and US Department of Defense procurement programmes. Current research and practice has led to WLCC, the principal differentiation being the movement from a static methodology to a dynamic asset management tool. Notwith-standing, the construction industry perception of WLCC is that of a politically driven albatross hung around the necks of practitioners, or of a nebulous concept with little value to procurement decision-making[1]. Clearly, further work is required to challenge this common perception.

The catalysts behind WLCC decision-making are numerous but dissipative; the sustainable design movement has irrevocably entrenched the need for a whole life or holistic approach to evaluating built asset procurement. Within the UK, government policy has aligned itself with whole life thinking through a tranche of policy and legislative mechanisms issued by the OGC, with the finer details of economic fiduciary covered by the Treasury Green Book. The current situation within the construction sector is confusing and variable. Prac-titioners have argued that increased take up of WLCC methods is frustrated by the lack of standardisation (save for BS 3843 (1992)) and no common cost breakdown structure (CBS). The latter issue would appear to be a central facet of work in ISO 15686-5: Whole Life Costing, which is yet to be published.

Since the publication of BS 3843 and ISO 15686-1, LCC has been the focus of significant research investigation, not just within the UK but internationally (particularly within the EU and USA). The various problems that were associ-ated with early LCC models, such as complexity, uncertainty (due to the nature of forecasting) and data collection, have been addressed in the new genera-tion WLCC models. These issues will be considered later in the chapter.

Presently, WLCC is not standardised either in BS or ISO format, although ISO 15686-5: Whole Life Costing is at the time of writing in draft. Conse-quently, there are various definitions and interpretations of WLCC in prac-tice, although these do not differ significantly from a theoretical perspective.

7.3 Definitions and disambiguation

BS 3843 (1992) defines LCC as:

The costs associated with acquiring, using, caring for and disposing of physical assets, including feasibility studies, research and development, design, production, maintenance, replacement and disposal; as well as all the support, training and operations costs generated by the acquisition, use, maintenance, and replacement of permanent physical assets.

In 2000, this definition was revised and incorporated into ISO 15686-8 Part 1: Service Life Planning, which cites LCC as:

A technique that enables comparative cost assessments to be made over a specified period of time, taking into account all relevant economic factors both in terms of initial capital costs and future operational costs.

At present, work is underway to standardise the definition of WLCC; this is due to be published in ISO 15686-5 in due course. Boussabaine and Kirkham (2004), in their work on WLCC defined it as:

. . . a dynamic and ongoing process, which enables the stochastic assessment of the performance of constructed facilities from feasibility to disposal. The WLCC assessment process takes into account the characteristics of the constructed facility, reusability, sustainability, maintainability and obsolescence as well as the capital, maintenance, operational, finance, residual and disposal costs. The result of this [risk-based] assessment forms the basis for a series of economic and non-economic performance indicators relating to the various stakeholders' interests and objectives.

The key difference between LCC and WLCC is the notion that the latter is a management tool that is used throughout the building life, rather than the static option appraisal tool that LCC is generally used for[2].

In the draft version of ISO 15686-5 (dated 2006), the definition of whole life [cycle] cost is given as:

The systematic economic consideration of all agreed significant costs and benefits associated with the acquisition and ownership of a constructed asset, which are anticipated over a period of analysis expressed in monetary value. The projected costs or benefits may include those external to the constructed asset and/or its owner (note may include finance; business costs; income from land sale).

So, whilst it is clear that the debate continues about the definition of the concept, the basic principles, which are described in this chapter, are undisputed. The key message is that best practice in cost planning now requires, where appropriate, a systematic consideration of the whole life costs of the asset, rather than simply the upfront capital costs.

7.4   Applications of WLCC – the public sector perspective

Within the UK public sector, WLCC should be taken into account in all business cases, which aim to justify capital investment in construction. This applies to projects financed by traditional public capital as well as through the PFI and PPP procurement routes. The tangible effects of this change in procurement can be seen, for example, in the NHS ProCure 21 strategy. ProCure 21 promotes the better use of NHS assets and resources to achieve the right buildings and equipment, in the right place, in the right condition, of the right type, at the right cost (from both capital and whole life points of view), at the right time whilst facilitating effective response to future needs of the service with minimal impact on the environment. The ProCure 21 programme incorporates WLCC models in the tendering process for its frameworks and requires specific models to be completed for each NHS scheme subsequently undertaken by the framework contractors in England. These models have helped the NHS to make significant steps forward in attaining better value for money in capital procurement.

The OGC has developed an established suite of guidance notes and tools encompassing project management and sustainability, through the Achieving Excellence in Construction Successful Delivery Toolkit[3]. The Achieving Excellence in Construction programme was launched in March 1999 to improve the performance of central government departments, their executive agencies and non-departmental public bodies (NDPBs) as clients of the construction industry. The strategy promotes a sustained improvement in construction procurement performance and in the value for money achieved by the UK government on construction projects, including those involving maintenance and refurbishment.

The Achieving Excellence initiative set out a route map with challenging targets for government performance under four headings: management, measurement, standardisation and integration. Targets included the use of partnering and development of long-term relationships, the reduction of financial and decision-making approval chains, improved skills development and empowerment, the adoption of performance measurement indicators and the use of tools such as value and risk management and whole life costing.

A central theme of Achieving Excellence is the delivery of construction products under best value for money conditions. This is not the lowest capital cost but the best balance of quality and whole life cost to meet the client's brief.

The National Procurement Strategy (NPS) sets out how local authorities can improve the delivery and cost effectiveness of high quality services through more effective, prudent and innovative procurement practices. The strategy illustrates the scope for potential cost savings through more efficient procurement procedures and partnership working and this is reinforced in page 50 of the report, which examines ways to 'achieve community benefits through procurement':

Adopt whole life costs and benefits as your contract award criteria. Procurement strategies and contract standing orders should establish 'the optimum combination of whole life costs and benefits to meet the customer's requirement' as the best value contract award criteria.

The summary checklist at the back of the report also reminds local authorities of the need to ensure that whole life-cycle costs have been taken into account during the various stages of the procurement process.

The Carbon Trust (CT) and the Energy Saving Trust (EST) have strongly advocated the use of WLCC methodologies to support a more sustainable environment, particularly in the construction sector. In January 2004, the EST published a report on the costs and benefits of community heating, which outlines the ideas and advantages behind the scheme. The guide also details the use of WLCC with a live example of how community heating compares to other forms of heating. The Perthshire Housing Association secured planning permission for a town centre site to be redeveloped into a four-storey single block containing 32 units, with a grant of £2,070 from the Community Energy programme which helped to finance production of a brief and report into the suitability and scope of a gas-fired community heating system. The development work undertaken indicated that community heating delivered the lowest WLCC option for the Association. A subsequent application for a Community Energy capital grant was successful.

Whilst the virtues of WLCC are now well versed, the implementation of thorough WLCC methodologies is still complex. This was recognised in a recent Low Carbon Buildings event organised by the Government Office for London[4] and DEFRA, which examined some fundamental issues with WLCC including:

- What are the main barriers to making more public procurement consider whole life costing principles? What action does the government need to take to ensure public funds are spent on developing high quality, sustainable buildings?
- What is the potential for improvements in the carbon management of public sector buildings at local and regional level? How can improvements best be ensured? What role is there for WLCC?

The key to success lies in clear briefing and involvement with all stakeholders from the outset.

7.5   WLCC beyond construction

Outside of construction procurement, WLCC has been applied successfully in the procurement of assets within the university sector under the Joint Procurement Policy and Strategy Group (now known as Proc-HE), where a basic spreadsheet model has been developed to help universities appraise the procurement of large assets such as laboratory equipment. The procurement of defence systems and equipment within the Ministry of Defence features the use of WLCC methods, particularly within the Acquisition Management System (AMS). Interestingly, CarCost is a motor industry whole life cost programme which allows consumers to compare the relative costs of car ownership; this has been exploited by Skoda UK who use a basic WLCC model on their consumer website to demonstrate the costs of various models against comparables. This final example is often used as the classic metaphor for engendering WLCC thinking; many purchases we make in everyday life are intuitively based on the concept. When we buy a car, for example, we don't think just about the capital costs, but also fuel consumption, maintenance and servicing costs, road tax, insurance, etc.

7.6   Private sector procurement

It is argued that developers and other speculative investors who account for a significant proportion of the construction industry output are not so readily accepting of the WLCC concept. The reasons given for this are based on the speculative investor's time horizon and profit maximisation. This is the standard neo-classical assumption that investors will seek to maximise profits by producing and then selling an output in the open market. In public sector procurement, this is often not the case as there are socio-economic, welfare and other intangibles that are seen as derivates from the investment, i.e. new hospital creates better healthcare, new prison reduces offending, etc. The difficulty lies then in convincing investors that the return on investment should be significantly increased by procuring buildings which are based on optimised WLCC – this element of future-proofing the building, as it were, adding value.

Of course, some larger private sector clients will require evidence of WLCC at the early stages of a project and will formally request it as part of a package of support options provided by the cost planner. Usually, risk and value management strategies take into account WLCC as part of a comprehensive cost planning service, particularly on multi-million pound schemes such as PPP/PFI.

7.7   Theory and methodology

It is important to understand that the fundamental theory underlying WLCC is not new, nor are the mathematical constructs used to calculate costs, cash flow and other financial measures (these techniques have been described in Chapter 6). Nevertheless, the key to successful application of WLCC lies in four areas:

- appropriateness of the methodology;
- quality of cost and performance data used;
- appropriate treatment of risk and uncertainty;
- involvement of the project stakeholders.

These caveats are important since purposeful WLCC cannot be achieved through the exclusive use of an off-the-shelf piece of software. Although in practice software does exist, it negates the importance of stakeholder involvement; such involvement is essential in ensuring that the models generate accurate results, and decision information that is informative and suitable. This briefing stage of WLCC can, depending on the complexity of the work, take anywhere between one day and twelve months to achieve. It is worth dedicating suitable resources at this point, since it reduces the possibility of reworking and model redesign later in the process.

Ferry and Brandon's three phases of cost planning, described in Chapter 3 of this book, give a useful guide to understanding the process of effective WLCC planning. In Figure 7.1 the WLCC methodology is characterised by the three phases of design detail, each level representing a decision stage (Kirkham 2005):

- strategic (concept) level (i.e. structure, envelope, services, etc.);
- system level (i.e. steel, concrete, timber frame, etc.);
- detailed level (i.e. concrete pre-cast or in-situ, RC grade, etc.).

The strategic-level stage involves WLCC modelling in the broadest sense, looking at the building design in its entirety and not detail-specific. This appraisal should be used initially to assess the substantially differing design solutions that are presented to the client at the briefing stage. It is at this juncture that the design team can begin to focus the client on WLCC, helping to deliver a cost-effective solution.

The results from this analysis, and any subsequent data extrapolation, form the basis to stage 2, where a more detailed system level analysis can be performed on the solution(s) identified in stage 1. Here, WLCC methods can again be used to assess the economic viability of various systems within the design, such as the WLCC comparison of steel, concrete and timber frames.

The data elicited from this stage should again provide the design team with the key information necessary to develop the design to stage 3, where component-specific WLCC analyses can be performed. At this stage, it is rare that WLCC methods are used on all design selections. A technique known as cost significance[5] (Saket 1986, Munns and Al-Haimus 2000) is proposed by some (in the context of traditional estimating and BQ preparation). It allows the planner to identify the cost items which contribute significantly to the overall cost (this can naturally be extended to WLCC), and this reduces the methodological effort by focusing on the major cost items and not, for example, on smaller insignificant items like doors and ironmongery.

This principle is quite similar to that of sensitivity analysis[6]. Usually, a sensitivity analysis will be performed in stage 2 to identify the most uncertain costs. This then allows the design team/analyst to focus detailed level WLCC analyses on the cost sensitive items. The complexity of the WLCC modelling required at stage 3 depends to a significant extent on the availability of cost and performance data, as well as the uncertainty attached to any assumptions.

Phase 1

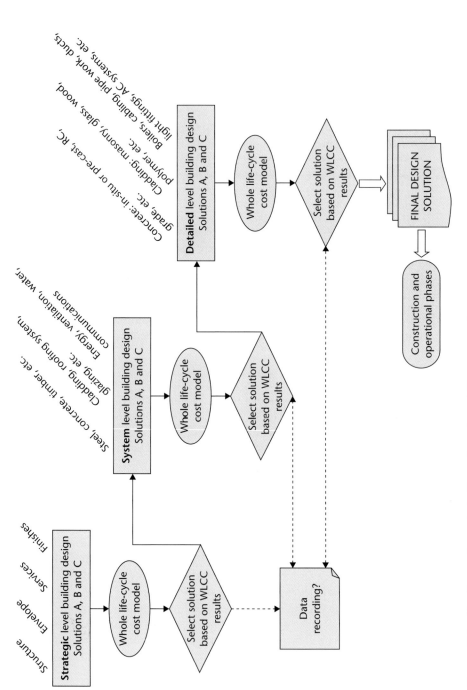

Figure 7.1 The three-phase whole life-cycle cost modelling approach.

Pareto's 80/20 law can often be applied to WLCC models in order to determine the focus. The law states that 20% of cost items are responsible for 80% of the total cost, and focusing on items that do not contribute significantly to cost is to be avoided.

7.7.1   Calculations and numerical methods

There is a variety of numerical approaches to the evaluation of WLCC, but the underlying theory is similar in most cases. In the most simplistic format, WLCC is given by the following formula:

$$WLCC = C_c + O_c + R_c \tag{1}$$

where C_c = capital cost, O_c = operational cost, and R_c = residual cost.

The cost centres identified in Figure 7.2 can be easily be grouped into the three categories described in equation (1), although this basic equation can tend to mask the complexity of the process overall. In complex models, the number of costs centres can be significantly high and it will be at the discretion of the cost planner to determine what should and should not be included within each cost centre. ISO 15686-5 is attempting to standardise

Phase 1

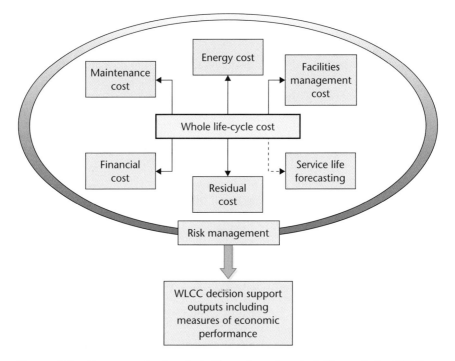

Figure 7.2   Components of a whole life-cycle cost analysis (Boussabaine 2004). Reproduced with permission from Blackwell Publishing.

this, although good WLCC models must be flexible enough to cope with a wide range of building types and structures. It is also worth remembering that this equation is constrained by the time value of money and as such, future operational and residual costs will have to be discounted to today's value. To demonstrate a very basic WLCC model, the costs of a cladding system are considered here.

Example

It is desired to compare the whole life costs of two types of cladding to a factory building, whose life is intended to be 40 years. The rate of interest allowed is 3% per annum compound.

Whole life-cycle costs of cladding A
Cladding A will cost £1,000,000, will require redecorating every 4 years at a cost of £120,000, and will require renewing after 20 years at a cost of £1,400,000.

			£
Capital cost			1,000,000

Present value at 3% of:

Redecoration after	4 years £120,000	at 88.8p =	106,560
Redecoration after	8 years £120,000	at 78.9p =	94,680
Redecoration after	12 years £120,000	at 70.1p =	84,120
Redecoration after	16 years £120,000	at 62.3p =	74,760
Renewal after	20 years £1,400,000	at 55.4p =	775,600
Redecoration after	24 years £120,000	at 49.2p =	59,040
Redecoration after	28 years £120,000	at 43.7p =	52,440
Redecoration after	32 years £120,000	at 38.8p =	46,560
Redecoration after	36 years £120,000	at 34.5p =	41,400
			2,335,160

Whole life-cycle costs of cladding B
The alternative cladding B will cost £1,800,000, and will last the life of the building without any maintenance, although a sum of £300,000 is to be allowed for general repairs after 20 years.

			£
Capital cost			1,800,000

Present value at 3% of:

Repairs after	20 years £300,000	at 55.4p =	166,200
			1,966,200

Saving by using Cladding B is therefore £2,335,160 minus £1,966,200 = £368,960. It would therefore appear justifiable to use the initially more expensive Cladding B, as this will prove much the cheaper in the long run. Note that the PV method of discounting has been used.

This principle can be applied to as many components as required, although it can easily be appreciated how complex the model would be for an entire building.

7.7.2   The effect of assumptions

In order to demonstrate the effect of quite small misjudgements of the future, the example relating to claddings A and B is now recalculated assuming some slight differences to the assumptions previously made. All these differences lie well within the range of error to be expected with careful estimates, and exclude either inflation or any other significant anomaly.

Revised example

The rate of interest allowed is 4% per annum compound, an increase of 1%.

Revised whole life costs of cladding A
Assume that cladding A is redecorated every 5 years instead of every 4 years, at a cost of £100,000, and lasts for 25 years instead of 20, costing £1,200,000 to renew.

			£
Capital cost			1,000,000

Present value at 4% of:

Redecoration after	5 years £100,000	at 82.2p =	82,200
Redecoration after	10 years £100,000	at 67.6p =	67,600
Redecoration after	15 years £100,000	at 55.5p =	55,500
Redecoration after	20 years £100,000	at 45.6p =	45,600
Renewal after	25 years £1,200,000	at 37.5p =	450,000
Redecoration after	30 years £100,000	at 30.8p =	30,800
Redecoration after	35 years £100,000	at 25.3p =	25,300
			1,757,000

Revised whole life costs of Cladding B
Assume that cladding B has to be repaired after 15 years and again at 30 years, instead of only once at 20 years.

			£
Capital cost			1,800,000

Present value at 4% of:

Repairs after	15 years £300,000	at 55.5p =	166,500
Repairs after	30 years £150,000	at 30.8p =	61,600
			2,028,000

Saving by using cladding A is therefore £2,028,100 minus £1,757,000 = £271,100.

This calculation gives a completely different result to the original; the initially cheaper cladding proves to be much cheaper in the long run also. It would still have some remaining life at the end of 40 years if it were decided to keep the building in commission for a longer period.

These simple examples demonstrate the use of WLCC techniques at the sub-element level. Typically, however, owing to the magnitude that WLCC models would take if they were to assume every single element in the building, a more generalist model is used, often based on the element level BCIS Standard Form of Cost Analysis (SFCA) structure.[7] WLCC is determined by the NPV method in this case (i.e. the sum of all PVs over the study period).

7.7.3   Uncertainty quantification in WLCC

Like almost any other type of future forecasting, WLCC has always been exposed to the criticism that decisions based upon the method are not reliable. This assessment manifests throughout earlier research (Bird 1987), and this was certainly the case in the earliest applications of the technique, particularly terotechnology, cost-in-use and latterly, LCC. The response to this has been clear and the latest generation of WLCC research demonstrates a clear metamorphosis from deterministic to probabilistic (stochastic) approaches. The benefits of a stochastic approach to WLCC lie in the treatment of uncertain parameters within the model. Where before, LCC models relied on deterministic 'point estimates' of future costs, stochastic techniques allow the analyst to model likely uncertain cost as probability distributions. The plethora of distributions available and the techniques used to construct them allow the analyst to represent the uncertainty of the variable, and perhaps more importantly, quantify it. A recent EPSRC project in the UK[8] investigated the development of a web-based risk simulation tool that can interact with whole life and risk data, with the intention of enabling designers to minimise the cost of uncertainty and achieve designs for optimum whole life cost and performance. One of the principal ideas behind the project was to develop a framework for effective data collection within a probabilistic environment, which could then interface directly into a WLCC model. Furthermore, the project investigated the interface between the tool and the designer to achieve a high level of acceptability. It is anticipated that the final WLCC model will assist in educating the design team about the need to manage risk more systematically across the whole life of a project.

7.7.4   Service life and performance modelling

Whilst a strong focus on uncertainty quantification has characterised WLCC research recently, the necessity for integrated building and building component service life prediction has been largely ignored. This is significant given the impact of ISO 15686, which focuses on the need for standardised service life prediction models. The requirement for service life prediction technologies lies predominately in the estimation of maintenance and life-cycle replacement times. It is here that service life models can provide information, either stochastically or deterministically, on likely deterioration and failure of components. The importance of service life assessment, coupled with the emphasis on whole life-cycle costs during the operational life, is driving the promotion of international standards in service life prediction.

7.8   Disadvantages of whole life costs assessment

We have discussed the benefits of WLCC within cost planning; indeed the advantages of the technique for comparing costs are self-evident: it enables us to consider the long-term implications of a decision and to provide a way of showing the cost consequences of short-sighted economies.

Unfortunately there are a number of fundamental disadvantages, which explain why this technique for comparing the cost of alternative materials and constructions has been seen more often in textbooks and in the university examination room than in real life. These disadvantages may be expressed through three lines of argument:

- initial and running costs cannot be equated;
- the future cannot be forecast;
- the process is heavily dependent on data and is overtly complex.

7.8.1   Initial and running costs cannot really be equated

This is because on profit developments:

- Where the building is to be sold, the maintenance charges will fall on the purchaser and so are of little importance to the developer, who is responsible for the construction costs alone.
- Where the building is to be let or used commercially, the repair and maintenance costs are deducted from the receipts in calculating profit for the year, and are therefore paid out of income before taxation. However, money spent at construction stage has to be raised as part of the capital cost and is eventually repayable. Repayment of, or interest on, capital expenditure is not normally deductible from receipts for the purposes of tax calculations.

In addition, on social developments:

Phase 1

- Even with publicly owned buildings, it is an advantage to pay maintenance out of running costs instead of incurring a heavier capital debt due to high construction costs.
- In some instances, such as schools, the bulk of the construction costs may be paid by one authority while another authority will be responsible for running costs, so that there will be little incentive to provide an unduly high standard of building with a view to subsequent saving.

On all developments:

- Money for capital development is normally more difficult to find, and is subject to more constraints, than money for current expenditure.
- Although a building may still be perfectly sound halfway through its planned life, it may be too old-fashioned in design and accommodation to do the job that is required of it in modern conditions.
- The same applies to expensive but durable finishes and fittings, which although still in good condition may give a very old-fashioned appearance to a building. Old joinery, shopfronts, tiling and other finishes or fittings may be ripped out long before they are life-expired, especially in the competitive world of commerce.
- In comparing figures of increased capital expenditure against future costs of repair and renewal, it must be remembered that once the money has actually been spent it is not possible to amend the decision in the light of future developments. As a domestic example of this, if very expensive finishes are chosen for a house to save repainting, and after a few years the owner becomes short of money, the annual interest on the expensive house still has to be met, whereas if the owner had opted for a cheaper house the redecorating could have been deferred for a year or two. (The house could also be redecorated in the latest fashion each time, if the owner so wished.) It is questionable practice to restrict the actions of future generations by committing them to high interest and repayment costs; this is just as bad as the other extreme of committing them to inflated running costs and maintenance costs by unduly low standards of design and construction.

7.8.2   The future cannot really be forecast

- While present-day capital costs can be estimated quite accurately, the cost of maintenance and other operational costs are not quite so straightforward.
- The amount of money spent on planned preventative maintenance of a building is determined far more by the current policy of the body responsible for the maintenance than by any quality inherent in the materials. Some owners will redecorate every few years, mend or replace worn or damaged work immediately and continuously carry out a policy of minor

improvements; others will spend the very minimum necessary to keep the building in operation.

- Major expenditure on repairs is usually caused by unforeseen failure of detailing, faulty material or bad workmanship, rather than by predicted overall ageing, and so is almost impossible to forecast. A well-designed and maintained piece of cheap construction might last much longer than its theoretical life, while some quite expensive work could require early renewal because of, say, entry of water at a badly designed joint.

- Interest rates, which in general tend to reflect the current minimum lending rate (MLR), cannot be forecast with any certainty, particularly over long periods of 20 years or more; remember that net interest rates are also affected by changes in taxation. Between 1988 and 1990 the MLR, although forecast by the Treasury to remain static or fall, actually increased from 8% to 15%; and that was during a period of only two to three years. Would you like to guess what the Bank of England (or the European Bank) will do in the year 2025?

- During the last 40 years there has been almost continuous economic inflation, and any building or maintenance work is likely to cost several times what would have been estimated, say, 20 years ago.

7.8.3   The process is heavily dependent on data and is overtly complex

- Clearly WLCC is most effective when good quality data is available.
- However, there are mathematical approaches that can be taken to evaluate the assumptions where no hard data exists.
- The use of probability distributions is appropriate here. For example, the Building Maintenance Information service (BMI) publishes service life data in the form of minimum, maximum and most likely values based on surveyors' experiences. Translating this information to a triangular distribution is quite straightforward.
- The complexity of the model should be managed by the analyst. The cost significance techniques that have been discussed are appropriate for this.

7.9   Software tools for WLCC

The momentum created by the WLCC paradigm has led to the emergence of various software solutions from the universities and professional practices. This section explores several systems. But it is important to remain mindful of the fact that good WLCC is not based simply on software, but also on close engagement with the client during the briefing stage; it is at this point that decisions are made which have a major impact on WLCC.

Phase 1

7.9.1   WLC comparator

Developed by the Building Research Establishment, WLC Comparator is a simple tool that has been designed to calculate the whole life-cycle costs of building components and elements. It provides a platform to assess the basic criteria for capital investments in an effort to ensure a reduction of future operating costs. The analytical approach enables WLCC to be performed at building component and element level. The model captures the economic analysis needed to improve the capital investment decision, optimise asset selection or design, and predict the best combination of interdependent systems. The model adopts the Treasury complaint standard and accepted accountancy rules for predicting the PV of future income streams, and supports ISO 15686. The tool is also recommended in OGC *Achieving Excellence in Procurement Guide No.7: Whole Life Costing* (OGC 2007).

7.9.2   EuroLifeForm

EuroLifeForm (European Life Performance), 'Probabilistic approach to life-cycle costs and performance of buildings and civil infrastructure', was a research project undertaken with the financial support of the European Commission and Taylor Woodrow Construction under the EC 5th Framework Competitive and Sustainable Growth Programme (G1RD-CT-2001-00497), concluding in 2004. This work focused on service life prediction within the WLCC model, recognising that existing models now provided the ability to quantify the risk in WLCC forecasts, but that other factors such as component deterioration were widely ignored. The model also responds to the problem that many WLCC approaches are used retrospectively and not as part of an iterative design process.

The EuroLifeForm approach is innovative in that it utilises a stochastic WLCC model in conjunction with a series of deterioration analysis algorithms and a decision support application to assist in optimising the WLCC design process. The integrated model enables the analyst to calculate WLCC results probabilistically, based on the forecasted effects of component deterioration on the maintenance and replacement intervention times. The decision support element facilitates the iterative application of the model throughout the entire design process, thus providing a repository of design decision-making information that can be used on a micro level to optimise the WLCC design, and also on a macro level to inform decisions in other similar projects.

7.9.3   WLC input tool – Glasgow Caledonian University

The Society of Construction Quantity Surveyors (SCQS) in mid-2004 commissioned this research project, which to date has involved the development of a framework document and WLC input tool for use in local government procurement decision-making. Based on an MS Excel platform, it provides a generic approach to the analysis of buildings and components.

7.9.4   4D cost model – Bucknall Austin

Bucknall Austin Cost Consultants have developed a '4 Dimensional Cost Model' (4DCM), which is advocated as a 'fully dynamic' whole life cost model where all variables within the model can be related and the impact on each other quantified. The information is stored in a format that allows similar projects or options to be analysed using the same template of logic and pricing. There are approximately 400 variables within the model that can be changed using different scenarios. The 4DCM model can also calculate capital allowances, CO_2, energy and water consumption in addition to the through life costs, and a detailed through life cost breakdown is available at the inception of a project. This supports design decisions at a very early stage with essential sustainability measures in mind. Evidence from the company suggests that early collaboration during the inception and feasibility stages of a project was vindicated with the Royal Ordnance Defence division of BAE Systems. The organisation saved an estimated £5 million over 25 years on a single new-build project.

7.10   Key points

- WLCC is a method of appraising the long-term cost implications of design decisions.
- It considers not only capital costs but also maintenance, operation, and energy costs.
- In order for the process to be most effective, it requires a significant element of client involvement at early stage, and this should be facilitated through the briefing process.
- It involves forecasting sums of money into the future and is thus governed by the time value of money concept.
- In certain procurement arrangements, WLCC must be carried out as part of the cost planning process.

Notes

(1)   Hunter, K. and Kelly, J. (2005) The development of a whole life costing tool for local government. *Proceedings of the RICS COBRA Conference*, Queensland University of Technology, Brisbane.

(2)　It should be noted that on an international basis, representatives from some countries do not accept the concept of WLCC and consider the current generation of methodologies to be a natural extension of LCC. This is reflected in the current version of ISO 15686-5 which is titled 'Life-Cycle Costing'.

(3)　http://www.ogc.gov.uk/sdtoolkit

(4)　Low Carbon Buildings Event: A Climate Change Programme Review Consultation Event, London 2005 http://www.defra.gov.uk/environment/climatechange/ccprog-review/pdf/lowcarb-summary.pdf

(5)　In Saket, and Munns and Al-Haimus, cost significant models are suggested as one way of overcoming the highly detailed and time-consuming approach to preparing such things as a traditional BQ. The work of these authors presents a methodology for selecting work packages, and recommends a refinement to the technique that reduces the variability in estimates produced using cost significance. Estimates are produced using both the traditional method of producing cost significant models, and a refined global cost methodology. Both techniques are tested against unpriced bills to measure the difference in results, with significant improvements being achieved with the new technique.

(6)　Sensitivity analysis determines how responsive the results of a model are to changes in the inputs and assumptions. It can be used to assess how robust the model is in dealing with uncertainties or assumptions about such things as cost and performance.

(7)　Please see http://pcwww.liverpool.ac.uk/~kirkham/FIG73.htm for a typical example and the calculation of the PVs for each year.

(8)　Engineering and Physical Science Research Council (EPSRC) project: Managing risk across the whole life of a facility: a design perspective (GR/N34024/01), February 2001.

References

Bird, B. (1987) Costs-in-use: principles in the context of building procurement. *Construction Management and Economics*, 5(1) Special issue.

Boussabaine, A.H. and Kirkham, R.J. (2004) *Whole Life-Cycle Costing: Risk and Risk Responses*. Blackwell Publishing, Oxford.

Kirkham, R.J. (2005) Re-engineering the whole life-cycle costing process. *Construction Management and Economics*, 23(1), 9–14.

Munns, A.K. and Al-Haimus K.M. (2000) Estimating using cost significant global cost models. *Construction Management and Economics*, 18(5), 575–585.

OGC (2007) *Achieving Excellence in Construction Procurement: Guide No. 7, Whole Life Costing*. OGC, London.

Saket, M.M. (1986) *Cost-significance applied to estimating and control of construction projects*. PhD thesis, University of Dundee.

Chapter 8
Procurement and the Relationship with Project Costs

8.1 Introduction

In the construction industry, procurement is the term used to describe the processes and procedures involved in the acquisition of an asset (building). Typically, the methods of contractual and organisational working are referred to as procurement systems or procurement strategies. Thus, procurement generally refers to the processes involved in designing, constructing and commissioning a new building. The concept of procurement is not unique to the built environment arena; procurement can be used to describe the purchase of anything from a small family car right through to a multi-million pound nuclear defence system. In the UK, the government has long-established agencies that deal specifically with procurement issues, including the Defence Procurement Agency (part of the Ministry of Defence) and the OGC. The latter has been instrumental in revolutionising public sector construction procurement in the UK through the Achieving Excellence in Construction programme. Procurement itself has become a discipline area (or indeed profession) in its own right, as exemplified by the increasing number of procurement managers employed specifically to govern, and provide advice on, purchasing protocols. In the built environment, it is rare to encounter a procurement manager; usually, this role is performed by various actors within the project, such as clients, project managers or specialist consultants. However, in some of the more modern procurement systems there is implicit evidence that such a role exists.

Procurement generally does not necessarily deal with the purchase of tangible 'hard' items such as buildings; it can also deal with 'soft' services such as social and healthcare. Good procurement should always try to encompass the different types of 'clients' in what are often complex organisations. For example, the procurement of a new school must take into account the requirements of the policy makers in the Department of Education, the procurement managers, the Local Education Authority and, of course, the end users themselves, i.e. the teachers and pupils. During this process, it is additionally important to ensure that the project demonstrates value for money throughout the project life-cycle. The emphasis from the public sector to review and enhance construction procurement has emerged through the more informed view that buildings should now be valued not in isolation, but as a means of delivering wider social benefits. In summary, procurement has been a key driver in the significant changes that have characterised UK construction industry since the mid-1990s.

This chapter will explore the various methods of construction procurement that are currently in use, and describe the appropriateness of each approach to the project environment. The 2005 edition of the Joints Contract Tribunal (JCT) guide to contract selection (JCT 2006) identifies three main systems: traditional, design and build and management procurement. The aim in this chapter is to give a contextual overview of procurement, rather than explain in detail each approach. The further reading list at the end of the chapter suggests some appropriate texts to support the concepts described here.

8.2   The traditional procurement system and cost planning

Historically, the growth of the quantity surveying profession is inextricably linked with the traditional procurement system, which has at its heart the BQ (Birnie and Yates 1996). This approach to procurement is perhaps the most well known and still continues to command a significant share of all construction project work. The discipline of cost planning as a process is derived from the primary skills of measurement and cost analysis, which perhaps underpin the traditional approach. However, changes to procurement have been significant, particularly since the last edition of this book, and the move away from the traditional BQ-centred procurement system has gained momentum. The next ten years may well continue to witness rapid change; the remainder of this chapter will highlight these changes and the impact on cost planning generally.

In the traditional system, the contractor agrees to build to the designers' drawings and specifications, and as the designers themselves do not have a direct link with the specialists, all communication is via the main contractor, who does not have a contractual design liability. This can result in a grey area of responsibility and liability, as information is passed from one to the other. In general, the designer (more often than not the architect) is the leader of the project and represents the client to implement the design process. The architect thus acts in *locum tenens*, as a sort of surrogate client, and takes the overall responsibility to ensure the project is delivered on time and on budget. In the traditional system, the client appoints these independent consultants who produce detailed designs and issue tender documentation. The documentation is used to invite competitive bids from tendering contractors, often on a lump sum basis. (Lump sums are considered later in this chapter.) The successful contractor enters into a direct contract with the client and carries out the work under the supervision of the design team.

Using a BQ, every aspect of the works is quantified to determine the contract price, as far as possible. It is possible to deal with uncertainty over the quality or nature of some work and a contractor can thus price the works without a full BQ; approximate quantities, schedule of rates, activity schedules, target cost contracts and cost plus/prime cost contracts are common examples of alternative methods of pricing. The traditional approach to procurement remains popular owing to the fact that most clients and contractors have experience of it. Price certainty (subject to the design being fully prepared at tender stage) is also a key benefit and gives the client greater control of the design (through management of the design team). Acceleration is also possible with the traditional method; design and construction can run in parallel using two-stage tendering or may be by negotiation on partial or notional information.

Another advantage is having an independent professional in the role of the contract administrator monitoring the project. However, the divorce between the construction and design stages has been recognised as a potentially significant deficiency. In many other manufacturing situations, such as automotive production, this divorce does not exist (or is of significantly smaller magnitude). Within the construction industry this separation can lead to disputes over variations and design changes.

In summary, traditional procurement is characterised by:

- Separation of the design and construction functions.
- Consultants are appointed for design and cost control.
- A contractor is appointed to construct the works, who:
    - is responsible for workmanship, materials, sub-contractors and suppliers;
    - usually has no design responsibilities;
    - is usually appointed through competitive tendering, supposedly based on complete design information.
- The contractor relies heavily on the design team for information; there is a possibility of claims if delay or disruption occurs.
- By using consultants, the client retains full control over design.
- The client obtains considerable certainty about cost before commitment to the contract/work begins.
- The employer, or designers, can select specialist firms to be used; contractors require certain safeguards. Care must be taken here though; in the JCT 2005 suite of contracts, the procedures for nominating sub-contractors by the client team have been omitted from the standard building contract

(SBC). However, the procedures for the architect or contract administrator to name an approved sub-contractor under the Intermediate Contract (IC) still exist.

8.3   Standard forms of building contract

Until the passing of the Housing Grants, Construction and Regeneration Act 1996 (HGCRA), building contracts and sub-contracts were governed only by the general law of contract. Early in the twentieth century the Royal Institute of British Architects (RIBA) had published a model form of building contract, which gained wide acceptance. Building contractors, however, felt that a contract drafted unilaterally by an organisation, however well meaning, which represented only one side of the industry was inherently unfair. An organisation called the Joint Contracts Tribunal (JCT) was set up to include contractors' representatives. The JCT took over administration and revision of the RIBA Form, and was gradually expanded to include representatives of all the bodies involved in procuring a building, including sub-contractors and clients. It now produces a whole range of forms to suit almost every conceivable kind of contract. The JCT forms do not have any kind of statutory authority, and there has never been anything to prevent clients or contractors from either using their own forms or altering the wording of the JCT forms to suit themselves. Although in theory a contract is mutually agreed between its parties, in practice a building contract or sub-contract is usually presented by one party to the other on a take-it-or-leave-it basis. This led to the imposition of unfair terms where either the client or the contractor was in a dominant commercial position, or had much greater experience than the other party.

The JCT 2005 suite of contractual documents[1] consists of 'contract families' made up of main contracts and sub-contracts, together with other documents that can be used across certain contract families. The suite is thus made up of:

- Minor Works Building Contract (with and without contractor's design)
- Intermediate Building Contract (with and without contractor's design)
- Standard Building Contract (SBC) – made up of Standard Building Contract With Approximate Quantities (SBC/AQ), Standard Building Contract With Quantities (SBC/Q) and Standard Building Contract Without Quantities (SBC/XQ)
- Design and Build Contract
- Major Project Construction Contract
- Construction Management Trade Contract
- Management Building Contract
- Housing Grant Works Building Contract
- Measured Term Contract
- Prime Cost Building Contract

- Repair and Maintenance Contract
- Generic Contracts
- Framework Agreement
- Adjudication Agreement
- Collateral Warranties
- Home Owner Contracts

8.4   Basic forms of building contract in the traditional method of procurement

Although there are many different forms of contract (including all the JCT variations shown above), these fall broadly into three different types.

Measurement contracts. The contract sum is not finalised until after completion, but is assessed on remeasurement to a previously agreed basis. This method is used when the contractor undertakes work that cannot be measured accurately prior to tender. The design is usually complete and the time/quality parameters specified. The contract of this type with least risk to the client is probably that based on drawings and approximate quantities. Measurement contracts can also be based on drawings and a schedule of rates or prices. A variant of this is the measured term contract under which individual works can be initiated by instructions as part of a programme of work, and priced according to rates related to the categories of work likely to form part of the programme (JCT 2006).

Cost reimbursement contracts. These are often referred to as 'cost-plus percentage/fixed fee' or 'prime cost' contracts and are used where the contracting firm agrees that all its expenditure on labour, materials, etc., will be met by the client, on top of which it will charge a fee on an agreed basis (e.g. a decorator will repaint the living room for the cost of the paint plus £10 per hour of his or her time).

Lump sum or 'price in advance' contracts. The contracting firm agrees to carry out its obligations for a sum of money agreed in advance (e.g. a decorator will repaint the living room for £200). Lump sum contracts with quantities are always priced on the basis of drawings and a full BQ. There are also contracts without quantities and these are priced on the basis of drawings and specifications rather than a BQ.

The last two types are rarely used in their pure form on large projects. The lump sum or price in advance contract involves too great a risk to the contractor (the building may need deeper foundations than was originally anticipated, or price inflation may substantially exaggerate labour and material prices). The cost reimbursement contract encourages waste and extravagant working by the contracting firm, which does not have to pay for anything that it uses. In fact, in the most primitive form of cost reimbursement contract

Phase 1

where the profit is a percentage of the cost, the contractor has a positive incentive to waste as much money as possible. From the point of view of cost planning, however, there is rather an interesting paradox. The price in advance contract gives the client almost no control over the details of methods, programming or expenditure, but gives an excellent forecast of the total cost. The cost reimbursement contract, on the other hand, allows the client to give orders about the acceleration or methods of work, and to obtain detailed allocations of actual site costs, but the total cost can never be forecast with the same degree of certainty.

8.5   JCT contracts for traditional procurement

This section looks at the various contractual methods of building, trying as far as possible to match them against the client's time and cost criteria as set out in Chapter 4. While this may suggest the best approach to the individual problem, it must be remembered that no form of contract is proof against things going wrong, and a keen combination of design team and building team will make a good job of things whatever the contractual arrangements. However, all other things being equal, a suitable form of contract will help. The reader is strongly urged to refer to the JCT publication *Deciding on the appropriate JCT contract* (JCT 2006), which provides a good, succinct reference to contract selection.

8.5.1   *Measurement contracts*

JCT recommends the use of measurement contracts (i.e. JCT 2005 Standard Building Contract with Approximate Quantities) for:

> . . . larger works designed and/or detailed by or on behalf of the Employer, where detailed contract provisions are necessary and the Employer is to provide the contractor with drawings; and with approximate BQs to define the quantity and quality of the work, which are to be subject to re-measurement, as there is insufficient time to prepare the detailed drawings necessary for accurate BQs to be produced; and where a Contract Administrator and Quantity Surveyor are to administer the conditions.

In section 8.4.1 the features of measurement contracts were briefly covered; the implication here is that these types of contracts are best applied where the contractor is required to design separate part(s) of the works (referred to as the contractor's designed portion). Due to this arrangement, measurement contracts are also ideal where the works are to be carried out in sections or stages.

The contract price is based on the initial tender figure submitted by the contractor. This is reassessed at practical completion, where on remeasurement the ascertained final sum (AFS) is obtained by valuation of all the work. Like many other forms of contracts and in accordance with the HGCRA, monthly interim payments are made to the contractor (unless otherwise stated in the contracts).

Measurement contracts require the design team to provide a set of drawings and approximate quantities at tender stage, and from this the contractor will quote the tender sum. However, due to incomplete design information and the use of approximate quantities, this tender figure can only be indicative. If the final quantity varies significantly from the approximate quantity in the bills, the contractor is usually entitled to renegotiate the rate. For example, if the bills state that 100 m^3 of concrete are required, but the remeasurement only indicates that 20 m^3 has been used, then the contractor would not have benefited from economies of scale and may be entitled to an increase in the rate for the item. Many contracts, especially in civil engineering works, state that a remeasurement of ±10% in the quantity would allow a renegotiation of the rate.

8.5.2 Cost reimbursement contracts

With cost reimbursement contracts the sum is arrived at on the basis of prime (actual) costs of labour, plant and materials, to which there is added an amount to cover overheads and profit. Within cost reimbursement contracts, there are three variants discussed here: cost plus percentage, cost plus fixed fee and a hybrid approach called target costing.

Cost plus percentage

This variant has several advantages:

- It is the most convenient contractual basis of all.
- The contractor can be selected, the contract placed and work started before the scheme has been finalised and without any estimates or quantities needing to be prepared.
- The contractor's management methods, in theory at any rate, can be used for the direct benefit of the client.
- The client also knows that the contractor will not make an exorbitant profit.

However, there are disadvantages to this approach, including:

- The drawback of poor cost forecasting facilities and low productivity arising from the fact that the project is not working to a price so can afford to do things 'properly' (or slowly and extravagantly).
- There is a positive incentive towards improvidence because the contractor is rewarded with a percentage of everything that is spent.

Phase 1

Because of these drawbacks it is not surprising that this type of contract has a poor reputation in terms of prudent cost control. However, because of the virtues mentioned previously, it is a widely used method for such projects as:

- emergency first aid and repair work;
- alterations and repairs to old buildings where the extent of the works cannot be foreseen until the contract has started;
- contracts where very high quality work is required, such as restoration to listed buildings and façade retention;
- contracts where cost may be important but where the client wishes to retain control over the method of working;
- contracts where a good long-term relationship exists between the client and the contractor.

Cost reimbursement approaches are also suitable for projects requiring an early start on site, where the works are designed – but not to the fully detailed design level – before the works commence.

Cost plus fixed fee

Cost plus fixed fee contracts are an attempt to overcome some of the negative aspects of cost plus percentage variants, in that paying the contractor a fee based on the estimated cost of the project instead of a percentage of the actual cost, mitigates the propensity for profligacy. This has the advantage (compared to the cost plus percentage variant) of reducing the incentive for wastefulness. However, there are two disadvantages:

- A fairly detailed scheme and an estimate have to be prepared before work can start, so losing some of the advantage of the 'cost plus' system.
- It is likely in practice that the so-called fixed fee will have to be renegotiated at the end of the job because of the major variations that are sure to arise in projects of the type for which such a contract would be used. This is likely to be based on the contractor's actual cost, so effectively returning to the cost plus percentage system.

In summary, cost reimbursement contracts create shared risks. The key issue though is that a contractor more used to the traditional forms of contract will have little incentive, other than repeat business from a large employer, to work efficiently and economically. Clearly, none of the standard forms of cost reimbursable contracts discussed here address the problem of incentivising the contractor to collaborate with the employer in forecasting the final costs, so that joint action may be taken to prevent any cost overrun.

Target costing

Target costing involves an attempt to get the benefits of cost reimbursement without the disadvantages described above. Usually, a BQ is prepared and priced, by negotiation, to arrive at a target cost; the contractor then carries out the work on an actual cost basis. In order to arrive at the final price, there are various methods available:

- If the actual cost is lower than the target cost, the contractor and client usually split the difference between them on a prearranged basis.
- If the actual cost is higher than target, the contractor has to be content with the actual costs plus a small overhead percentage.

As an aside, the philosophy of target costing is encapsulated in the pain/gain contracts which are now becoming common in many procurement scenarios. Several public sector projects have been procured under the Option C of the New Engineering Contract/Engineering and Construction Contract (NEC/ECC) forms as a target cost contract with a negotiable pain/gain share. This means that both contractor and client share either the reward or cost of deviation from the target cost. Unfortunately, many clients, or their advisers, insist on the 'pain share' being 100% to the contractor, whilst the 'gain share' is weighted significantly in favour of the client. This clearly is contrary to the philosophy of target cost contracting and effectively creates a guaranteed maximum price (GMP) contract.

The extolled virtues of target costing are:

- The contractor has an incentive to do the job as cheaply as possible and the client gets a direct benefit if this is achieved.
- The system retains many of the benefits of cost reimbursement, especially the ability of the contractor and client to work closely together in the management of the project.
- It can be used with particular success on a 'continuation' basis where the client has a succession of projects in which the same contractor can participate.

The main disadvantages of target cost are:

- The target cost is subject to revision in respect of the many likely variations (so that one might as well have an ordinary lump sum or remeasurement contract).
- The QS's fees are likely to be high because there is dual documentation (cost reimbursement accounts to be checked and BQs to be prepared, priced, agreed and updated).

8.5.3   Lump sum and price in advance contracts

This is probably still one of the most widely used variants of the JCT forms of contract, in spite of various difficulties. One of these difficulties is of course inflation. Many large building contracts can run over several years;

consequently, it is unreasonable to ask the contractor to bear all the risk of inflation during this period.

The example of Wembley stadium in Chapter 1 demonstrates this and also reveals the potentially devastating effects it can have on the project. Therefore, most of the contract methods set out in this chapter have two versions:

- for use on small to medium projects in times of economic stability where the contractor bears the inflation risk;
- for larger projects or at times of economic instability, where the risk is borne in whole or in part by the client.

It should be noted that while the version where the contractor bears the risk gives the client a firm cost forecast, the alternative approach may well be cheaper, even on short-term jobs. This is because contractors do not like this type of risk, and thus will tend to overprice it if they are given a choice of tendering on the two methods. There are various types of price in advance contracts, which will now be considered.

Lump sum contract (with firm or approximate BQs)

This is often seen as the best form of contract, since drawings and detailed specifications are prepared which include sufficient information to allow a detailed BQ to be produced, all of which is then issued to contractors for pricing.

Bills can either be firm or approximate (i.e. where the level of detail to prepare a firm bill is lacking, but there is sufficient information to proceed with the project).

Lump sum contract with quantities

Although the relative importance of the BQ is changing in UK procurement, a good deal of major work in the UK and elsewhere is undertaken on this basis. Once the scheme has been designed, a number of contractors are asked to submit lump sum tenders based on the pricing and totalling of a BQ prepared on behalf of the client. The successful contractor's tender BQ then becomes the instrument for financial administration of the project, so that the one document provides a simple means of contractor selection, price commitment and contract management. This method of contract pricing probably gives the lowest price, which is deemed best value by many clients and consultants. However, although superior to any of the previous methods in regard to cost forecasting and budgeting, it has a number of disadvantages in this respect, including a requirement for the scheme to be fully designed prior to tender. It is this type of contract, with its competitive BQ rates, which provides most of the raw data for elemental cost analyses, and indeed acts as a control against which the cost of buildings erected under less competitive arrangements can be judged. In practice, the provisions of the JCT standard forms of contract (which attempt to be fair to everybody) make the total price and time commitments rather less firm than they appear to be in theory.

Lump sum contract with approximate quantities

This is used as an alternative where the scheme has not been fully designed, the BQ being only a notional but weighted representation of the finished product. The work is measured as executed, and a final price agreed using the items and rates in the BQ as far as possible. This method has the advantage of permitting an earlier start to be made; however, control, both of time and money, is much weaker than in the previous case. A contracting firm that has underestimated the cost of its commitments when tendering is sure to be able to find some way of recouping its actual costs unless supervised by experienced consultants.

Lump sum contract without quantities

Where, for whatever reason, a BQ is not provided, drawings and specification/work schedules are provided at tender stage instead. The contractor either prices in detail the specification or the work schedules, thus determining the contract sum, or states the lump sum required for carrying out the work shown on the drawings and specification. In the latter case, a contract sum analysis or a schedule of rates will normally be provided. The priced documents will be used as a basis for the valuation and control of variations.

Negotiated tendering

This is frequently used where the early involvement of a contractor's specialist skills may be recognised as essential to project success. Normally, a contractor is selected early in the design stage, and as the design develops, the BQ is prepared and priced jointly with the contractor, who will be involved in the detailed design stage as well.

Advantages of this style of tendering include:

- The system assists cost forecasting and budgeting, as there is some contractual commitment to the cost assumptions made during design.
- An early start can also be made on site.
- Because everything is negotiated, the contract terms and the arrangement of the BQ can be tailor-made to suit the actual requirements of the parties.
- This method of contracting is very suitable for use on a 'continuation' basis, where the same client, professional advisers and contractor can establish a good working rapport.

Phase 1

Disadvantages of this style of tendering are:

- It is unlikely to produce competitive prices.
- It requires some degree of special expertise on both sides to obtain the best results.

For this last reason, the QS may well be the most influential member of the professional team and may be appointed first.

Prime contracting (serial contracting)

This is a variation on a theme of the above. The serial contracting approach, now commonly referred to in the UK as prime contracts, is a procedure where the contractor will tender for a series of similar contracts (usually related to the estate of a particular client such as the Ministry of Defence or National Health Service). This method is open only to clients with a continuing work programme, however. Contractors are asked to competitively price a typical BQ, on which basis they agree to carry out any similar work that the client may require over a fixed period of possibly two or three years. Usually the contractor is asked to take the rough with the smooth, and to quote average rates that will apply on both difficult and straightforward projects. Serial contracting provides an ideal platform for cost forecasting, as contractually binding detailed prices are available for future schemes before they are even designed. However, there are several disadvantages to the approach including:

- The system relies heavily on mutual trust and works only where the contractor has confidence in the integrity and fair play of the client, for which reason it has mainly been used in the public sector.
- As already stated, it does not usually produce competitive prices.

Prime contracting is unlikely to be appropriate for clients that procure buildings only occasionally. Importantly, prime contracting approaches require there to be a single point of responsibility (the prime contractor) between the client and the supply chain. The prime contractor needs to be an organisation with the ability to bring together all of the parties (consultants, contractors and suppliers) necessary to meet the client's requirements effectively. There is nothing to prevent a designer, facilities manager, financier or any other organisation from acting as the prime contractor. A key part of the prime contracting[2] procedure involves the development of WLCC models; these should be developed as early as possible in the design stage.

Schedule of prices or measured term contract

These are used mainly for repair and maintenance work contracts. Increasingly, local authorities, housing associations and other public sector agencies contract out the repairs and maintenance of their buildings to private contractors under instruments like measured term contracts (MTCs). These have something in common with the serial contract, as the contracting firm binds itself to a price list at which it will undertake work as required for a period of years, subject to an agreed discount or premium as the case may be. The most commonly used version is the *National Schedule of Rates for Building Works* published by The Stationery Office. This type of contract is sometimes used as a matter of convenience for small new works.

Framework agreements

These are similar to MTCs in that they are often used for rolling maintenance work. They can be defined in a contract using the JCT 2005 Framework Agreement (both binding and non-binding versions). Framework agreements are becoming increasingly popular within the public sector for driving continuous improvement under the requirements laid out in the National Procurement Strategy. One of the most well-known framework agreements in the UK is the NHS ProCure21[3] system. NHS ProCure21 was originally developed by the then NHS Estates following consultation within the NHS and the private sector, industry and academia. ProCure21 is intended to promote more efficient capital procurement in the NHS by developing a partnering programme using pre-accredited supply chains engaged in a long-term framework agreement (framework partners).

Within the ProCure21 collaborative framework, NHS clients select a framework partner at an early stage; the partner will help the client develop the full business case (thus expertise should in theory be available to inform the cost planning process). The ProCure21 framework agreement uses the NEC option C, which stresses the importance of early warning on project issues and collaboration to solve problems, and includes provisions for pain/gain sharing.

Recently, the largest ever ProCure21 scheme was completed on Merseyside: the £80 million hospital redevelopment at Broadgreen, part of the Royal Liverpool and Broadgreen University Hospitals NHS Trust.

Other major clients involved in framework agreements include the British Airports Authority (BAA); the organisation is currently involved in the procurement of the redevelopment of Heathrow Airport, including the new Terminal 5 development.

8.6   Design and build procurement

> . . . in no other important industry is the responsibility for design so far removed from the responsibility for production. (Emmerson 1962)[4]

Earlier in this chapter, the divorce between design and construction – the signature of traditional procurement – was discussed. Emmerson's statement truly exemplifies this, and perhaps acted as a catalyst to the design and build procurement route. Under this system, the contractor acts directly for the client, filling the roles of both the professional design team and builder. Typically, the process of design and build would involve the following stages:

1. Client employs designers to produce outline design.
2. Contractor tenders price to complete design and perform construction.
3. Tendering usually competitively selected and best overall submission should win – price, design, programme, etc.
4. Winning contractor carries out design and construction through employed design consultants and sub-contractors.
5. Client pays price in monthly instalments as traditional route.
6. Significant shift in risk to contractor compared with the traditional route.

The design and build approach has gained increasing popularity over the years. The reasons why are encapsulated in a Chartered Institute of Building report[5], which reported completion times for projects being 25% faster than for those using a traditional approach. The report also identifies that up to 80% of design and build projects in their study were completed on time, compared to 56% of those using a traditional approach. Finally, the report also indicated that design and build projects were 15% cheaper than equivalent traditionally procured schemes.

There are a wide range of advantages to design and build procurement, including:

- single point responsibility;
- speed of construction;
- overlap of design and construct;
- better communication;
- more timely completion;
- improved financial control;
- all-in lump sum;
- simpler;
- reduced finance;
- competitive design fees;
- good relationships.

Because of the lack of a competitive tender, the client usually appoints a QS to act as the client's representative in order to control costs. The client may also ask a consultant architect to provide advice on design issues. However, the latter role is so limited and frustrating that not every architect is keen to take it on. The design and build firm may have its own in-house design team or may employ outside consultants; as the latter are responsible to the design and build firm and not to the client, this is not much more than a domestic detail.

Other possible disadvantages, which the client should be aware of, include:

- The drawbacks of production-oriented design.
- The lack of competition. Even if two or more tenders are obtained it is difficult to evaluate the best buy, as what is being offered by each tenderer will be different.
- Loss of control by the client over the end product if the offered price is to be maintained.

As a general rule, the design and build contractor will be dependent on a clear brief from the client, since the lack of the traditional architectural role in translating the brief into design is removed. The design and build route is also suitable for work of a standard or routine nature or where certain contractors can demonstrate technical expertise and efficiency. Complex projects, and schemes where high levels of architectural quality and specification are required, are not particularly suited to design and build methods of working. However, the reduced tendering costs of this procurement method have contributed to an upward trend in the volume of design and build work, and continue to do so.

Design and build procurement has also found itself associated with the concept of GMPs. These featured as part of the Scottish Parliament enquiry where the client investigated the possibility of moving to a GMP contract for the remainder of the scheme. It is basically a self-explanatory concept, i.e. the cost of the project should not increase beyond the GMP, but a recently published internet blog by David Lewis, the Principal of Lawbuild Solicitors, London, provides a fascinating and informed insight into the concept of GMP. Part of it is reproduced below, with permission of the author.[6]

> ... The buzzword is 'Guaranteed Maximum Price' or GMP, and it is often used in the context of a design and build contract ...
>
> ... There was an immediate meeting of minds between my QS friend and me, when we found that neither of us had been able to discover any difference between a GMP and the contract sum under a standard JCT Design and Build Contract. The contract sum under a design and build contract is a guaranteed maximum price for all practical purposes.
>
> When someone talks about GMP, ask them if a client under a building contract with a GMP can require the contractor to (say) build an extra storey without an increase in the contract sum. They will obviously have to admit that a significant variation must entitle the contractor to an increase in the contract sum, GMP or not.
>
> Then try to find out what else might distinguish a GMP from an ordinary contract sum under a design and build contract. Is it that the contractor bears the risk of adverse weather conditions or other 'neutral' delaying events? When they gratefully seize on this, point out that such a transfer of

Phase 1

risk wouldn't affect the contract sum; instead it would require the contractor to pay liquidated and ascertained damages for the resultant period of delay.

Is it that a GMP contract sum remains the same even if the contractor finds difficult ground conditions which cost him money to overcome? Perhaps, but under a design and build contract adverse ground conditions are normally at the contractor's risk anyway.

So what is a GMP, precisely?

I discovered the apparent answer to this question in Cockram's *Manual of Construction Precedents*, which contains a 'Price and Payment Schedule (Target Cost/Guaranteed Maximum Price) for use with JCT 2005 SBC/XQ'[7].

The learned author of this work, in a footnote, says that the principle behind this form is to convert the contract sum into a prime cost arrangement, under which the contract sum has three elements: the amounts payable by the contractor to subcontractors and suppliers (the 'work cost'); the contractor's site overheads ('prelims cost'); and a percentage mark-up on works cost and prime cost for the contractor's head office overheads and profit ('fee'). And the form also allows for provisional sums, i.e. elements which cannot be priced before the contract is awarded.

The Cockram form provides for a Target Cost, which is the estimated total of the works cost and the prelims cost and is stated in the contract. The Target Cost can be adjusted for variations or provisional sums. And then you have an incentive adjustment, so that the contract sum is reduced if the actual cost exceeds the Target Cost, or increased if the Target Cost exceeds the actual cost. So the contractor (apparently) has a monetary incentive to keep his costs down.

It strikes me that a contractor who is incentivised to keep his costs down might be equally incentivised to cut corners. But be that as it may, a contractor under an unamended lump sum contract is just as incentivised to keep his costs down because (since the contract sum is fixed) he can thereby increase his profits.

It was at this point in my researches – and possibly while pondering the absence of His Imperial Majesty's new clothes – that I was reminded of Voltaire's famous remark about the Holy Roman Empire: that it was neither holy, nor Roman, nor an empire.

Could one not likewise say that a Guaranteed Maximum Price (according to Cockram) is neither guaranteed, nor maximum, nor a price?

8.7 Management-based procurement

In the early 1990s, management-based projects enjoyed considerable popularity, owing in part to the distinct advantages these methods enjoyed over other methods of procurement, such as contractual relationships based on collaboration instead of confrontation, and availability of the contractor's construction expertise to the design team. But experience has led to some reversion to more traditional methods, partly because of a number of high-profile failures together with concerns over the forms of contract which are available. In Chapter 1 two examples were highlighted, at the Scottish Parliament Building in Edinburgh and the British Library project at St Pancras, London. There are two variants of management-based procurement: management contracting and construction management.

Management contracting

Management contracting is the closest to the traditional practices and structure of the contracting industry. In its most basic form, it is arguably little different to the older prime cost plus fixed fee contract, although its protagonists beg to differ. The main difference in theory is that the managing contractor firm does not undertake anything other than site management with its own resources, all direct works being sub-contracted by it. However, this is now so much the general pattern of the industry that it is not easy to discern much real difference in practice. The typical process involved in this method of procurement is:

- Client appoints design team as for traditional route.
- Early in design process, once the scope of the project is reasonably clear, a management contractor is appointed.
- Usually, competitive tendering is used to obtain a percentage or fixed fee bid for management and sometimes a GMP.
- Management contractor works as member of client project team:
  - □ role, mainly, is to manage the execution of the works
  - □ may lead the team on programme issues
  - □ may be directly involved in preliminaries activities: setting up site accommodation, fencing, services, etc.

The criteria for selecting the MC route and some of the key features include:

- Best suited to large, complex, fast-moving projects where early completion is desirable.
- Depends on a high degree of confidence and trust: there is no firm contract price before work actually begins on site; decision to proceed usually has to be taken on the basis of an estimate.
- Management contractor is the agent of the client, and should therefore put the client's interests first throughout the job.
- MC expertise available to design team in crucial pre-construction period.
- Detailed design work can proceed in parallel with site operations of some early works packages.

Potentially, MC offers a considerable degree of flexibility in design matters as early work proceeds, and effective cost control is possible provided good QS services are available. Due to the contractual arrangements, a competitive element is retained in works packages but it does demand high levels of management/co-ordination skills.

Construction management

Construction management (CM), on the other hand, is more well known, and perhaps a more notorious too! The principal differential between the two variants is the contractual arrangements. In CM, the client contractually employs each work package contractor directly, unlike in MC. The key advantage of this system is that the client can exercise a high degree of control over the entire procurement process. However, the client must have the necessary expertise and resources in order to exercise this control effectively. Where the client is naive and ambiguous, and where multiple stakeholders characterise the client group, CM can lead to major problems.

Advantages

The advantages claimed for both the MC and CM approaches, compared to the traditional standard JCT-type contract, are considerable:

- Fast-tracking, i.e. making the design and construction processes concurrent instead of consecutive, becomes possible.
- Construction-based expertise becomes available to the designer at an early stage.
- The collaboration of client and designers with the builder, instead of the usual confrontation, is especially valuable on very large and complex time-sensitive projects or in refurbishment work.

These advantages have not always been realised in practice however. Many traditionally trained contractor staff (and contractor organisations) experience difficulty in adapting their methods to a new client-oriented role. The removal of adversarial relationships occurs only at the top management level. Down where the action is, the direct works contractors are normally employed on the usual lowest-price-for-the-work basis and the usual confrontations apply.

It is sometimes said that these confrontations are worse than usual, because the trade contractors feel that the builder, who used to be a fellow sufferer, is now insulated from cut-throat tactics whereas they are not. However, the advantages of MC and CM have undoubtedly led to their employment on large and difficult projects where any system would find itself in trouble. On such projects traditional methods would probably have got into even greater difficulties (even if it had been possible to use them at all). However, for all their new-found status, the management contractor and the construction manager are still acting largely in the builder's traditional role. In particular, they are dependent for their success on satisfactory performance by the architect, consulting engineers and QS, none of whom they control. Their ability to guarantee the client an on-time and on-cost project is therefore limited. Sometimes, therefore, the management contractor may seek to control the designers, particularly regarding the timing of information flow. But this is an unhappy compromise and the client looking for a manager who will accept total responsibility for performance would do better to consider a full project management service.

It has to be remembered that at the end of the day total costs are likely to be greater with managed projects than with more traditional methods. Apart from anything else, there is usually an extra level of management to be paid for. Also, in the UK the traditional lump sum/BQ approach brings two major benefits which are lost when newer management methods are employed: all procedures are so well known that the parties do not have to go through a learning process (at the client's expense) when they are running the project. The provisions of the standard forms of contract are well tried in case law, unlike remodelled or bespoke forms. So it is probably a mistake to employ management methods on moderately sized straightforward jobs which are in no particular hurry.

Finally, it is of the essence of professional project management, either at construction level or at total scheme level, that the management organisation must not carry out any of the other professional or construction tasks for which it is responsible to the client; an MC firm should not do any of the building work. There are some very grey areas at present involving works contractors which are group subsidiaries or group associates of the MC firm. Also, a quantity surveying firm that is acting as project manager should not do the cost planning themselves. Similarly, where project management is in use, the project manager should not attempt also to act as construction manager – the roles are quite different. It is perhaps for this reason that the role of scheme project manager has not often proved attractive to construction firms, in the UK at any rate. The usual cost reimbursement advantages and disadvantages apply to all managed projects, although the management team will almost certainly let out most of the work to sub-traders and specialists on a price in advance basis. As mentioned previously, where the CM division of a building contracting firm is acting as the construction or project manager, there is often not much difference between this and the more traditional cost plus fixed fee contract. However, it is more usual for the cost of site staff and facilities to be included in the fee than is the case with the traditional contract.

8.8 Partnering

Partnering agreements emerged from recommendations made in the Egan Report and *Trusting the Team*, the Reading Construction Forum's (now Constructing Excellence) initial publication on partnering in the team.

Partnering involves: '... a management approach used by two or more organisations to achieve specific business objectives by maximising the effectiveness of each participant's resources. The approach is based on mutual objectives, an agreed method of problem resolution, and an active search for continuous measurable improvements'.

Partnering is essentially a structured methodology for organisations to set up mutually advantageous commercial arrangements, either for one-off projects or in long-term strategic relationships. The three basic components of partnering are:

- establishment of agreed and understood mutual objectives;
- methodology for quick and co-operative problem resolution;
- culture of continuous, measured improvement.

The benefits of partnering relationships are cumulative, so strategic alliances produce significantly more advantage than single project arrangements; and the benefits are increased further if partnering is applied throughout the entire supply chain. By definition, it is clear that the success of partnering arrangements is to some extent reliant on the premise that all parties are committed to working together for mutual benefit. Although the construction industry is good at creating relationships when things are going well, this must also apply when things aren't running so well!

Partnering features 'open book' working practices and relationships, which attempt also to introduce systematic approaches to problem resolution rather than seeking parties to blame. There is also an emphasis on customer-focused working practices and adding value through the elimination of waste, and harnessing best practice.

This concept of partnering has now evolved further still and evidence from several PFI and commercial partnerships has shown that the relationships between partners have become more sophisticated; in other words, a 'second generation' of partnering has evolved.

PPC2000 was the first standard form of project partnering contract, again a direct result of the Egan Report; the contract provides the tangible foundation to the project partnering process. The latest version, PPC2000/3, can be applied to any type of partnered project in any jurisdiction, with the support of an experienced partnering adviser or with appropriate legal or other professional advice on its implementation. Other partnering-based contracts include the NEC Partnering Option and the ICE Partnering Addendum. beCollaborate (now Constructing Excellence) produced the Be Collaborative Contract[8], a new form of contract for construction projects that underpins collaborative behaviour. Be (Collaborating for the Built Environment) was an independent association for companies across the construction supply chain in the UK, but merged with Constructing Excellence. The JCT has also issued a non-binding Partnering Charter, for use with its 2005 suite of contracts.

8.9 Public sector construction procurement and the Private Finance Initiative

In 2003, £33 billion was spent on public sector construction in key sectors such as schools, hospitals, roads and social housing. This capital investment is set to continue expanding over the next decade.[9] The government has spearheaded procurement reform in the UK through various reports such as Egan and Latham, and more recently through the OGC Achieving Excellence in Construction Programme (OGC, 2007).

Results of the Achieving Excellence in Construction Strategic Targets in 2005 demonstrated that significant improvements had been achieved since the introduction of the initiative back in 1999. In comparison with 1999 figures, the results showed that:

- 65% of projects were being delivered on time, compared with 34%;
- 61% of projects were being delivered to budget, compared with 25%.

Moreover, the NAO report, *Improving Public Services through better construction*, published in March 2005, highlighted that an £800 million overspend on construction projects had been avoided through the adoption of the Achieving Excellence in Construction best practice principles. The same report estimates that further value gains of up to £2.6 billion in annual construction expenditure is possible if good practice was applied across all the public sector. Clearly the reforms are working, but a more controversial aspect of the government approach to procurement has been in the form of the PFI.

PFI is fundamentally different to all the other methods of procurement described in this chapter in that:

- It is exclusively used for the delivery of public buildings.
- The procurement involves not only the design and construction of the building, but also the provision of services within it over a predetermined period known as the concession period.

The PFI was introduced by the Conservative government in 1992, by the then Chancellor of the Exchequer, Norman Lamont. The primary objective of PFI was to encourage private investment in major public building projects like schools, prisons, hospitals and roads. The first examples of PFI in the UK included the construction of HM Prison Parc in Bridgend, South Wales, and Prison Altcourse in Liverpool, Merseyside, (the latter procured under a variant of PFI known as DCMF – Design, Construct, Manage and Finance). PFI is advocated as a method of risk transfer in capital procurement; the private investment implies that the level of government borrowing falls and that risk is transferred from the public to the private sector. The concept of risk transfer is of significant debate at present and should the reader be interested to learn more, they are referred to the further reading section at the end of this chapter. The procurement procedures for PFI are long and complex and fall

Phase 1

out of the scope of this book. However, to demonstrate simplistically how PFI works, the following stages are typical of the procurement on, say, a new hospital building for an NHS trust:

- A private consortium pays for a new hospital, where the consortium usually consists of a construction company, a bank or financier, a facilities management contractor and consultants.
- The local NHS trust then pays the consortium a regular fee for the use of the hospital, which covers construction costs, the rent of the building, the cost of support services and the risks transferred to the private sector.
- Thus in essence, most new NHS hospitals will be designed, built, owned and run by a consortium or grouping of companies.
- The NHS will employ some of the staff, mainly doctors and nurses, and will rent the building and other facilities from the consortium for at least 25 years.
- The deal is constructed in such a way that the consortium is guaranteed a full return on costs, including interest on the capital borrowed, plus an element of profit.

8.10 Key points

- There is a variety of client requirements, and many contractual methods of satisfying them.
- The method of contracting for a particular project or group of projects should be chosen with the individual client's needs in mind, not just on the usual basis. It is, for instance, pointless to prepare a BQ for a client who is concerned not with money but with convenience or time.
- We must also remember that the client organisation may not really need a building at all; perhaps it should be changing its distribution methods instead of building a new warehouse. It may therefore be important that it should not go initially to somebody who has a vested interest in putting up buildings.

Notes

(1) The full electronic versions of these documents can be obtained from the JCT website at www.jctltd.co.uk or through the subscription-only Construction Information Service (CIS).

(2) Client Pack, Construction Works Procurement Guidance, Scottish Executive, HMSO.

(3) NHS ProCure21 is a partnership framework where NHS construction projects costing from £1 million–£20 million can bypass Official Journal of the European

Union (OJEU) procurement requirements by working with principal supply chain partners (PSCPs) and pre-accredited supply chains.

(4)   Emmerson, H. (1962) *Survey of Problems Before the Construction Industries: A report prepared for the Minister of Works*. HMSO.

(5)   *How To Use The Construction Industry Successfully: A Client Guide*. The Chartered Institute of Building, Ascot, UK.

(6)   'GMP and the Holy Roman Emperor's new clothes', by David Lewis, the Principal of LawBuild Solicitors, London, on Sat. 14 Oct 2006 06:18 PM BST.

(7)   JCT 2005 SBC/XQ refers to the JCT 2005 Standard Building Contract Without Quantities.

(8)   The beCollaborate contract can be obtained via the Constructing Excellence web portal: www.constructingexcellence.org.uk/sectorforums/buildingestatesforum/bcc/index.html

(9)   National Audit Office (2005) *Improving Public Services through better construction*. HMSO, London.

Further reading

OGC (2007) *Achieving Excellence in Construction Procurement: Guide No. 3, Project Procurement Lifecycle*. OGC, London.

Boussabaine, A.H. (2006) *Cost Planning of PFI and PPP Building Projects*. Taylor & Francis, London.

Hackett, M., Robinson, I. and Statham, G. (2006) *The Aqua Group Guide to Procurement, Tendering and Contract Administration*. Blackwell Publishing, Oxford.

JCT (2006) *Deciding on the appropriate JCT contract*. Sweet & Maxwell, London.

Masterman, J. (2001) *Introduction to Building Procurement Systems*. Spon Press, London.

Reference

Birnie, J. and Yates, A. (1996) Procurement Choice – Implications for the Quantity Surveyor. http://www.rics.org/NR/rdonlyres/FDE6FE4F-CB86-4B57-845D-0FF2B9530106/0/procurement_choice_19960101.pdf

Phase 1

Phase 2
Cost Planning at the Design Stage

Chapter 9
The Design Process and the Project Life-cycle

9.1 Introduction

The previous chapters have explored the key elements and processes required to understand appropriately the budget and scope of the building project. Without this, it is folly to expect the design team to be adequately equipped with the necessary information to produce a robust response to the brief. This chapter will explore the building design process and will draw from this the role that the cost planner assumes. It will also briefly touch on the concepts of process mapping; this is an academic theory that helps us to understand how good design is accomplished.

9.2 The building design process

The building design process is a complex interaction of skills, judgement, knowledge, information and time, and has as its objective the satisfaction of the client's brief. The optimum solution is that which is obtained within the constraints imposed by such factors as:

- statutory obligations;
- technical feasibility;
- environmental standards;
- site conditions;
- cost.

Problems arise, however, in establishing which of the alternative options available is the 'best', as some factors such as personal comfort or aesthetics are difficult to measure (these concepts in economics are considered under the theory of utility). Neither is it easy to translate all attributes into a common unit of value, for example excessive noise levels compared with higher initial cost. In practice, compromises in the client's demands are nearly always necessary to keep within the constraints. The role of the design economist/cost planner is to provide information with regard to initial and future costs so that the design team can make decisions knowing the cost implications of those decisions. It is not usually the building economist's responsibility to provide 'value' as this must be the province of the team as a whole, of which, however, the building economist should be a contributing member. In theory the team will pool their combined knowledge for the benefit of the client, whose representative should wherever possible be a member of the team and play a part in the corporate decision-making process. As with any group activity, the composition of the team should be carefully planned to avoid vociferous members unduly influencing the final decisions.

A design economist who is to be an effective member of the team must:

- understand when the major decisions of cost significance are to be made, so that information can be provided at that crucial time;
- acquire techniques, knowledge and experience to provide answers to questions of cost that will be posed as the design is refined;
- appreciate the manner in which the design team, and in particular the architect, thinks and operates, commonly referred to as the design method.

Each designer adopts a different approach, but as there are many ways of solving the same problem, it is possible to identify some of the common techniques, which are very often incorporated in a typical approach to the task of achieving good design. This knowledge can be used to select the form of cost advice most appropriate to a particular problem.

9.3 The design team

It is not possible to place all the responsibility for establishing a successful cost solution solely on the shoulders of the building economist/cost planner or QS. The architect must co-operate in, and contribute to, the cost planning process, cost being one of several constraints that have to be faced in varying degrees, depending on the type of project and the client's financial resources. In this respect the architect needs to have a reasonable grasp of the factors affecting the cost of the project and the options that are available within that constraint. This knowledge will often be gained by experience, especially if the firm specialises in one particular type of contract, e.g. housing, factories or hospitals. Even without this experience the architect should be aware of the very basic design and cost relationships.

The advantage of this triangular set of relationships is that any two of the factors can be seen as functions of the remaining one. For example, if the size of the building is fixed, together with the form and specification, then a certain cost will be generated. Conversely, if the cost of the building is established together with its size (as is the case with some government yardsticks), then this constrains the form and specification that can be chosen. Alternatively, if the shape and quality standard of the specification is declared, together with an established sum of money, then the amount of accommodation is the design variable, which is limited. Since one factor must be the result, it is never possible to declare all three in an initial brief.

This is an oversimplistic view of the cost system but it is a starting point in the understanding of the complex relationships that exist between design and cost. It is the skill of the design team in achieving the right balance between these factors that makes a project a success or a failure.

9.4   The RIBA Plan of Work

Recognition for the team approach to design is included in what has come to be known as the RIBA Plan of Work (RIBA 2000). This is a model procedure dealing with some basic steps in decision-making for a medium-sized project, and is included in the RIBA *Handbook of Architectural Practice and Management* (Vol. 2). In this procedure the responsibilities of each member of the design team at each stage of the design process are identified. A design–tender–construct procedure is envisaged and from the QS's point of view it anticipates and incorporates some of the cost planning techniques described in later chapters. The pre-tender procedure includes the following stages:

Stage A	Inception	Appointment of design team and general approach defined.
Stage B	Feasibility	Testing to see whether client's requirements can be met in terms of planning, accommodation, costs, etc.
Stage C	Outline proposals	General approach identified together with critical dimensions, main space locations and uses.
Stage D	Scheme design	Basic form determined and cost plan (budget) determined.
Stage E	Detail design	Design developed to the point where detailing is complete and the building will work. Cost checking carried out against budget.
Stage F	Production information	Working drawings prepared for tender documents.
Stage G	Bills of quantities	

It is of course recognised that a certain amount of overlap is bound to occur in practice. The RIBA Plan of Work has come under fire from a number of quarters because of the inherent inflexibility of its procedures, and because it

Phase 2

tends to delay the letting of the contract, which in times of rapid inflation can severely penalise some types of client. However, it was never intended as a rigid set of rules and the procedures adopted for any contract should be those which best suit the prevailing economic climate and the particular interests of the client. The plan has been included at this point in order to give an overview of the complete design process, but the emphasis in the rest of the chapter will be on the very early design period, prior to sketch design, when the critical decisions affecting cost are usually taken.

9.5   Comparison of design method and scientific method

To understand the concept of design it is useful to compare the design method with the traditional approach of the scientific method. To establish a scientific law:

- An observation is made in nature and an inference drawn, e.g. light passing through a prism breaks down into several colours.
- A hypothesis is set up (e.g. white light is a combination of several colours).
- Tests are applied to establish whether the hypothesis is true.
- If the results of the tests conclude that the hypothesis and original inference are correct then a scientific law can be established based on the hypothesis.
- If not, a new solution must be set up and tested until the tests corroborate the hypothesis.

Design can be said to follow a similar pattern:

- The brief is observed and some inference obtained.
- A hypothesis is set up in the form of a model (e.g. a drawing).
- This is tested by evaluation to see whether it works.
- A check is made to see whether it complies with the interpretation of the brief, and if it does then the design is accepted.

There are, however, some very important differences. The design team cannot loop round the system producing new hypotheses (i.e. designs) ad infinitum. The team works within the constraints of time and cost, and they must produce a final solution within the design period set by their client and within the fee structure by which they are paid. In addition, much of their creative work can only be evaluated in terms of social responses (such as aesthetic appeal), which at the present time do not have a satisfactory quantitative measure by which they can be tested.

Consequently, the team's objective tends to change from 'What is the best solution for our client?' to 'How can we produce a design which satisfies as many of the demands of the brief as possible in the time available?' To arrive at a satisfactory solution, some kind of strategy, very often incorporating rules of thumb learned from previous experience, is used to narrow down a particular range of alternatives. The responsibility of the design economist/ cost planner in this search for a satisfactory solution should be to indicate to the team where it should look in terms of form, quality, spatial standards, etc., to solve the cost problem. This will avoid wasting time on abortive designs that will not meet the cost criteria, and will thereby increase the chance of arriving at the 'best' solution in the time available. In other words, the building economist should contribute to the overall design strategy.

9.6   A conceptual design model

A good deal of research work has been undertaken to establish a general pattern for design. The Building Performance Research Unit at Strathclyde University has put forward the following view, which is shared by several other writers. They suggest that design consists of three stages:

1. *Analysis*. Where the problem is researched in order to obtain an understanding of what is required.
2. *Synthesis*. Where the information obtained in analysis is used to converge on a solution at the level being investigated.
3. *Appraisal*. Where the solution is represented in some form, which is then measured and evaluated.

QSs have traditionally been involved almost exclusively in the appraisal process, and yet the real decision-making role is in the analysis and synthesis stages. New techniques are therefore required to enable the design economist to contribute to understanding and solving the design problem. It was assumed by the Strathclyde team that decision-making became more detailed and refined as time went on (from, say, concepts of building form and spatial arrangement through to the eventual choice of ironmongery), and that at each stage different methods of problem solving were adopted. Any cost technique must therefore follow a similar pattern, which may involve coarse measurement and evaluation in the very early stages and a more reliable measurement and cost application when more information becomes available. It is with these thoughts in mind that cost models have been constructed that can input information at each level of refinement.

One of the criticisms that can be made of traditional cost planning methods is that they do very little to contribute to the pre-sketch design dialogue, where all the major decisions of form and quality tend to be taken. Current research suggests that there is a heavy commitment of cost prior to a sketch design being formalised. This may amount to over 80% of the final potential building cost, leaving perhaps only 20% available actually to be controlled. Figure 9.1 illustrates the point in a hypothetical diagrammatic form. With an

Phase 2

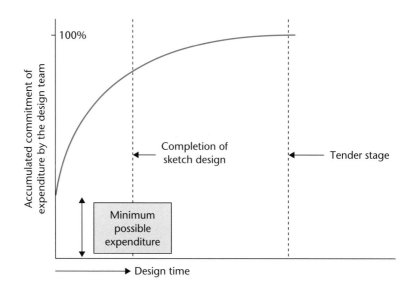

Figure 9.1 Accumulated commitment of expenditure by the design team.

improved understanding of design method it may be possible to input information prior to sketch design, which will reduce the number of abortive design solutions needing to be produced.

9.6.1 Research into design process mapping

Earlier chapters referred to the EuroLifeForm project. This project investigated a process mapping of design in the context of WLCC[1]. The rationale was based on understanding this approach to building costing as a synergy with the design process. If building design can be conceived as a systematic framework of activities comprising transformation, flow and value creation (Koskela 2000), then it is appropriate to model the design process in a logical way. Such a model, it must be remembered, is a model of the design decision-making process (when decisions are made and by whom and how they are linked, but not the actual technical decisions) and the information flows relating to the decision (input data, knowledge transfer, verification, recording and monitoring). Two issues arise as a consequence: the actual model of the process and the modelling medium (e.g. data flow diagrams). When reviewing the research into these areas, particularly when related to construction design, these two issues become interrelated but with the consensus view that the established modelling methods are applicable. The debate centres on the complexity of the models and how a generic design model can be produced from these formal modelling methods.

Phase 2

A number of design process models have been developed for engineering manufacture (automotive and aerospace in particular) and these are often referred to as new product introduction process models as they attempt to involve activities in the supply chain outside the strict design activity and also may encompass more strategic issues. (The establishment of the Innovative Manufacturing Research Centres within schools of Construction Management and Quantity Surveying in the UK Higher Education Institutions demonstrate this 'Egan thinking'.) In addition, a number of models have been developed specifically for construction design. None of the models available extends to any degree beyond the start of manufacture or start of construction phase, although a few make some reference to 'operations' without going into any detail. The approach in all cases is to assume that design is effectively a linear process (i.e. following the 'transformation' concept), although with some integration into the manufacturing processes. The models are all very similar but differ in detail, primarily because of the original aim in identifying and defining the model. The issue of design iteration, whilst considered, is not usually a central feature of these models. Macmillan *et al.* (2001) produced a summary of an extensive review of design process models, which indicated, as expected, broad similarities between the different models but some significant differences at the conceptual design stage. These differences relate to whether they are engineering based (very prescriptive) or architect based (generalised descriptions of stages).

As already mentioned, the RIBA Plan of Work is a well-known design model. It is designed to identify the main steps from client instruction to commissioning, primarily as a contractual aid. Architects' fees can be paid against achievement of the various stages. It implies a particular procurement route (competitive tender) and, in reality, the gates between the stages are 'fuzzy'. However, it provides an easily understood, widely used and simple model of the process and has been implicitly used in the development of more sophisticated design models. The British Airports Authority guide to the construction project process (BAA 1995) is a process map but with some features which relate to the specific needs of the authority. Inception is decoupled from feasibility and there is no tendering stage. BAA operates via a partnership mechanism (see Chapter 8) and has its own internal arrangements for project definition. The 'gates' in this model are more pronounced and are used in an active way as part of the project management process. There is an attempt to cover operations and maintenance within this model, but it seems to have been added almost as an afterthought. Network Rail operates a similar approach called GRIP (Guide to Railway Infrastructure Projects), which is a structured project delivery model with design included. The UK Defence Procurement Agency's CADMID system is an asset acquisition life-cycle, used primarily in the purchases of ordinance rather than buildings.

Perhaps the most comprehensive and ambitious attempt to model the design process is the 'process protocol' (Kagioglou *et al.* 1998), which is a

Phase 2

truly generic model although, again, it is effectively restricted to the design of the building rather than including maintenance and operational activities. The process protocol model includes gates both 'hard' and 'soft' to accommodate the fuzzy issues discovered by previous researchers. Current work is aimed at trying to bridge the gap between the high level abstraction and the detailed design tasks. Since briefing and the client/design team interaction are so critical in construction design, there have been a number of studies that have concentrated on this aspect. Latham (1994) and Blyth and Worthington (2001) have modelled briefing as an iterative process including feedback from previous projects. Macmillan *et al.* (2001) consider a fresh approach in the search for a generic framework for conceptual design. It is instructive to note that a framework rather than a model is proposed. What is described as a categorical framework with five levels is demonstrated, but extensive discussion with design professionals indicated that a three-level approach was perhaps the most useful, and that a framework rather than a prescriptive model was preferable.

The EuroLifeForm process map proposed in Figure 9.2 considers design (from identification of client requirement through to completed design) as a

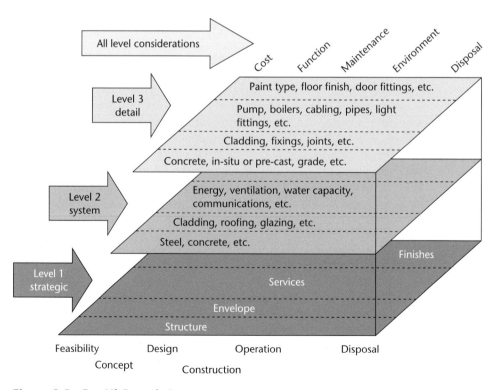

Figure 9.2   EuroLifeForm design process map.

linear, sequential process along the time axis and indicates the issues/stages where decisions are made (and the relevant data and information needed) in relation to WLCC performance. It is not a map designed for the purposes of project monitoring and control. Note that it is of course recognised that the design is not easily described in detail by a formal model.

Furthermore, the detailed models of the design process that have been suggested do not necessarily reflect design practice. Consequently this is a map of the decision stages to enable the correct WLCC data and information to be available at the appropriate point in the design process. The map is designed to be at different levels (three are proposed rather than the five or six proposed in some other design models), to preserve the generic nature of the map but to allow for varying levels of design detail.

By having different levels in this way, it should be possible for performance and cost data to interact with the design map at different levels depending on the precision and detail of the data and the level of detail decision. It is envisaged that, as design proceeds, the cost, performance and environmental data can become progressively more precise and can then be entered into the lower levels of the design map. Another advantage of this multi-level approach is that it may be able to handle the inevitable design interactions. In practice, the interaction between the three levels is likely to be at the later stages of design (after the feasibility and concept stages) so the three-dimensional map will be of this general nature.

9.7   Design techniques

It is not possible to describe fully in a chapter or book of this nature, the vast range of techniques applicable to design. Those interested in developing their understanding in this area of study should consult other publications (see the further reading section at the end of the chapter). It is possible to identify some decisions that form the basis of most design approaches. In very simple terms these can be illustrated by some general questions relating to matters of principle affecting the design. These include the following.

9.7.1   'What are the constraints within which I have to work?'

The constraints on a project can be of three kinds:

■ *Constraints imposed by physical factors.* These very often relate to the site and include the position of boundaries and easements, the method of access, the nearness to service supplies, any visual aspects and views, soil and environmental conditions and adjacent structures. In addition to the site, the physical performance of materials acts as a limitation and may exclude the use of a particular specification. Constraints imposed by physical factors are usually fixed and therefore provide very clear boundaries to the design problem.

■ *Constraints imposed by external bodies*. These, however, can often be the subject of negotiation. Planning requirements are usually the result of the policy of the local planning committee and are often open to interpretation, hence the facility for appeal when these interpretations conflict. Some building regulations can be waived if the appropriate authority can be convinced that an alternative construction form or arrangement is satisfactory. Resolving these problems can be a time-consuming business, and until they are settled the solution to the design problem has to remain flexible. However, once they are defined they again provide a clear boundary to the design problem.

■ *Constraints imposed by the client body and its advisers*. These tend to be far less stringent than the previous two categories, and can sometimes be compromised more easily. Even here some may be inflexible due to a specific demand, which takes precedence over all other needs. An example may be the requirement to design a form which envelopes an expensive manufacturing process; because the plant is so expensive the form of the building must take second place. Another example is a cost limit which is subject to the financial standing of the client; this may, however, be imposed by an outside body such as a government department or finance company, which will define the cost that it sees as realistic. Most self-imposed constraints can be reconsidered in the light of experience and gradual evolution of the design solution.

Definition of the constraints is of enormous assistance in containing the design solution. It helps in narrowing down the range of possible solutions, which for practical purposes are almost infinite without constraints. It is the responsibility of the design economist to account for these controlling factors in the budget and, therefore, to set a realistic strategy of cost. Ignorance of these issues will possibly result in abortive effort and a less than satisfactory service to the client.

9.7.2 'What are the priorities of the scheme?'

In a sense this is the question that sets the self-imposed constraints and whose solution should provide value for money. If priorities can be ranked and given their due importance in solving the design problem, then it should be possible to spend the client's money in accordance with these requirements. This would be the ideal design and would provide the optimum solution. Unfortunately, ranking all the priorities is an extremely difficult task. For example, should maintenance-free windows take priority over an improved reduction in noise levels between rooms in a commercial office block? It is difficult to compare the two, and it is even more difficult to award a satisfactory weighting. However, this kind of decision is at the root of good cost budgeting and, if possible, it should be made in conjunction with the designer. The theoretical aim should be to spend money in the same order and importance as the priorities.

Whether this can ever be achieved is a debatable point, as an item which is not given a high priority by the client may yet be essential, for example easy-to-clean windows on a multi-storey block. To provide this item may be more expensive than providing an item of greater ranked importance. This does not mean that the higher ranked item is of less value to the client, but just that because of external economic forces and the nature of buildings, the client has to pay more to achieve the desired objective. This emphasises the difference between the two concepts of value put forward by the economist Adam Smith: value in exchange or value in use. In most buildings of a public or social nature, it is value in use that is being considered, whereas in a speculative housing or commercial development, it is value in exchange (which is much more dependent on scarcity factors) that is the major concern. There is, of course, a link between the two because the greater the degree of user satisfaction, the more likely it is that the client will be prepared to pay more for the building. Value for money is achieved when the priorities of the client, for example profit, symbolism, welfare, religious worship, etc., have been successfully balanced with the initial and future costs allowable for the development.

9.7.3   'How much space is required?'

The purpose of building is usually to provide space for a particular activity, within which the climate is modified so that the activity may function more efficiently. Other considerations, such as the environmental and social impact of the building and its activity, arise out of this prime need. It is therefore usual for the client's brief to give an indication of the usable area required, but even where this is stated there is considerable flexibility, for example the circulation area (corridors, waiting areas, lifts, public areas, etc.) is not usually included in the list. In addition, the multiple use of space (e.g. assembly halls doubling up as dining halls in schools) may allow a more efficient use of a certain area and reduce the overall requirement. Part of the design problem is to discover the most efficient use of the spaces required to satisfy the client's brief, and to arrange the areas in such a way that circulation is kept to a minimum.

As there is a strong correlation between the area of a building and its cost, the design economist should be an active participant in the discussion of areas and their spatial arrangement. Statutory requirements with regard to means of escape in case of fire, disabled access and disability requirements, health and hygiene, etc., in conjunction with the client's requirements for efficient movement in the building, will provide an indication of the minimum areas allowable for circulation and ancillary purposes. A careful study of these needs will provide the constraints for the area and thereby indicate the balance between quantity and quality of building that can be achieved within a given cost limit.

Phase 2

9.7.4   'What arrangement of space is required?'

This question is heavily influenced by the amount of space needed. In answering it, a good designer's knowledge can be exercised to enormous advantage. Indeed, determining spatial organisation has been recognised as one of the most fundamental of the architect's range of skills.

Despite the advent of computer programming techniques which attempt to optimise the positioning of space, the ability of the human architect to take into account a large number of factors and bring them into a suitable relationship has not yet been surpassed.

A number of techniques have been developed to assist the architect in this important task, and perhaps the most common is the association matrix. In this design aid the relationship between spaces is identified in a table, rather like a 'mileage between towns' indicator in a road atlas. The figures in the table, however, will relate to the 'cost of communication or movement' related to a unit of distance between spaces, rather than measured distance. For example, the salary cost of the managing director spending time in walking one metre may be five times that of the administrative clerk. It follows that the MD's office should be given priority in being closer to the centre of communication.

Problems arise in this technique when the subject has no direct wages cost, for example the casualty patient in a hospital who does not cost the hospital administration any salary. Without an artificial weighting factor, hospitals might be designed with the patient always taking the longest route!

The simplest form of the matrix is that in which a simple weighting system is used to define the ease of movement between spaces. To illustrate the use of such a table, let us take the example of a new administration/accommodation wing added to the existing appliance garage of a fire station. The spaces required by the fire authority are as follows:

Blankstown: extension to existing fire station

Space		Area (m²)
(1)	Administration area including watchroom	120
(2)	Lecture room	80
(3)	Firefighters' dormitory	140
(4)	Television room	25
(5)	Recreation room	150
(6)	Mess and officers' lounge	200
(7)	Visitors'/administration toilet and washroom (male and female)	25
(8)	Visitors' entrance	10
(9)	Firefighters' entrance	20
(10)	Firefighters' ablutions area	40
(11)	Appliance garage (existing)	820

The first step is to identify a suitable scale for the ease of movement required between spaces. The following may be considered reasonable:

Weighting	Degree of ease of movement
5	Essential
4	Desirable
3	Tolerable
2	Undesirable
1	Intolerable

By considering the relationship of each space to the other spaces, an association matrix can be established using the above key.

It can be seen from the developed matrix (Figure 9.3) that the crucial space is the existing appliance garage. This is fairly obvious as the efficiency of the station is centred around the ability of the firefighters to reach their equipment quickly when an emergency call is received. The next step is to arrange the spaces in a diagrammatic form to show the desirable clusters of accommodation. This is usually shown by means of a bubble diagram identifying the spaces and their required links. In arranging the groups of spaces, consideration needs to be given to important constraints, such as the site not allowing ideal groupings to be made. For example, the parking and practice area required behind the fire station may restrict the usable site area and force a two-storey solution. The resultant bubble diagram incorporating the site constraint and association table may be as indicated in Figure 9.4.

The bubble diagrams reinforce the relationship between spaces and emphasise the view that the appliance room is the controlling factor in any arrangement. In addition, the size of the spaces is illustrated, and a visual indication of the likely room clusters is given ready for incorporation into a suitable building form. The efficiency of spatial arrangements in building is not a field that has involved the cost adviser to any great extent in the past, apart from commenting on the ratio of circulation space to usable floor area. In buildings in which ease of movement between spaces is a priority, such as those housing industrial processes, this area of study may prove fruitful for cost research and investigation as the layout has a major effect on the overall economy of the scheme.

9.7.5 'What form should the building take?'

Answering this question involves another of the essential design skills of the architect, who should be able to translate the functional spatial arrangement of the bubble diagram into a building form that will reflect the relationships determined. In doing so the architect will also be aware of the constraints which usually reduce the design options considerably, the largest constraint in many cases being the site itself. In other cases, planning requirements or cost limits may be paramount.

In the case of the Blankstown fire station, the available site area is considerably reduced by the need for a practice yard at the rear. The orientation is also fairly limited because of the desirability of having a frontage on to the High Street. The resultant plan is an attempt to segregate administrative

Phase 2

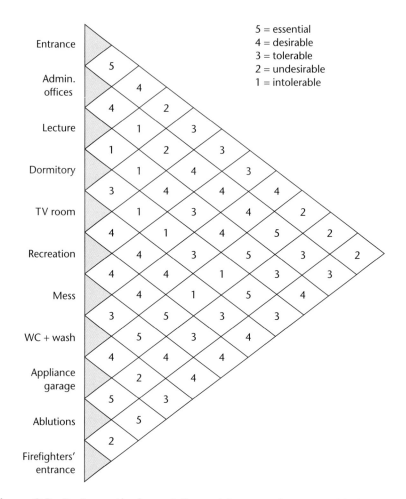

Figure 9.3 Design method association matrix – ease of movement between spaces, extension to Blankstown fire station.

areas from recreation/work areas and at the same time provide speed of access to the appliances from any part of the building or surrounding areas.

Figure 9.5 illustrates the first sketch of a design that may be suitable. Note that from force of circumstances the architect has had to use open space in rooms for through access to the appliance garage. The visitors'/administration toilet facilities have had to be split between ground and first floor to obtain male and female facilities. In addition, an attempt has been made to isolate noisy areas from the quiet/rest rooms, and the recreation block has been set back, thus providing passing motorists with a better view of the garage doors – very necessary in an emergency.

Phase 2

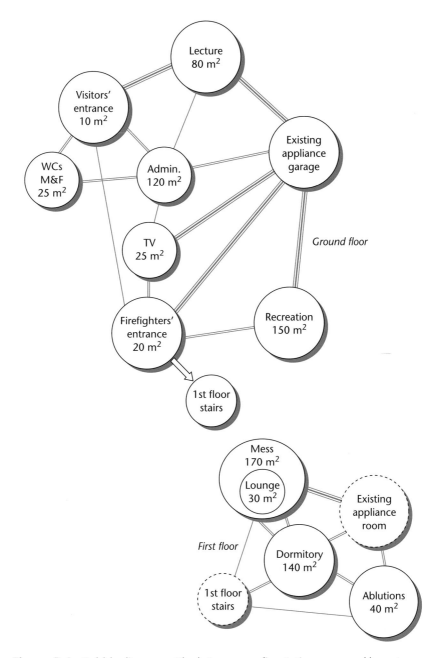

Figure 9.4 Bubble diagram – Blankstown new fire station proposed layouts.

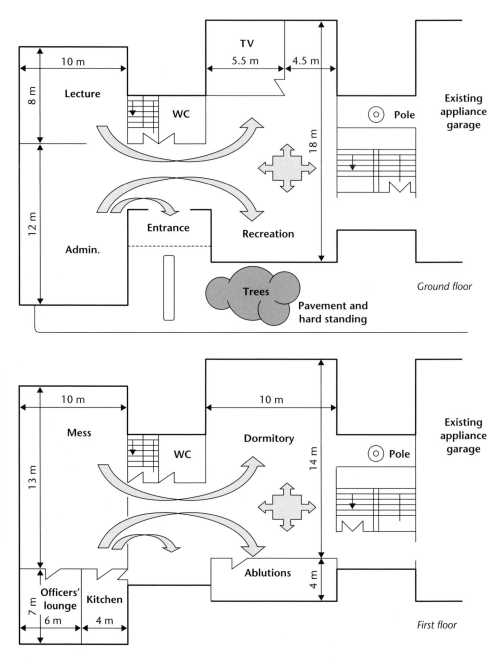

Figure 9.5   Sketch plans for Blankstown fire station.

It is interesting to note that this is the stage at which the cost adviser is usually first asked for an estimate, yet it is clear that there is already a heavy commitment to space, circulation and form standards. In addition, the specification standard will probably have been indicated in the brief and therefore the degree of cost control that can be exercised from this point on is severely limited. This reinforces the view that cost advice is most effective if given prior to the sketch design. A request to design a more compact form once the architect has already produced a solution is likely to involve considerable effort – and also resentment. One of the aims of good cost budgeting must be to avoid abortive effort and therefore to advise where cost-acceptable solutions are likely to be found.

9.7.6 'What is the level of specification?'

This is very often the decision that has the least external constraint. It is also the design decision that suffers most when cost reduction is required at a late stage of the design process.

At present there are very few numerical techniques that attempt to measure quality. It is therefore a decision that is based largely on intuitive judgement arising out of the previous experience of the design team.

The client's brief will have attempted to give an indication of need, perhaps in the form of a requirement for soft or hard finish in certain rooms, maintenance-free exterior, and so on. Standards of environmental comfort and the need for prestigious public areas, etc. will also have been conveyed by the client.

In these circumstances the selection by the designer has to be narrowed to those materials that will fulfil the function required, and a selection made based on comparative performance. An important aspect of that performance will be the economic considerations of installation and durability that will generate capital, maintenance and operational costs.

One role of the cost adviser should be the production of alternative estimates for each major solution, in order that the cost implications of a decision can be known.

9.8 Generally

The above series of questions is a gross oversimplification of the design problem and the text has been orientated towards those factors which have cost implications. The object has been to give the student, in disciplines other than architecture, a grasp of the early stages of the design problem and the potential for economic advice, and in particular to dispel the common misapprehension that the architect starts work by sketching a building form neatly divided up into rectangular rooms and passages.

Phase 2

However, it should not be assumed that a linear chronological order has necessarily been implied by the sequence of the above questions. In some cases a linear sequence may be the method adopted for the solving of a particular problem, but in the vast majority of cases there will be a good deal of overlap. There may be simultaneous decisions covering all the points discussed, and possibly reversal of the process.

The above procedure has been largely a process of designing from the inside out, from internal space requirements to external form. In buildings where external form is important or the site constraints are very tight, then it may be necessary to design from the outside in.

Each designer's methods and techniques will have been developed to suit that particular person's approach; not every architect, for example, will use numerical methods to assist in deciding on circulation priorities.

There are, of course, a large number of design decisions still to be made on Blankstown fire station and these include:

- the working and arrangement of the building services;
- the sizing and choice of the structure;
- the fixing of components;
- the detailing of the specification;

and so on. Some of these decisions will be made in conjunction with those already outlined, and in fact may depend on those decisions, and vice versa.

Design cannot be neatly contained in watertight compartments of sequential decisions, and this does make the input of information by other consultants more difficult. An understanding of design method will, however, help the cost adviser to know when cost advice is likely to be most helpful in its contribution to solving the design problem. It will also assist in determining the degree of refinement in cost advice that will match the stage of detail in the design process.

9.9   Recognition of design methods in cost information systems

It is generally recognised that any cost information system must be compatible with the design method. However, unless cost is of paramount importance, it should not dictate the approach that the designer takes; if the financial tail wags the project dog this will not normally be conducive to satisfying all the client's requirements.

The cost adviser should also be aware that striving to achieve optimisation in cost is only a part of the total objective. While all consultants would like to achieve optimisation in their own subsystem of activity, it is the performance of all systems acting together that will determine the degree of client satisfaction.

Two examples will suffice (in the following sections) to demonstrate recent attempts that have been made to marry cost advice to design method in the pre-sketch design stage, when the major cost significant decisions are made. The systems are not fully developed, but they give an indication of current trends. Since these are the first of many cost control systems at different levels, it is necessary to highlight that systems such as the following represent overall strategies within which a cost control process operates. It is very rare in practice, however, for the cost adviser to have sufficient resources available to be able to carry out every single step of the strategy in respect of every aspect of the design, and it is doubtful that this would be a very economical thing to do in any case.

Just as a general commanding an army cannot afford to attack all along a front but must concentrate its efforts at the most sensitive points, so the cost adviser's resources must be used where they will be most effective. This level of intuition and understanding is probably one of the most important and difficult parts of the cost adviser's job, and is one of the things that makes it a truly professional task.

9.10   An evaluative system

In this system the cost adviser attempts to evaluate the cost implications of design knowledge and decisions immediately they are postulated. Design is considered to be the arrangement of spaces into a building format. The procedure could be as follows.

Designer	Cost adviser
Receives brief with floor areas given.	Evaluates cost of alternative finishes.
Organises rooms into groups and location in a building shape and form.	The form generates a type of production method and site transportation which the cost adviser evaluates.
Looks at the structure needed to support the chosen form.	Evaluates cost of alternative forms of support.
Selects envelope strategy.	Evaluates alternative specification of external walls and roof.
Organises circulation and services layouts, etc.	Evaluates alternative arrangements of lifts, service runs and corridor space.

There are two problems with this method:

- It assumes a linear design method, with each event occurring after the other, although in practice these decisions are made on an iterative process. This concept is closely related to the WLCC process, which again should be iterative and conducted simultaneously with design.

Phase 2

■ It depends on the generation of information by the designer before evaluation can take place, and therefore gives no indication as to where to look for a good cost solution prior to a committed decision being made on building form.

It is, however, a significant step forward in providing an evaluation of the decision-making process prior to sketch design.

9.11   A strategic cost information system

By use of this method the cost adviser attempts to explore a range of possible solutions available to the designer using a structured search process in which the designer is involved. The object is to identify a cost strategy which the designer can employ in the synthesis of building form and which will avoid abortive redesign at a later date. The procedure may take the following form.

Design team	Cost adviser
Receives brief with accommodation area and quality standard given.	Explores the cost of different components of the building according to changes in the major design variables (e.g. the area and shape of a bay in a structural frame).
Selects the specification and parameters for each component from the explorations undertaken.	Explores the use of these components and parameters in buildings of different shape and height according to a predetermined series of building descriptions, e.g. plan shape, number of storeys, density of partitions. A cost table is produced of the feasible solutions available within the identifiable constraints.
Identifies the lowest total cost from the table and uses it as the point of reference for selecting any other alternative.	Contributes to the discussion of an alternative solution.
Chooses a solution from the table.	Provides a breakdown of the major component costs, which is then adopted as the point of reference for the initial cost plan for use in traditional budgeting. The design variables incorporated in the selected solution are communicated to the designer.
Uses the parameters and descriptions as guidelines in the preparation of the design solution.	Uses traditional budgetary procedures to maintain control.

The descriptions conveyed to the architect for the design strategy should not be so rigid that a straitjacket is imposed, resulting in only one possible solution. Rather, they should be coarse measures which still allow reasonable scope for using the designer's skills of modelling and spatial organisation within the strategy adopted by the team as a whole. The choice of descriptors is important in making sure that the creativity of the designer is not stifled.

The advantage of this method is that it attempts to undertake a systematic search through possible solutions in order that the design team may identify the preferential options based on cost. It identifies a 'least cost' solution, which can then form the point of reference for any alternative selection. It is, therefore, possible to obtain a gauge of value by comparing a specific choice with the lowest cost.

There are, however, quite a number of problems. The setting up of models that will cope with the large number of components and design variables is an onerous task, even though the models need not be very refined at this stage of the design process. It is also extremely difficult to produce a satisfactory range of descriptors that will be simple enough for the design team to incorporate into their thinking and also comprehensive enough to allow a reasonable evaluation of the building as a whole.

Both the evaluative system and the strategic approach depend heavily on the use of computers to provide the information sufficiently quickly for decision-making purposes.

9.12   Key points

An understanding of design method enables the cost adviser to know:

- the time when cost advice will be most effective;
- the type of advice that needs to be given;
- the objectives of the design team additional to minimisation of cost.

It would appear that the earlier the advice is given, the greater the chance of the advice being incorporated into the final design solution. The key is the need for techniques that contribute to the analytical understanding of the problem and assist in the convergence of the best solution within the shortest possible time. The development of cost models and IT-based solutions appear to offer a way forward in achieving these objectives.

Note

(1)   The process map was published in the Final Report of EU TG4: *Life-Cycle Costs in Construction*, 2003.

Further reading

Broadbent, G.H. (1995) *Emerging Concepts in Urban Space Design*. Spon, London.

Kirkham, R.J., Alisa, M., Pimenta da Silva, A., Grindley, T., Brøndsted, J. (2004) Rethinking whole life-cycle cost based design decision making. *Proceedings of the 20th Annual Conference of the Association of Researchers in Construction Management*, Heriot Watt University, Edinburgh. ARCOM: www.arcom.ac.uk

References

BAA (1995) *The Project Process: A Guide to the BAA Project Process*. Internal publication, BAA, London.

Blyth, A. and Worthington, J. (2001) *Managing the Brief for Better Design*. Spon Press, London.

Kagioglou, M., Aouad, G., Cooper, R. and Hinks, J. (1998) The Process Protocol: Process & IT Modelling for the UK Construction Industry. *Proceedings of 2nd European Conference on Product & Process Modelling in the Building Industry*, October 1998, Salford. Salford University, Salford.

Koskela, L. (2000) An exploration towards a production theory and its application to construction. *VTT Building Technology*, VTT Publications. www.inf.vtt.fi/pdf/publications/2000/P408.pdf

Latham, M. (1994) *Constructing the team*. Final report of the government/industry review of procurement and contractual arrangements in the UK construction industry. Department of the Environment, HMSO, London.

Macmillan, S., Steele, J., Austin, S., Kirby, P. and Spence, R. (2001) Development of a generic conception design framework. *Design Studies*, **22**(2), 169–191.

RIBA (2000) The RIBA Plan of Work. Royal Institute of British Architects, London.

Chapter 10
Standard Methods of Cost Modelling

10.1 Prototypes

When a manufacturer, say Audi for example, produces a new motor vehicle, the process of product development that they adopt will involve the construction of a prototype model. Often a number of prototypes will be developed before a final design solution is attained. Audi will do this for a number of reasons, including:

- To identify and solve three-dimensional problems that are not apparent in the drawings.
- To identify the tools required for production.
- To help estimate the cost of production.
- To test and evaluate its functional performance.
- To test its marketability.
- To provide a sample for a customer as evidence of the quality standard to be achieved.

By building, manipulating and testing the prototype, the manufacturer can iron out and avoid future problems when the product is actually manufactured, supplied or sold. However, it is not always feasible to build a prototype, and in the case of buildings there are particular problems:

- Buildings are very large and prototyping would be prohibitively expensive.
- They are intended to last for a very long time, and it would be difficult to simulate this in prototype testing.
- Normally, only one production model is to be constructed and the prototype costs cannot be written off over a large production run.

Therefore, because of the time needed for construction and the great expense and the individual nature of buildings, it is not realistic to construct trial examples of a whole project to see whether or not it will work satisfactorily. This concept is interesting since the Egan report advocated the merits of understanding processes such as manufacturing and production and applying these to the construction industry. This contention elicited some criticism from the industry, particularly given the significant variance between the manufacturing scenario and the traditional construction environment.

It may, however, be possible to build and test samples of major components, and this possibility may need to be taken into account in cost planning a building which is to incorporate some measure of innovation or a great degree of repetition.

It might be interesting at this point to mention the Sydney Opera House, a building of great prestige and enormous innovation, where the construction and testing of prototypes and mock-ups was undertaken on an exceptional scale. The result of this was that very few problems were encountered in the use and maintenance of the building, considering the potential for trouble in a project of this nature, but of course a considerable penalty was paid in terms of construction time and cost and the political fallout that ensued (see the further reading section for an excellent text on this iconic building).

10.2   Other types of model

If physical prototypes are not normally possible, the design of a building must still be assessed in some way to try and ensure that the demands of the client are going to be satisfied. For the sake of expediency and cost it is, therefore, necessary to construct models that represent the real situation in another form, or to a smaller scale, so that a realistic appraisal of performance can be made. There are many approaches to doing this including:

- physical (such as a typical architectural model);
- three-dimensional (i.e. CAAD systems);
- mathematical;
- statistical (where data is used perhaps to forecast or identify trend or variance);
- simulation (using mathematical algorithms to evaluate building performance – Hevacomp[1] for example).

Today, Computer Aided Architectural Design (CAAD) is the most common approach to modelling building designs. The power of modern day CAAD systems in allowing designers to produce three-dimensional, fully rendered designs represents a significant step forward. The first generation of CAAD and the geometric modelling that is associated with it were utilised widely in manufacturing industries. Realising the potential, designers in the construction industry readily adapted to this new technology.

Cost is one of the measures of function and performance of a building and should therefore be capable of being modelled in order that a design can be evaluated. In recent years a considerable effort has been made to construct models that will help in the understanding and prediction of the cost effect of changing design variables.

Cost modelling may be defined as the symbolic representation of a system, expressing the content of that system in terms of the factors which influence its cost. In other words, the model attempts to represent the significant cost items of a cash flow, building or component in a form that will allow analysis and prediction of cost to be undertaken. Such a model must allow for the evaluation of changes in such factors as the design variables, construction methods, timing of events, etc.

The idea is to simulate a current or future situation in such a way that the solutions posed in the simulation will generate results which may be analysed and used in the decision-making process of design. In terms of quantity surveying practice this usually means estimating the cost of a building design at an early stage to establish its feasibility.

10.3   Objectives of modelling

Models for estimating and planning costs have evolved gradually. Their adoption by the profession at large has led to the establishment of what might be called traditional techniques. These will be discussed and referred to in this chapter, together with some more recent developments.

At this stage it is useful to consider what a good cost model should be attempting to achieve. Traditional and future models can then be tested to see to what degree they comply with this set of requirements. The broad objectives can be listed as follows:

- To give confidence to the client with regard to the expected cost of the project, i.e. economic assurance.
- To allow the quick development of a representation of the building in such a way that its cost can be tested and analysed.
- To establish a system for advising the designer on cost that is compatible with the process of building up the design. This should be usable as soon as the designer makes the first decision that can be quantified, and should be capable of refinement to deal with the more detailed decisions that follow.
- To establish a link between the cost control of design and the manner in which costs are generated and controlled on site. This involves dealing with the cost of resources at as early a stage as possible to aid communication between the design team and those responsible for managing the construction process.

Phase 2

Linked to these objectives will be some guiding principles that can be applied to the way we approach and verify the model. These may be summarised as follows:

- The degree of refinement of the model should be tailored to suit the stage of design refinement, and should call on as much design information as is available at the time when the model will be used. The cost data applied to this information should represent the degree of reliability that can reasonably be expected from an estimate at that stage.
- The cost data in the model should be capable of updating and evolution in the light of changes in external market and environmental conditions, without too much time being involved.
- The representation of the building or component in the model should bear an understandable relationship to design method (e.g. the arrangement and use of space) and if possible to the manner in which the costs are incurred (e.g. the production method). Ideally the model should show the relationship between the client's objectives expressed in the brief and the subsequent cost of resources used to achieve these objectives. This is, however, an extremely difficult task as many of these objectives will be of an intangible nature.
- The model should cope with constraints imposed on design and be able to test the feasibility of a proposed solution within these constraints in order that definite decisions can be made.
- The results of the model should enable this knowledge to be incorporated by the designer into the drawings, specification and quantities, so that they form part of the strategic decision-making process.

10.4 Traditional cost models

If the above definitions are understood, it becomes clear that QSs have been using a form of modelling technique for a number of years. In their measurement for BQs they have been representing the building in a form suitable for the contractor's estimator; and when prices are applied to the measured quantities, the BQ becomes a representation (or model) of the cost of the building. By altering the quantity of the measured items or changing the price according to variations in specification, it would be possible to evaluate the effect on cost of manipulating certain design variables.

However, the BQ has to be prepared at a very late stage of the design process, and any information obtained from changing the quantities or price rates would come too late to avoid abortive design effort. An estimate obtained from a BQ would also be too late to give an indication of the client's likely cost commitment at the outset of the project or to allow any cost control to be exercised.

10.5   Horses for courses

There is, therefore, an obvious need to employ much simpler models at an earlier stage of design to overcome these problems. The complexity of the model will depend largely on the amount of information the designer can provide to the cost consultant since:

- There is little point in using a complex model that takes into account shape and layout of the building if all that has been determined is an idea of the approximate area of accommodation.
- Conversely, it is wrong to use an oversimplified costing technique when the building form and specification are known and sufficient time is available to do a more thorough job.

As is so often the case, the selection of the most appropriate technique is a case of selecting the right 'horse for the course'.

10.6   The pyramid

Figure 10.1 shows some of the more traditional models that have been developed over the years to suit various stages of the design process. The pyramid is an attempt to show that more detail is required in the structure of the model as one descends the list.

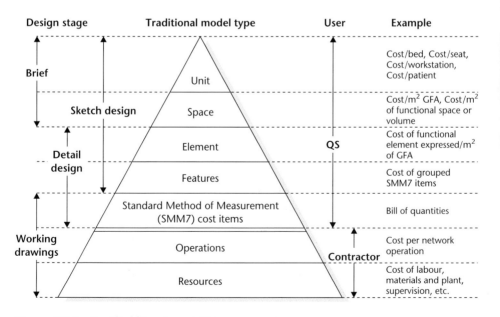

Figure 10.1   Traditional cost modelling.

10.7   Single price rate methods

The first two model types, unit and space, are basically single price rate methods of estimating cost, and those of practical application have already been dealt with in Chapter 5.

10.7.1   The storey enclosure method

This single price rate system is now of historical interest only, but it was the first attempt to compensate for such factors as the height and shape of buildings. In this method the areas of the various floors, roof and containing walls were measured, each then being weighted by a different percentage and the resultant figures totalled to give the number of storey enclosure units. The rules were:

- The area of the lowest floor multiplied by 2 (or by 3 if below ground level).
- The area of the roof (measured flat).
- The area of the upper floors multiplied by 2 and plus a factor of 0.15 for the first floor, 0.30 for the second, 0.45 for the third, and so on.
- The area of the external walls, any part below floor level being multiplied by 2.

It was recommended that lifts and other engineering services should be excluded from the calculation and worked out separately. This is common practice in single price rate estimating and applies equally to the cubic and superficial methods.

The system depended heavily on the weightings, which were permanently built into the rules, and these were unlikely to apply equally to every building. In addition, the measured units tended to be abstract, not relating to the physical form of the building or to the client's accommodation requirements. Therefore the technique suffered from deficiencies similar to those of the cube method. The paper presenting the method (in the early 1950s) claimed that in tests it had performed far better than the other single price approaches, but lack of use meant that it was never possible to verify this claim in practice. It never achieved wide acceptance because it was quickly superseded by the elemental estimate.

10.7.2   Comparison of single price methods

Figure 10.2 shows each of the cube, superficial area and storey enclosure methods applied to a multi-storey office block. The actual cost of each building is known and the analysis of building A is used in each case to forecast

Phase 2

Dimensions

Office A

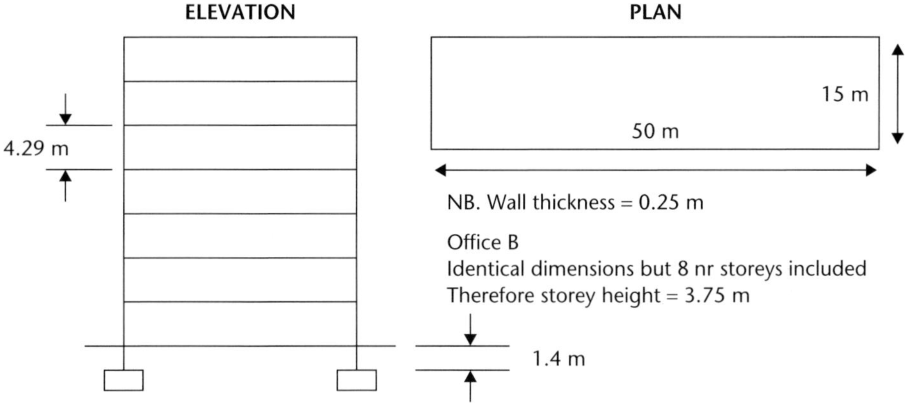

ELEVATION   PLAN

4.29 m

15 m

50 m

NB. Wall thickness = 0.25 m

Office B
Identical dimensions but 8 nr storeys included
Therefore storey height = 3.75 m

1.4 m

Office A					
Actual cost	= £1,758,000	Cube	= 24,000 m³	Storey height	= 4.29 m
Area	= 5,024 m²	Nr storeys	= 7	Foundation dep.	= 1.4 m

Office B					
Actual cost	= £1,995,000	Cube	= 24,000 m³	Storey height	= 3.75 m
Area	= 5,742 m²	Nr storeys	= 8	Foundation dep.	= 1.4 m

Cubic metre
Office A analysis

$$\frac{£1,758,000}{24,000} = £73.25/m^3$$

Office B forecast
£73.25 × 24,000 = £1,758,000

Error
Underestimate of ≈ 12%

Square metre
Office A analysis

$$\frac{£1,758,000}{5,024} = £349.29/m^2$$

Office B forecast
£349.29 × 5,742 = £2,005,600

Error
Well within acceptable range

Storey enclosure
Office A analysis   m²

Lowest floor	717.75 × 2	= 1,435.50
First floor	717.75 × 2.15	= 1,543.16
Second floor	717.75 × 2.30	= 1,650.83
Third floor	717.75 × 2.45	= 1,758.49
Fourth floor	717.75 × 2.60	= 1,866.15
Fifth floor	717.75 × 2.75	= 1,973.81
Sixth floor	717.75 × 2.90	= 2,081.48
Roof	750.00 × 1	= 750.00
External walls	717.75 × 1	= 3,900.00
		= **16,959.42**

$$\frac{£1,758,000}{16,960} = £103.65 \text{ storage enclosure unit}$$

Office B forecast
As above   16,959.42
Add
Extra floor (7th) 717.75 × 3.05   2,189.14
Total   **19,148.56**
19,149 × £103.65 = £1,984,800

Error
Well within acceptable range

Figure 10.2 Comparison of single price estimating models using a cost analysis of office block A with forecast for office block B.

building B. The buildings have exactly the same dimensions, but A has seven storeys and B has eight storeys within the same envelope.

■ It can be seen that the cubic method has not taken the extra floor into account and there is, therefore, a shortfall of approximately 12% on the estimate.
■ The superficial method, on the other hand, has overcompensated for the extra floor by not taking into account the fact that the envelope area remains the same. It has, however, produced a more reasonable figure.
■ The storey enclosure method, by considering both envelope and floor, has certainly produced an even closer estimate well within acceptable limits for this kind of early exercise.

However, in each case a change in specification, site or location would result in additional problems and call for the skill of the cost adviser's judgement to compensate for the changes. In addition, changes in plan shape and storey height would affect the cubic and superficial methods to a greater or lesser degree. These alterations, or variables, are enormously complex and difficult to assess.

Although adequate for establishing a budget, it will also be remembered that such an estimate suffers from the defect that during the development of the design it is not possible to relate the cost of the work shown on the working drawings to the estimate.

It is for these reasons that single price rate methods now tend to be rejected except for the very earliest estimates where little is known about the building form.

10.8   Elements

The model by which design cost planning has been best achieved is that of dividing the building into functional elements (level 3 of the pyramid in Figure 10.1). Elements are defined as major parts of the building that always perform the same function irrespective of their location or specification. For example:

■ Internal partitions always vertically divide two internal spaces.
■ A roof encloses the top of a building and keeps the weather out and the heat in.
■ The substructure transmits the building load to the sub-soil.

These elements have some relation to the design process, can be readily measured from sketch drawings and are easily understood by all parties including the client, thus aiding communication.

Example

Suppose an external wall element costs £80,000 and the area of the wall is 1,000 m². The cost per square metre of wall (known as the elemental unit rate) is £80.00. If there are 1,500 m² of the same type of wall on the proposed project, then the cost will be:

$$\frac{£80,000}{1,000 \text{ m}^2} \times 1,500 \text{ m}^2 = £120,000 \tag{1}$$

A cost index (see Chapter 12) is then used to update the prevailing price levels to the proposed tender date, and the new estimate for that part of the building has been established. Having set out a series of estimates, one for each element (known as cost targets), it is then possible to consider each element in turn instead of trying to cope with the whole building. Costs can therefore be monitored as the design develops.

10.9   Features

Cost models based on features (groups of abbreviated quantities) rather than elements were popular for a period among some major local authorities, but this non-standard approach was never widely used and disappeared with the decline in local authority building programmes.

10.10   Standard Method of Measurement (SMM7) – the BQs as a cost model

Level 5 of the model pyramid shows the position of the BQ in terms of the detailed modelling used in traditional techniques. There has been an attempt in the current revision of the rules for measuring BQs (SMM7) to orientate the measurement of the building to the way costs are incurred on site. However, the vast majority of items are still largely measured 'in place', that is to say they are measured as fixed in the building with no allowance for waste and no identification of the plant required during installation.

Consequently the BQ remains primarily a document for obtaining a tender in a short time, rather than a document for management and cost control of construction on site. It is not possible, for example, to ascertain from this document the lengths of timber required for roof joists (although this must have been known by the measurer) or the hoisting requirements for materials and components, unless the document has been annotated for the purpose.

In the case of levels 1 to 4 of the pyramid in Figure 10.1, the data used is actually based on an analysis of a BQ, which is somebody else's view (the estimator's) of what the firm needs to charge for the resources, plus profit and overheads. Because of the large number of assumptions on which a BQ price is based, it is most unlikely that the real mix of resources that will be used for the project has been determined by the estimator. The factor that allows cost techniques based on the BQ to work is the knowledge that the overall cost of

Phase 2

the job, which has been analysed to provide data for the model, is the going rate for that building.

10.11   Operations and resources

The last two levels, 6 and 7, are more closely related to construction.

A contractor's network outlines the activities or operations in the order that they will be undertaken on site. For example, external cavity wall brickwork, which may be one item in a BQ, will be broken down into floor levels and possibly into zones marked on the floor plan. This aids site management and the organisation of labour, plant and material.

These basic resources form the most detailed level of modelling as they are the ingredients of the production process. In fact, all the other models used for cost forecasting entail the assumption that the resource requirements of the building will happen to work out all right (because they usually do).

10.12   Spatial costing

Some 30 years ago a variation on the superficial and elemental methods was developed under the title of spatial costing. In this method the basic concept was the cost of a room of a certain type, taking account of its floor area. There is a high correlation between the cost of room finishes, fittings, ceilings and walls, and the activity for which the room is designed.

If, at the brief stage of design, all that is available is the accommodation schedule (i.e. the list of required rooms), then it should be possible to forecast the cost of these items quite reliably as a function of the areas and, if possible, the shapes of each room or space. The other major items of structural frame, foundations, external envelope and main services would then be considered separately.

The wall costs attaching to a room would be the cost of the finishings plus 50% of the structural partition; similarly with the floor and ceiling. The extra cost of external walls over and above the 50% partition costs, and the extra cost of roof over the ceiling costs, are added on a measured basis, as are the other exclusions which have been mentioned.

A major problem in using this system, or any other system which depends on BQ analysis, is the chicken-and-egg situation whereby:

- no system is going to become widely used until there is a considerable body of data to support it; but
- no-one is going to spend money on analysing BQs to provide data for a system which nobody is yet using.

The cost and difficulty of preparing detailed analyses have been contributory causes of the lack of data available for elemental cost planning. The room concept requires even more complex analysis and therefore would be even more expensive to use. The system also lacks the simplicity of the elemental form and the clear communication of what is meant and included.

10.13   Synthesis

Although single price rate estimates have their drawbacks, these simple models of building cost are very useful for budgeting and other very early estimates where little information is available about the configuration of the proposed building. Of these the most commonly used is the superficial floor area method, and there is a good amount of data available for it. At later stages more detailed methods, such as elemental cost planning, should be used. However, a cost plan without subsequent cost control is still really nothing more than an approximate estimate.

10.14   Design-based building cost models

Building cost models are basically of two kinds:

- A product-based cost model is one that models the finished project.
- A process-based cost model is one that models the process of the project's construction.

It is often argued that since it is the process that actually generates the costs, the second of these two models must be the more accurate. However, the construction process cannot be modelled until the form of the building has been postulated, and therefore the process-based model has little place in the scheme of things in the early design stage. In fact, an attempt to model the construction process at too early a stage can have the effect of overriding the design process in order to arrive at a bricks-and-mortar solution before the user criteria have been properly worked out. Construction process modelling therefore has its place at a later stage in the cost planning process, and will be dealt with in due course. For the moment, therefore, we must be concerned with modelling the product. Such a model has to be based on data relating to finished work.

The simplest form of such a model takes no account of the configuration, or details of design, of the building but is based simply on one of the following: the floor area of the proposed project (gross or net); the volume of the proposed project; and some user parameter such as the number of pupil places for a school or number of beds for a hospital. It is customary to exclude site works from the single price rate calculation and to estimate them separately, since their cost obviously has no relationship to the size of the building. The engineering services are also sometimes treated in the same way (with rather

Phase 2

less justification). After World War II, and before cost planning systems became established, single price rate estimates increasingly acquired a reputation for inaccuracy, but this was because there was no follow-up process for keeping the estimate and the design in tune as the design developed. Such an estimate merely attempts to forecast that a building of a certain size can be built for a certain sum of money. It cannot analyse whether a particular design is going to meet that cost. Of course, it is possible to weight the estimate subjectively on the grounds that the proposed solution looks to be at the expensive end or the low-cost end of the market, but here we are well into the realms of guesswork. However, although it cannot say that a particular design will achieve the required result, the single price rate approach has an important role as the first stage of a system which in the end will do that. It has the great advantage that it can be used before any design has taken place, whereas even the most simple process-based models require a tangible design as their basis.

10.15   The bill of quantities

The traditional BQ is a very good example of a product-based cost model. Because of its insistence on measuring finished work in place, the BQ has often been criticised by production-oriented people (although it never purported to be a production-control document), but it does have the virtue of providing a total cost model within a single document. There are two important points in favour of the product-oriented approach to this definition of the project:

- What the client will be contracting for is finished work in place, not a production process. The model is therefore couched in contractual terms.
- The design team can define and categorise the finished work according to an industry agreed convention (SMM7). This is not the case with the production processes, which depend as much on the methods of the particular constructor as on the details of the design, and which cannot easily be analysed or categorised outside the context of the particular project – there is no agreed basis for doing this.

Although the BQ is a useful cost model, it is not usually available until the design of the project is completed and it is therefore of little use for cost control of that design. Its great importance, however, is as a source of cost data for subsequent projects.

Since the BQ is a major source of cost data it is important to be quite clear about its role. It is essentially a marketing document, not a production control document, and the rates in it are prices not production costs. Although there clearly has to be a relationship between the builder's prices and the builder's costs, this really only applies at the level of the total project. The building firm cannot afford to undertake the project for less than its costs, and competition will usually ensure that it cannot charge an unreasonable profit on top of them.

This does not apply to individual pieces of work, however, because the rates in the BQ are not separately offered prices in the sense of the supermarket shelf. You can only buy the brickwork at so much a square metre if you also buy the carpentry at whatever price is charged, so these rates are nothing more than notional breakdowns of the total price, and are made with commercial rather than cost control objectives in mind. Very often the client's QS will object if the rate for a particular item is very different from that usually charged, but there is very little sanction to enforce an objection, except the rather impracticable one of advising the client not to accept the tender.

10.16   Elemental cost analysis

Elemental cost analysis is perhaps the most common product-based cost model and provides the data on which elemental cost planning is based. This technique is currently used by the quantity surveying profession at large and, in spite of a number of failings which have become apparent over the years, has made possible a degree of control over costs which was previously unknown. It is the experience gained with this system, and the attitude of mind which it has engendered, which has led the way to more fundamental approaches to cost control. Because elemental cost analysis was developed in order to provide data for the preparation of a design cost plan, it is first necessary to look at the requirements of the latter. This is done in the context of the traditional competitive tendered type of contract, which is still popular but which at the time the system was developed was almost universally used. What are the essential features of a cost plan? The obvious requirement (common to any form of estimate prior to tendering) is that it should anticipate the tender amount as closely as possible, but there are two particular requirements that must be fulfilled before the estimate becomes a cost plan:

- It must be prepared and set out in such a way that as each drawing is produced it can be checked against the estimate without waiting until the whole design is complete. This enables any necessary adjustments to be made before the drawing is used for the preparation of quantities.
- It should be capable of comparison with other known schemes in order to see whether the amount of money allocated to each part of the building is reasonable in itself and is also a reasonable proportion of the whole. From the point of view of economical building, this second requirement is possibly even more important than the first.

It can be seen that this philosophy is especially suited to work in the public sector where, as we have seen, the only real cost criterion is comparison with

Phase 2

what has been done before. At the time the system was developed such work dominated the building programme; it is possible that a totally different approach might have been used had profit development then been dominant. However, once a system has become established it tends to form the basis for future development as circumstances change, and this is what has happened with elemental cost planning, still widely used today. How can the costs of one building project be compared with another? The first suggestion that would occur to a QS would be to look at the priced BQs for the two contracts. Suppose we look at the summary pages of the BQs for two different schools:

Cost item	Clay Green School (£)	Woodley Road School (£)
Preliminaries	57,300	129,000
Excavation	24,240	28,980
Concrete work	129,915	272,925
Brickwork and blockwork	154,305	101,145
Masonry	10,950	—
Asphalt work	21,150	—
Roofing	74,655	39,045
Woodwork	228,675	290,895
Steelwork and metal work	189,660	197,850
Plumbing, engineering and electrical work	319,500	399,825
Floor, wall and ceiling finishes	136,725	133,275
Glazing	14,640	25,530
Painting and decoration	39,135	37,170
Drainage	43,260	33,495
Site works	109,890	79,470
	1,554,000	1,768,605
Insurances and summary items	5,250	37,950
Contingencies	30,000	30,000
	1,589,250	1,836,555

These trade totals tell us very little. We can see that the second school is dearer than the first (it is in fact larger), but it is difficult enough to try and compare the rather varied trade breakdowns without having to make repeated mental adjustments for the difference in size between the two buildings. So a first step is to divide each trade total by the floor area of the respective schools in order to obtain comparative prices per square metre for each trade:

Cost item	Clay Green School (2000 m²) (£/m²)	Woodley Road School (2500 m²) (£/m²)
Preliminaries	28.65	51.60
Excavation	12.12	11.59

Concrete work	64.96	109.17
Brickwork and blockwork	77.15	40.46
Masonry	5.48	—
Asphalt work	10.57	—
Roofing	37.33	15.62
Woodwork	114.34	116.36
Steelwork and metal work	94.83	79.14
Plumbing, engineering and electrical work	159.75	159.93
Floor, wall and ceiling finishes	68.36	53.31
Glazing	7.32	10.21
Painting and decoration	19.57	14.87
Drainage	21.63	13.40
Site works	54.95	31.79
	770.00	707.44
Insurances and summary items	2.63	15.18
Contingencies	15.00	12.00
	794.63	734.62

We now have the costs on a more truly comparable basis, and it turns out that the first school is in fact the more expensive of the two. Clay Green has:

- a steel frame;
- timber pitched and tiled roofs;
- mainly brick-faced external walls with window holes;

while Woodley Road has:

- a reinforced concrete frame;
- felt flat roofs on composite decking;
- large wall areas of metal windows with concrete panel in-filling.

As we would expect, Woodley Road shows a far higher total for concrete work and a lower figure for brickwork, roofing and steelwork. At Clay Green there are some asphalt-covered concrete flat roofs and the crawl-way floor duct is tanked in asphalt, whereas at Woodley Road the floor ducts are water-proof rendered. The glazing figures allow for the larger window area at Woodley Road, while the drainage and site works figures are consistent with Clay Green being on a larger and more rural site with a consequently greater extent of external works.

The differences in floor, wall and ceiling finishes and decoration are simply due to a higher standard of specification at Clay Green. However, while the effect of these differences in specification can be traced to some extent in the trade costs per square metre, the inclusion of parts of several elements of the building in one trade section makes it impossible to carry comparisons very far. For instance, the figures for Woodley Road for concrete work are affected by:

Phase 2

- the reinforced concrete frame in lieu of structural steelwork;
- some concrete walls in lieu of brickwork;
- the concrete facing panels in lieu of faced brickwork.

We do not know how much of the difference is due to any one of these causes. Similarly, the steelwork figures are affected by the omission of the steel frame and the increased area of higher quality metal windows at Woodley Road. These are compensating differences and in fact the two figures do not reflect the scale of the variations between one school and the other in this section. Other sections are similarly affected; the rates for woodwork are almost meaningless without detailed breakdowns.

Although the figures we have obtained are interesting, they do not enable any really valid cost comparisons to be made. We do not know whether the steel and the concrete frames are competitive in cost, and we cannot tell how much is saved by a felt roof instead of a tiled one. To be able to make such comparisons the BQ has to be split up in a different way – to divide the work into elements. An element has been defined as 'that part of the building which always performs the same functions irrespective of building type', and (we might add) irrespective of specification. What list of elements should be used? The possibilities are nearly endless. As few as six could be used, for example:

- substructure and ground floor, complete with finishes;
- external and internal walls, complete with finishes;
- upper floors including staircases, complete with finishes, and proportion of frame;
- roof, complete with finishes and proportion of frame;
- engineering and electrical services;
- site works.

Alternatively any greater number could be used, within reason; some authorities have used over 40. Points to bear in mind in arriving at a decision are:

- The definition of an element as stated previously. Any element chosen must be capable of being defined exactly, so as to ensure uniformity between the elemental breakdowns of any number of contracts, even if the breakdowns are done by different people.
- The element must be of cost importance.
- The element must be easily separated, both in measuring from sketch drawings and in analysing BQs.
- The list of elements chosen should be capable of being reconciled with those used by others, for comparison purposes.

A cost planner who does not have access to the records of a very large firm or other organisation will frequently need to make comparisons with analyses produced in *Building* (such as those by Davis Langdon) and elsewhere, or with analyses obtained from the BCIS of the RICS. As an example of the need for standardisation, in most published forms of cost analysis the cost of parapet walls and copings is included under 'roof'. A cost planner might prefer to include it under 'external walls' (it would make the analysis of a traditional BQ easier), but would then be unable to compare the figures for these two elements with the published information, as the basis of measurement would be quite different.

10.17   The Standard Form of Cost Analysis (SFCA)

As the use of elemental cost planning increased, differing forms of cost analysis were developed by those independent authorities and firms who were building up their own cost records, and by journals, which published cost information for the benefit of their readers. The weekly magazine *The Architects' Journal* was one of the pioneers in this field, and produced a detailed set of rules for their published analyses, while the BCIS and others used somewhat different rules.

As previously mentioned, it was obviously desirable that a uniform set of rules should be established so that users could benefit fully from cost data prepared outside their own organisations. The RICS therefore set up a working party to standardise cost analyses. This proved to be difficult because, in practice, it is not possible to define a set of totally independent functional elements that can be related to a BQ. Any standard cost analysis therefore becomes a compromise between independent functions on the one hand and ease of producing the data from traditional documentation on the other.

In December 1969 the first SFCA was published by the BCIS. In addition to its sponsorship by the RICS, the SFCA was also supported by the chief QSs of all the main government departments concerned with building (and this was at a time when these bodies directly controlled a major part of the non-housing building programme). It was this wide measure of support that gave the SFCA such importance, as anybody using a different format would soon have become isolated from the cost experience of the rest of the quantity surveying profession. Thirty years later the SFCA is little changed since it works reasonably well and any alteration would have the effect of making the comparison of old and new projects unnecessarily difficult. As an example of a practical elemental breakdown, the headings in the current SFCA are as follows[2]:

1   Substructure
2   Superstructure
      2.A   Frame
      2.B   Upper Floors
      2.C   Roof
            2.C.1   Roof structure
            2.C.2   Roof coverings
            2.C.3   Roof drainage
            2.C.4   Roof lights

Phase 2

2.D   Stairs
    2.D.1   Stair structure
    2.D.2   Stair finishes
    2.D.3   Stair balustrades and handrails
2.E   External wall
2.F   Windows and external doors
    2.F.1   Windows
    2.F.2   External doors
2.G   Internal walls and partitions
2.H   Internal doors

3   Internal finishes
    3.A   Wall finishes
    3.B   Floor finishes
    3.C   Ceiling finishes
        3.C.1   Finishes to ceilings
        3.C.2   Suspended ceilings

4   Fittings and furnishings

5   Services
    5.A   Sanitary Appliances
    5.B   Services Equipment
    5.C   Disposal Installation
        5.C.1   Internal drainage
        5.C.2   Refuse disposal

etc.

Most other forms of elemental analysis are basically similar, but in order to be successful they must incorporate the same sort of hierarchical principle. It will be seen that the SFCA can be used at different levels of generality, i.e.:

- the element groupings: Substructure, Superstructure, Internal finishes, etc.;
- the elements themselves: 2.A, 2.B, 2.C, etc.;
- the sub-elements: 2.C.1, 2.C.2, etc.

Since the detail is grouped in this hierarchical way, an analysis at level 1 into six items only will be quite compatible with a fully detailed analysis at level 3 of another project.

The construction of the list therefore required a careful selection of elements, each of which was significant on its own but which could form part of a larger significant group. Note, however, that the third level of detail is no longer used by the BCIS in its published analyses – a tacit admission of the unreliability of sub-elemental analysis referred to later in this chapter. One or two forms of analysis which have been used differ by including finishes, windows, and external doors in the external walling element, and including finishes, internal doors, and partitions in the internal walling element. This is more logical but means splitting finishes between external and internal walls, which is difficult to do when analysing a traditional BQ.

10.17.1   Using elemental analyses

If the costs per square metre of Clay Green and Woodley Road schools could be expressed in terms of these elements, it would enable the answers to be found to the questions: Which is the cheaper frame? How do the two roofs compare in cost? These answers would then enable the cost planning of a third school basically similar to Woodley Road but with a steel frame.

10.17.2   The Standard Form, CI/SfB and Uniclass

It was a source of disappointment to many people that in the BCIS form the elements themselves, their grouping and their coding were not the same as in the CI/SfB classification system used by the architectural profession for coding and classifying design information. It would have been very convenient for the architect to have a record of typical costs filed with his design information. However, the incompatibility of the systems was not quite such a disadvantage as might appear. We have seen that the so-called elemental costs obtained at present are not true costs at all, but are only a breakdown of a BQ in which money may have been allocated to the various parts of the work in a fairly capricious manner, quite apart from the overall level of pricing of the BQ itself. It would therefore be dangerous to detach these costs from the analysis and index them for the use of an architect as though they were scientific data like thermal insulation values. One day it may be possible to do this, but at present it is vital that this cost data should only be used by a qualified cost planner who has the experience and knowledge necessary to assess and manipulate it.

The united classification for the construction industry (Uniclass), published in 1998 to replace CI/SfB, contains an element table much closer in form to the SFCA and is a compromise between the needs of librarians and those of the cost planner. It has a flexible form which allows the elements to be ordered in different ways for different purposes. There is one-to-one mapping between the SFCA elements and the Uniclass elements, which will allow elements to be coded in both systems.

10.17.3   Preambles to the SFCA

In the SFCA rather more information is required to be given about the size and nature of the building than was previously customary. The gross floor area is measured in the normal way, that is the overall area at each floor level within the containing walls, but the basement floors (grouped together), the ground floor and the upper floors (grouped together) are each required to be shown separately.

Phase 2

The definition of 'enclosed spaces' means that open entrance areas etc. are excluded from the gross floor areas, although even a light enclosing member such as a balustrade will suffice to include the area. It is possible that doubt might arise when a wall becomes so pierced by blank openings that it no longer acts as an 'enclosing wall'; obviously a series of columns does not meet the definition.

Although 'lift, plant, tank rooms, and the like above main roof slab' are to be included in the gross floor area, there is the option of excluding these if one is prepared to allocate their costs to 'builder's work in connection'. If the plant room is only required because of the existence of a particular service, it seems quite logical to adopt the second course.

10.17.4   Roof and wall areas in the preambles to the SFCA

Roof and wall areas are both measured gross over all openings etc., the roof being measured on plan area. As it is 'walls of enclosed spaces' which are required, parapets, gable ends of unused roof spaces, etc. would not be included in the wall area. The roof, on the other hand, is measured across overhang and would presumably include roofs over open entrance porches and other areas that do not count as 'enclosed spaces'. If there are substantial areas of open covered way it would probably be better to exclude them from the elemental analysis altogether and deal with them under 'site works'. The wall area is not shown on the form, except in calculating the wall-to-floor elemental ratio, which is the wall area divided by the floor area. The lower the ratio, the more economical the design. For this purpose the wall area is normally measured across openings.

10.17.5   Interdependence of elements

Although it is quite easy to define a cost-planning element, it is more difficult actually to divide up a BQ into elements that comply with this strict definition. A major difficulty is that an indivisible building element may have several functions (some of which it shares with other elements). For example, 'external walls' may have all or some of the following functions:

- keeping out the weather;
- thermal insulation;
- sound insulation;
- supporting themselves (dead loads, wind loads);
- supporting floors and roofs;
- transmitting light and ventilation (curtain walls).

If an external wall only performs a few of these functions, then it is obviously unreasonable to compare its cost with that of a wall that performs a greater number. It is therefore usually necessary to refer to the frame and window elements in order to arrive at a true indication of the wall's cost performance, and similar cross-references may be required when costing other elements.

10.17.6   Preliminaries and insurances

These may be shown either as separate costs per square metre (treating them, in fact, almost as extra elements) or allocated proportionately among other elements. This is an important point and deserves some consideration, first with regard to preliminaries and insurances. The cost of any or all of the following items (mostly at the contractor's discretion) may be included in the preliminaries or insurances sections of a priced BQ for a new project:

- huts, temporary buildings, latrines;
- canteen, mess-rooms, site catering staff, welfare;
- huts for clerk of works;
- site architect, consulting engineers, QS, and attendance on these people;
- huts for contractor's own supervisory staff and (on a large project) sub-contractors' supervisory staff;
- mechanical plant, including tower cranes, excavating and concreting plant, lifts, dumpers, etc.;
- scaffolding;
- non-mechanical plant and small tools;
- water for works;
- temporary electricity supply;
- consumable stores;
- temporary fencing and hoardings;
- temporary roads and standings, car parking spaces;
- health and safety requirements;
- cost of agent, foreman and other site supervisory staff;
- cost of timekeeper and site clerks;
- cost of security staff;
- security lighting;
- heating of building for winter working and for drying out;
- temporary weather protection;
- attendance on sub-contractors and artists;
- National Insurance payments;
- superannuation;
- guaranteed week (wet time);
- travelling time and expenses, subsistence and lodging;
- redundancy payments;
- training levies;
- anticipated increases in costs of labour or material;

Phase 2

- bonus or other supplementary payments;
- making good damage and defective work;
- fire insurances, third party insurance and any other insurances required by the client;
- public liability or contractor's all-risk insurance;
- head office expenses (overheads);
- profit.

However, almost any of these items (and certainly any of the major ones) may be included in the rates for the building work instead of being shown separately. Some contractors do not price preliminaries at all, while others do but in such a way that it is impossible for the QS to find out what is or is not supposed to be included. Sometimes the contractor may have second thoughts about a tender at the last minute, and may adjust it by adding a lump sum to preliminaries or taking one off. Thus any large differences between the amounts of money inserted against the preliminaries items on one project and on another are less likely to be caused by genuine contractual differences (site conditions, access, etc.) than by the different pricing habits of the two contractors; remember that the allocation of costs within a contract BQ is done largely for commercial purposes.

Some of the principal items, such as profit, supervision, scaffolding, plant and overheads, could affect the level of pricing of the work sections by 15 to 20%, according to whether or not they are included in preliminaries. As these pricing habits are to some extent regional, it may be possible for the QS's department of a local authority, or for a firm of cost planners whose work is confined geographically, to consider preliminaries and insurances as a separate element with some degree of consistency. However, it would be safer on the whole to add preliminaries and insurances to each element as a percentage, in order to give a common basis of comparison. Referring back to the summaries of the two schools, Clay Green and Woodley Road, we can see that the work section prices for the latter appear low by comparison because preliminaries and insurances have been priced more fully.

This advice, of course, relates only to the analysis of BQs and to the early stage estimates for a new project based on such data. When preparing a more detailed estimate for a major new project, preliminaries and insurances have to be considered on their merits. If the new project has some abnormal feature such as a difficult site or an uneconomically short contract period, it will be necessary to give special consideration to these matters from the start.

10.17.7 Contingencies

Contingencies also have to be considered. Unlike preliminaries, the contingency sum is an arbitrary amount decided by the client or the design team. It is not really part of the contractor's tender but is an amount the contractor is instructed to add to his tender in order that there may be a cushion to absorb unforeseen extras. It normally has no effect on the level of pricing of the BQ, and is better treated as a separate element rather than as a percentage on the remainder of the work.

10.17.8   Analysis of final accounts

On first consideration it might seem a good idea to analyse the final account instead of the tender, since this will give a more accurate picture of the actual cost of the building. There are two objections: it would be much more difficult to analyse both BQ and variation account than to analyse the BQ alone, and the analysis would not be available until perhaps three or four years after a tender analysis and so would only be of historic interest. However, these difficulties do not seem to be insurmountable, and the convention that it is the tender that is analysed probably owes a lot to the fact that the practice of cost analysis started in the days of public sector building, when attention was focused on forecasting tenders rather than final accounts. Although the differences between the tender and final account are not usually great enough to invalidate an analysis obtained from the BQ, this is not always the case.

10.17.9   Cost analysis of management contracts

Management contracts and the like pose a difficult problem. There is little point in analysing the master estimate since there is no contractual commitment to this; it is itself part of a cost planning system and may well have been prepared on an elemental basis. On the other hand, by the time all the costs are known we would be effectively analysing a final account, with all the problems just mentioned. However, management contracts imply a production orientation to the project, and it is perhaps the resource-based cost information obtainable from these projects that is more useful than the product-based costs.

10.17.10   System

Whatever methods of cost breakdown and of cost planning are chosen, they must be adhered to rigidly, otherwise not only are the figures futile for reference purposes, but the whole idea of working to a system is lost. The technique of cost analysis depends on working in accordance with a fixed method; this is why standard forms are preferable. There is plenty of scope for rough working, but the forms must be used at all vital points so that there

Phase 2

is no possibility of preliminaries (for instance) being left out because the cost planner who did the previous estimate believes in showing them separately, whereas the person using this estimate for reference supposes that they are included in the rates. In most cases the forms and instructions issued by the BCIS can be used, thus ensuring compatibility with information obtained from other sources.

10.17.11   Preparation of an elemental cost analysis

Cost analyses are ordinarily carried out today using software tools such as those introduced in Chapter 2, but many cost planners find that commonly available spreadsheet packages enable them to construct their own cost analysis system very easily. The aim of cost analysis is to provide data for use in elemental cost planning; as little time as possible should be spent on it consistent with obtaining a fair degree of accuracy. Meticulous allocation of trivial sums of money, or the identification of insignificant changes in specification, should be avoided.

If there is no time to prepare a full analysis, an outline analysis will be better than nothing and may take less than an hour to prepare. For the preparation of an elemental cost analysis the following are required: a priced BQ, a drawing showing plans and elevations, and a list of elements. Each item in the BQ has to be allocated to one or more elements until every item has been dealt with and the elemental totals will equal the total of the tender. Once an office has adopted a certain form of analysis it will be possible to prepare the BQs with subsequent analysis in mind; this will ease the task of the analyser considerably and will make it unnecessary to refer back to the taking off. It is not necessary to depart radically from the usual order of billing as long as the main elements can be kept separate within each trade or section of trade. For instance, the sawn softwood section of woodwork could be billed under headings of roof timbers, upper floors, and stud partitions; and the reinforced concrete section of the concrete work could be separated into substructure, frame, upper floors, roof, and staircases. This should not involve lengthening the BQ greatly as there will be very little duplication of items between different elements; the concrete work is the only section where this should occur to a significant extent.

A BQ prepared in this manner will not only be useful for analysis but will be convenient for interim valuations and for the contractor's use in site organisation generally and in calculating performance-related payments to operatives. The so-called Northern system of taking off, where the BQ is written straight from the dimensions, lends itself to the preparation of a subdivided BQ of this sort. Any further breakdown of the tendering BQ into a completely elemental format is very unpopular with builders' estimators, since design cost planning elements are of little significance to them and a good deal of rearrangement into trade order has to be carried out by them when sending it out to sub-contractors and suppliers.

10.17.12   Definition of terms

Before going any further there are a number of terms to be defined in order that the processes of cost analysis and cost planning may be understood:

- *Elemental cost* is the cost of the element expressed in terms of the superficial area of the building.
- *Elemental unit quantity* (sometimes also called quantity factor) is the actual quantity of the element, expressed in square metres for such elements as floors, roof, walls, finishes or in terms of number of elements where this is not practicable.
- *Unit cost* is the cost of the element expressed in terms of the element unit quantity, e.g. 1,000 m^2 of internal walling costing £20,000 gives a unit cost of £20.00/m^2.
- *Elemental ratio* is the proportion that the unit quantity of one element bears to that of another. A commonly used example of this is the ratio of external wall area to gross floor area.

When analysing a BQ there are three different ways in which the analyser may be helped to deal with the items:

- The description of the item may indicate the element to which it belongs, in which case there is no necessity to refer to the original taking off.
- The item may be too trivial to spend time on, in which case it may be allocated as seems most obvious or as is most convenient.
- It may be necessary to refer to the taking off in order to allocate the item correctly. Since it would take far too long to do this for every item in the bill, analysers must use a good deal of discretion about this.

In cases of doubt they would be influenced very much by the cost importance of the item.

10.17.13   Element unit quantities

Unfortunately a simple analysis of building cost into elements will not satisfy all the requirements. It will give a money total for each element, and each total can be divided by the floor area to get the elemental cost per square metre, but it does not give the unit quantities or the unit costs. These are needed if the analysis is to be of much use. Some commonly used element unit quantity factors are set out below as a guide; where 'none' is marked, the elemental cost per square metre is the only basis of cost comparison.

Phase 2

Work below lowest floor finish	Area of lowest floor
Frame	Area of floors relating to frame
Upper floors	Area of upper floors
Roof	Area on plan of roof measured to external edge of eaves, but excluding area of rooflights
Rooflights	Area of structural opening measured parallel to roof surface
Staircases	Number of total vertical rise staircases or area on plan
External walls	Area of external walls excluding window and door openings (basement walls to be given separately)
Fenestration	Area of clear opening in walls
External doors	Area of clear opening in walls
Internal walls and partitions	Area of internal walls excluding openings
Internal doors	Area of clear opening in walls
Ironmongery	None
Wall finishes	Area of finishes
Floor finishes	Area of finishes
Ceiling finishes	Area of finishes
Decorations	None
Fittings	Often none, but where appropriate the total length of benches, number of tables or other details may be given
Plumbing and hot water services	Number and type of sanitary fittings, number of hot and cold water draw-offs
Heating services	Heat load in kW, cubic capacity of accommodation served
Gas services	Number of outlets
Electrical services	Number of points, total electrical load
Special services	Such information as will indicate the extent of each service (e.g. for lifts the number, capacity and velocity of each and number of stops)
Drainage	None
External works	None

Example

Assume a building of 2,400 m^2 of total floor area
Element: Upper floors

> Total cost of element: £85,525
> Cost/m^2 of total floor area (elemental cost) £35.63
> *Unit quantity and cost*
> Element unit quantity 1,000 m^2
> Unit cost/m^2 £85.00
> *Sub-division*
> 550 m of 150 mm rc slab at £82.50   = £45,375
> 350 m of 225 mm rc slab at £105   = £36,750
> 100 m of 25 mm softwood boarding on
> 175 × 50 mm joists   at £34   = £3,400
> = £85,525

While these sub-divisions are very useful as explanations of how the total cost is affected by specification, it must never be forgotten when using them that the greater the detail in which a priced BQ is analysed, the less reliable are the results. It is more than likely that a BQ for the same job priced by another builder would give completely different figures at this level of breakdown, although quite similar in total. The overall unit costs will be found particularly useful when preparing early estimates of cost before specification details are available.

10.17.14   Obtaining unit costs from elemental analyses

In order to obtain unit costs as well as elemental costs, the analysis has to be more elaborate than if elemental costs alone are being recorded. This is especially so if one wants to keep separate costs for the different forms of structure or finish within each element. There are two problems to solve:

- the separation of the costs within the element;
- the recording of areas and other quantities.

Such things as areas of walls, floors and finishes can be obtained most easily from the BQ. However, sizes of window and door openings may be difficult to get from this source (particularly where there are fanlights, sidelights or windows glazed directly into frames) and it may be necessary to refer to the taking off dimensions or the drawings. There is also the difficulty of areas that occur more than once, for instance:

- The areas of concrete in floors will be duplicated by the formwork areas and these must not be added in again.
- However, the separate areas of concrete floors and hollow pot floors will require to be added together.

Thus either the analysis needs to be done by somebody technically qualified, or very clear procedures have to be laid down. No attempt need be made to separate constructions that differ only in detail (such as 100 mm, 125 mm and 150 mm floor slabs), because the aim is to obtain overall unit rates for basic constructions, not 'bill rates' for individual items.

10.17.15   Final form of the analysis

This is one of the places where a standard form, preferably the SFCA form, should always be used. It is not necessary to fill in too much specification detail if the analysis is for office use, as it should be possible to refer to the contract papers if anything more than a very broad outline is required. There should be a reference number for the analysis so that a list of analyses can be kept, and there should also be a cost index value so that allowance can be made for changes in market prices when comparing with past or future jobs. Note that elemental costs continue as a running total but unit costs cannot be carried forward, as a total of them would be meaningless.

10.17.16   BCIS Online

As well as providing printed elemental cost analyses, the BCIS offers a comprehensive database whereby analyses can be examined and any interesting examples downloaded to the cost planner's computer for use in the cost planning of new projects. This service also provides the facility to amend the BCIS analyses for the cost implications of a change in tender date or UK region of construction, so enabling comparison between jobs carried out at different times or in different places.

10.18   Design cost parameters

According to the *Concise Oxford Dictionary*, a parameter is 'a quantity which is constant in a particular case considered, but which varies in different cases'. It is now necessary to look at the factors which influence the components of the building in terms of area, number and size, as well as their quantity in terms of cost. Unfortunately insufficient research has been undertaken to date to give clear indications of the degree to which changes in the parameters of the building (or by implication its model) will affect the cost of that building. There is, however, a very great depth of knowledge gained by practitioners, which provides us with some general rules of thumb. In some cases we can be quite specific about how cost varies. For example, if we change the shape of a single-storey building so that the area of the external

Phase 2

brick cavity wall is increased, then we can be sure that, all other things being equal, the wall cost will probably have increased in direct proportion to the increased area. Similarly, if the quality of facing bricks is increased and the shape of the building remains fixed, then the wall cost will have increased by the extra material cost of providing the better specification.

Whilst we can probably rely on this type of simple wisdom for small brick buildings, it may not be adequate for dealing with more complex multi-storey framed and curtain-walled structures. If we change the shape or height of the building, it may not be just the extra quantity and quality that we have to pay for, but also indirect costs such as:

- different lifting equipment;
- improved fixings to deal with increased exposure;
- access, and manoeuvrability and dispersal of plant on site.

A particular difficulty lies in producing rules of general application, rather than in relation to one constrained set of circumstances, and indeed there is no real agreement that such rules exist. Very often the answer seems to depend on the methods that a given builder normally uses. Using our existing knowledge we can, however, establish some starting principles which could be the foundation for any further cost research, but which meanwhile can be drawn on in developing a design. The parameters can be viewed at two levels:

- the form of the building itself (morphology), where the effect of shape and height on cost can be studied;
- the major components of the building and the factors that influence their size, quantity and cost.

10.19 Building shape

The building shape has a major impact on:

- areas and sizes of the vertical components, such as walls, windows and partitions;
- perimeter detailing such as ground beams, fascias and the eaves of roofs.

It would seem obvious that the building that has the smallest perimeter for a given amount of accommodation will be the cheapest as far as these items are concerned. However, the shape that has the smallest perimeter in relation to area is the circle, and this does not very often produce the cheapest solution for the following reasons:

- The building is difficult for the constructor to set out.
- Curved surfaces, particularly those incorporating timber or metalwork (e.g. in joinery or formwork) are expensive to achieve.

- Circular buildings seldom produce an efficient use of internal space, as inconvenient odd corners are generated between partitions and external walls.
- There is a tendency for circular buildings to generate non-right-angled internal arrangements. Standard joinery and fittings are based on right angles and will not fit against curved surfaces or into acute-angled corners.
- The circle does not normally allow efficient use of site space.

In these circumstances it would appear that the right-angled building with the lowest perimeter will provide the best answer. This shape is of course the square. But although a compact square form is generally recognised as the most economical solution because of its reduction in cost of external vertical elements (and the lowest area of external wall for heat loss calculations), there are some important qualifications to be made:

- Where the square plan produces a very deep building requiring artificial ventilation, air-conditioning and lighting which would not otherwise have been needed, then the shape may become uneconomic. It is only the single-storey building with its opportunities for natural top-lighting and ventilation which may be regarded as a partial exception to this rule, although really satisfactory arrangements for top-lighting and venting are likely to be expensive in first cost and (particularly) maintenance.
- Where there is a high density of rooms the use of the external wall as a boundary to the room helps to reduce the amount of internal partitioning required. It is therefore sometimes preferable to elongate the building so that rooms can be served from either side of a spinal corridor, rather than have a deep building resulting in a complex network of corridors to serve all rooms plus the possibility of artificial ventilation to those that are internal. A real-life case concerns a high-tech building where the internal offices with no natural lighting were so claustrophobic that no-one would work in them and they were used as stores for rubbish; a small saving in cost/m² of floor area had produced a lot of floor area that was in fact unusable, and a poor bargain.
- A given amount of accommodation housed in a square multi-storey block may be much more expensive than the same accommodation housed in a less compact two-storey block, for reasons discussed later in this chapter.
- On a sloping site involving cut-and-fill, it may be more sensible to provide a long building running with the contours rather than a square building, which would cut more extensively into the site.

These are just a few of the qualifications that need to be made when talking about the efficiency of shape. Like a number of rules of thumb, this one was developed for traditional buildings in the UK, and two exceptions show how dangerous it is to regard such rules as universal laws:

- Modern high-tech buildings normally require large floor areas, air-conditioning and artificial lighting whatever their shape.
- Part of the reason why external walls are expensive in countries like Britain is because they need to have a good thermal performance. In warm countries this is not the case, and the ability to obtain natural ventilation cheaply may lead the cost planner to try to maximise the perimeter of the building.

In any case, the shape of the building is often dictated by the site boundaries, topography and orientation, and the degree of choice is therefore rather limited. If the national construction programme tends towards redevelopment rather than the exploitation of greenfield sites, then ideals of building shape can become fairly meaningless. There have, however, been a number of attempts to measure the cost efficiency of a building shape, and some simple examples are listed here:

- *Wall/floor ratio*. This is perhaps the most familiar of all the efficiency ratios but it can only be used to compare buildings with a similar floor area and does not have an optimum reference point such as those below

- $\dfrac{P - Ps}{Ps} \times 100\%$ (J. Cook), where P = perimeter of building and Ps = perimeter of square of same area. This formula relates any shape to a square which would contain the same area, thus providing a reference point for shape efficiency

- *Plan compactness or POP ratio* (Strathclyde University).

  $\dfrac{2(\pi A)^{\frac{1}{2}}}{P} \times 100\%$, where P = perimeter of building and A = area of building.

  In this case the point of reference is the circle (a square would have a POP ratio of 88.6% efficiency and yet is probably the best cost solution in initial cost terms).

- *Mass compactness ratio or VOLM ratio* (Strathclyde University).

  $2\pi \dfrac{\left[(3V/2\pi)^{\frac{1}{3}} \right]^{2}}{S} \times 100\%$, where V = volume of hemisphere equal to volume

  of building and S = measured surface area of the building (ground area not included). This formula chooses a hemisphere as the point of reference for considering the compactness of the building in three dimensions.

- *Length/Breadth Index* (D. Banks).

  $\dfrac{P + \sqrt{(P^2 - 16a)}}{P - \sqrt{(P^2 - 16a)}}$, where p = perimeter of building and a = area of building.

  In this index, any right-angled plan shape of building is reduced to a rectangle having the same area and perimeter as the building. Curved angles

Phase 2

can be dealt with by a weighting system. The advantage here is that the rectangular shape allows a quick mental check for efficiency. As these formulae are for guidance purposes only, this index is probably sufficient for early stage advice.

- *Plan/Shape Index* (D. Banks).

$$\frac{g + \sqrt{(g^2 - 16r)}}{g - \sqrt{(g^2 - 16r)}}$$ where g = sum of perimeters of each floor divided by the

number of floors and r = gross floor area divided by number of floors. This is a development of the previous index to allow for multi-storey construction. In effect, the area and perimeters are averaged out to give a guide as to the overall plan shape efficiency.

While the above indices are useful, they obviously have severe limitations as they consider only those elements that comprise the perimeter of the building, or in the case of VOLM the perimeter and roof. However, the repercussions of shape on many other major elements are considerable. For example, wide spans generated by a different plan shape may result in deeper beams, which in turn demand a greater storey height to give the same headroom, and thus will affect all the vertical elements. These implications need to be represented in any advanced model of building form, and an awareness of knock-on cost effects of this kind must be part of the cost planner's knowledge. Size of building is another important factor in cost efficiency. The larger the plan area for a given shape, the lower will be the wall/floor ratio. The next section takes the example of a single-storey square building with a height of 5 m.

10.20   Height

Here it is possible, and indeed desirable, to be dogmatic. Tall buildings minimise land costs in relation to floor area, but are invariably more expensive to build than low-rise buildings offering the same accommodation, and the taller the building the greater the comparative cost. The only partial exception to this rule is that the addition of a further storey or storeys to a tall building in order to make the best use of lifts or other expensive services may slightly decrease the cost per storey, but this does not invalidate the general rule.

10.20.1   Cost problems of tall buildings

The reasons for the high cost of high buildings are:

- The cost of the special arrangements to service the building, particularly the upper floors. Apart from the necessity of providing sufficient high-speed lifts, it is necessary to pump water up and to break the fall of sewage and other rubbish coming down. Complete service floors often have to be provided at intervals of 10 or 15 floors to deal with these problems.

- Special ventilation and lighting arrangements are needed because of the impossibility of providing adequate light wells in a tall building.
- A high standard of fire-resistant construction and practicable escape arrangements are required.
- The necessity for the lower part of the building to be able to carry the weight of the upper storeys, which obviously makes it more expensive than if it were carrying its own weight alone.
- The structure of the building and its cladding will have to be designed to resist a heavy wind loading, a factor which hardly affects a low building at all. Experience with many of the tall buildings of the last 30 years has shown how demanding is the required standard of windows, wall panels, etc. at high levels, and how expensive is the failure to meet these standards. One is talking about a very different price range indeed from similar components for low-rise construction.
- The cost of working at a great height when erecting the building; costs of:
  □ Hoisting all materials and operatives to the required level.
  □ The time spent by operatives going up to their work and down again at the beginning and end of each day and at break times.
  □ The extra payments for working at high altitudes and all the safety requirements which this entails.
  □ The bad climatic conditions for working at many times of the year.
- The increased area occupied by the service core and circulation. As the height of a building increases, so it needs more lifts, larger ducts, wider staircases, etc. and these installations take up more and more of the lower floors, so cutting down on the usable area. It is possible to imagine a building so tall that the whole of the ground floor is occupied by vertical services; adding extra floors to such a building would produce no increase in usable area at all!
- The cost of dealing with the effects on neighbouring properties, such as rights of light, and the considerable costs of overcoming planning objections.

Many of the above factors will also influence the running and maintenance costs; such items as window cleaning, repainting and repairs to the face of the building will all be much more costly than similar work to a low-rise structure. Therefore high building should never be considered favourably on cost grounds unless the saving in land costs, owing to the smaller site area in relation to accommodation, will pay for the considerable extra building costs. Land values must be high for this to occur, since a tall building needs a lot of space around it and the reduction in land requirement is not proportional to the reduction in plan area.

Phase 2

An increasing problem today is the near impossibility of dealing with the car-parking needs of a tall building within the plan area of the site, except at an excessive cost.

10.20.2   Costs of single-storey buildings

Just as high building is not usually economical, neither is single-storey building, but the exceptions here are more numerous and important, for instance:

- Where large floor areas free from obstruction by wall or columns are required, it is more economical to build horizontally rather than vertically, since to provide load-bearing floors over such areas would be far more difficult and expensive than a roof.
- Similarly, where very heavy floor loadings are required, it is cheaper to build floors resting on the ground than high-performance suspended floors over other storeys.
- Single-storey temporary or sub-standard buildings, where the low-cost foundations and structure cannot be made capable of supporting a further storey, can be an economical solution.
- Both for retailing and for manufacture there are so many user advantages in single-storey accommodation that this type of building is becoming very common. One of the problems in some areas is in finding a reasonably level site for a large building of this kind.

10.20.3   Cost advantages of low-rise buildings

The reasons for the relative economy of two- or three-storey buildings compared to single-storey are:

- One roof and one set of foundations will be serving two or three times the floor area.
- The walls or frame will be capable of carrying the extra load with little or no alteration.
- In domestic construction it will be possible to use cheap timber-framed upper floors, which will help the comparison still further.

Once the building exceeds this number of storeys, various factors make it difficult to attain the low costs possible with two-storey construction:

- It is less often possible to dispense with a separate frame.
- The frame itself must be more substantial.
- Lifts, fire-resisting construction and other expensive measures are required.

10.21   Optimum envelope area

The envelope of a building, that is the walls and roof which enclose it, forms the barrier between the inside and the outside environments. The greater the difference in these environments, the more expensive this envelope will be, but it is always at least a significant factor in the cost of constructing and running the building.

We have already seen that a square building is inherently economical in wall area, but the total envelope/floor area ratio will also depend on the number of storeys chosen for the accommodation. For example, if the accommodation in a large single-storey building was instead arranged on two floors, the roof area would be reduced by more than the consequent increase in wall area, so the total envelope area would be reduced.

The same might happen if the same accommodation was arranged on three floors, but if the building is made higher and higher while retaining the same gross floor area, the process eventually reverses and the increase in wall area becomes greater than the roof area saving. This reversal happens quite slowly, so the envelope area changes very little over a range of several possible storey arrangements close to the optimum. It is obviously useful to know what this optimum is as a design guideline. The following formula can be used to calculate this optimum for a square building thus:

$$N\sqrt{N} = \frac{x\sqrt{f}}{2s} \qquad (2)$$

where N = optimum number of storeys, x = roof unit cost divided by wall unit cost, f = total floor area (m^2) and s = storey height (m).

If the desired width in metres (w) is known, the formula for a rectangular building is:

$$N^2 = \frac{xf}{2sw} \qquad (3)$$

More complex formulae involving several elements have been developed to optimise the shape of the whole building, but although interesting as research tools these result in oversimplification, which renders them of little practical help.

10.22   Further cost modelling techniques

We have looked at the way in which simple traditional cost models have developed, noted their deficiencies and put forward a list of criteria for judging cost models. It is now intended to look at the possibilities of adopting other models, currently extensively used in other disciplines, in the search for better cost information. When we refer to better cost information what do we really mean? There are probably five major ways in which it is possible to advance. We can:

- provide cost information more quickly;
- provide more information so that a more informed decision can be made;
- provide more reliable cost information, which will introduce more assurance into the decision-making process;
- provide information at an earlier stage in the design process;
- provide information in a more understandable form.

By harnessing the power of modelling techniques it is hoped that each of these objectives can be achieved, or at least a step can be taken to improve the chances of achieving them. The traditional models evolved in the way that they did because it was expected that manual labour would be employed to do the calculations. This resulted in an oversimplification of the models, at all levels, for the sake of expediency. A new view of what is needed in terms of cost information, without reference to the constraints of manual computation, is required if the full potential of the new technology is to be tapped for the benefit of the client. However, almost by definition, all models are simplifications of the thing they seek to represent and are consequently imperfect.

10.23   Classification of models

There are a number of ways of classifying models. For example, classification may take place according to the function they perform, e.g. evaluating, descriptive, analytical, optimising. On the other hand it may be according to the form of construction:

- iconic (physical representation of the item under consideration);
- analogue (where one set of properties is chosen to represent the properties of another set, for example, electricity to represent heat flow);
- symbolic (where the components of what is represented and their interrelationships are given by symbols).

Perhaps the most important knowledge concerning any model is an understanding of its limitations within the context of its use. It is dangerous to ignore the simplifications that are inherent in the construction of cost models in particular. A considerable amount of research work has been carried out in the universities in developing models during the past 20 years. However, few if any of them have had any impact on the practice of cost planning in the real world. This is almost certainly because the vast expense of developing experimental computer systems into robust, foolproof and fail-safe commercial packages has not been seen as worthwhile in the context of potential use and profitability. Potential users of computer-based cost models are therefore very limited by what is available as commercial packages.

10.24   Key points

Building-cost models are basically of two kinds:

- a product-based cost model that models the finished project;
- a process-based cost model that models the process of its construction.

This chapter deals with the first of these. The simplest form of such a model takes no account of the configuration, or details of design, of the building but is based simply on one of the following:

- the floor area of the proposed project (gross or net);
- the volume of the proposed project;
- some user parameter such as the number of pupil places for a school or number of beds for a hospital.

More complex cost models include the BQ and elemental cost analysis. An element has been defined as 'that part of a building which always performs the same functions irrespective of building type and specification'. The most commonly used list of elements is that published by the RICS as the Standard Form of Cost Analysis (SFCA). There are a number of useful rule-of-thumb generalities about building form and cost, which tend to favour low-rise buildings with a low wall/floor area ratio, but there are some exceptions to these.

Notes

(1)   Hevacomp is a software option that is used by building services engineers to simulate the effects of such phenomena as heat gain and loss in buildings.
(2)   The 2004 version of the BCIS Detailed Form of Cost Analysis can be obtained from the RICS and is available for download, free of charge, at the institution's website: www.rics.org.uk. The standard form of cost analysis is also produced in full in Appendix A of this book (with notes).

Phase 2

Further reading

Flanagan, R. and Tate, B. (1997) *Cost Control in Building Design*. Blackwell Science, Oxford.

Fortune, C., Moores, N. and Lees, T. (1996) *The Relative Performance of New and Traditional Cost Models in Strategic Advice for Clients*. RICS Research Paper series. Royal Institution of Chartered Surveyors, London.

Murray, P. (2003) *The Saga of Sydney Opera House: The Dramatic Story of the Design and Construction of the Icon of Modern Australia*. Spon Press, Abingdon.

Chapter 11
Cost and Performance Data

11.1   Introduction

In reviewing traditional cost models it was suggested that the techniques used are merely the structure around which the professional cost adviser's judgement is applied in order to make them work. The model itself is ill-equipped to produce a reliable answer on its own, and reliability at any stage of refinement is entirely dependent on the costs applied to the measured quantities in their various forms. If the right cost figure is applied to any of the single price methods, then they will give better results than more sophisticated techniques with the wrong cost data applied to them.

The factor that makes detailed cost planning techniques more satisfactory than traditional methods is the control that is exercised as the design develops. Even this control, however, is based on unit rates applied to abbreviated quantities (another model) and is therefore dependent on the reliability of cost information. At the end of the day any estimate is dependent on the prevailing market conditions at the time of tender rather than conditions at the time of making the estimate, and this again requires information in order that market trends can be detected.

So at the root of all this forecasting and control activity is the need for cost data to supplement the numbers, areas, volumes, etc. which have been used to describe the building. It is this data that is critical in determining whether or not an estimate is reliable.

11.2   The ambiguous problem of software

One of the major problems with more recent models, in which software is used, is the need to ensure that the cost information held within the database is reliable and relevant for all conditions of use of that model. Very often, human intervention is not envisaged in the use of IT systems and professional judgement is left until the results have been produced. This is very often too late, because the cost consultant has little chance of knowing why a particular figure has been generated, without considerable investigation and interpolation. And if the consultant is to adjust a result to one that is assumed to be more reasonable, then a traditional model, using the consultant's judgement from the start, might as well have been used.

There is a common phrase in computer parlance: GIGO (Garbage In – Garbage Out). The above amply demonstrates this problem, and reinforces the view that incorrect data used in any form of modelling will produce inaccurate results.

11.3   Types and origins of cost data

The study of 'what information should we use?' and 'where can this be obtained?' is essential for the correct understanding of the application and development of costing techniques. First, however, we need to understand what the information is used for.

11.3.1   The applications of cost data

The work undertaken in a typical quantity surveying office would use information for four main purposes:

- The control and monitoring of a contract, for which the contractor has already been selected, through interim and final valuation procedures.
- The estimation of the future cost of a project and the control of its design to ensure that this figure is close to the tender figure.
- The 'balancing' of costs in a cost plan to ensure that money has been spent in accordance with the client's priorities.
- The negotiation of rates with a contractor for the purposes of letting a contract quickly.

Arising out of these activities are other uses, which might well be the concern of a special section in a quantity surveying practice or in the research department of a university. These uses relate to a deeper understanding of the external political and economic trends, and the relationship between design decisions and the degree to which they affect cost. These studies usually require a more detailed level of investigation, involving the analysis of large quantities of data or the classification and structuring of the data in a particular way to assist in the development of evaluation models.

To incorporate all uses of data we can probably classify them under four main headings:

- forecasting of cost;
- comparison of cost;
- balancing of cost;
- analysis of cost trends.

11.3.1.1   Forecasting of cost

Under this heading would come such information as:

- cost per square metre for various types of building;
- elemental unit rates;
- BQ rates;
- all-in unit rates applied to abbreviated quantities.

This information would almost certainly arise from an analysis of past projects, or 'historic costs'. The figures would be updated by the use of a building cost index, itself a form of cost data, and would be projected forward to the proposed tender date by an intuitive or calculated prediction technique. In recent years the problems associated with historic costs (see later in the chapter) have spurred researchers and practitioners to develop techniques that rely on current resource costs (labour, plant and material) to forecast the cost of a project. This certainly helps in the negotiation of contracts and also in the reliable modelling of the building process. Resources are where the cost is generated, and the adviser is therefore dealing with the origins of cost.

However, the problem lies in the acquisition of information on resource costs by people who are not members of a large building organisation, and in the time-consuming task of synthesising costs from detailed resource inputs. In fact techniques employing resource costs to forecast building costs have various ways of simplifying the problem to avoid the user being overwhelmed by too much data. These usually involve concentrating on major items and quantities at the expense of those with less cost significance. However, it should be noted that even these techniques rely heavily on 'historic' information regarding the normal time and quantity requirements for the resources required for a particular item or building.

11.3.1.2   Comparison of cost

In this use of data the need is not so much to discover what the building or component will actually cost at the time of tender, but to make a comparison between items with similar function or buildings of different design, to decide which is the better choice. The problems of the tender market are not so critical here, unless there is evidence to suggest that there may be a change in the cost relationship between the alternatives by the time the decision takes effect.

The criterion in choosing data for this task should be that it is structured in such a form that, if the design or specification of an item is changed, the use of the cost data will reflect the true change in the cost of the item.

11.3.1.3   Balancing of cost

In determining a budget for the cost control of a building, it is necessary to break down the overall cost into smaller units. These smaller units are used not only for checking purposes but also to allow a cost strategy for design to be developed. This strategy will attempt to spend money in accordance with the client's requirements, by allocating sums of money to the various major components of the building.

Data for this use is usually obtained from past projects, and it may be in the form of actual costs or as a proportion of the total cost of a similar project. At the stage that this information is used, the design will not have been developed (and may not even have commenced), and therefore the categories used for this breakdown will tend to be broad and probably not exceed 40 items for any one project. The data can therefore be described as coarse as opposed to refined.

11.3.1.4   Analysis of cost trends

Of paramount importance in any prediction technique which seeks to project costs into the future is information, which tells us what is happening to costs in the industry over a period of time. By looking at the way in which costs for different items are changing in relation to one another, or changing between one point in time and another, it is possible to have a better chance of selecting the specification which will suit the client's requirements over the short and long term. It will also allow us to obtain a more reliable prediction of what the market price to the client will be when the job eventually goes out to tender.

These trends may be shown as the change in the cost of materials and labour, or they may be a detected change in the total cost of a particular type of building or component. When data is used for the detection of cost trends, the cost is very often related to a base-year cost, in which case the presentation of the information is in the form of a cost index. It should be noted that the detection of a cost trend does not necessarily imply that it will continue into the future. Indeed, it is very unwise to extrapolate a cost movement without taking into account all the political, social and economic factors that contribute to a change in cost levels. Just extending a straight line or a curve without thinking about it is a recipe for trouble.

The number of variables involved in such an evaluation is enormous, and consequently the establishment of what future costs will be is nearly always left to the professional experience of the cost adviser, who will probably take account of economic reports – which often differ widely in their predictions. It has been said that when you get two different economists discussing a particular economic problem you are likely to get three different opinions!

These are the major uses of cost data and we will now look at the problems in retrieving and storing the information for these particular applications.

Phase 2

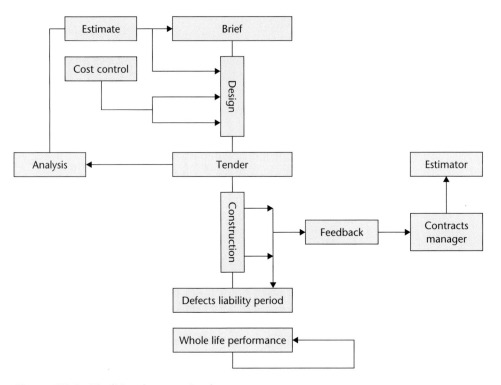

Figure 11.1   Traditional cost retrieval.

11.4   The reliability of cost data

Nearly all data used in cost planning techniques has been processed in some way, very often with a corresponding loss in accuracy and certainly with some loss of context. The published information received by the QS/cost planner usually relates to a 'typical' building in a 'typical' location, with only a brief summary of the contributing factors to such a cost (i.e. the explanatory variables). The feedback of such cost information focuses principally around the BQ. Figure 11.1 shows a diagrammatic representation of traditional cost retrieval and planning practice.

In theory, information on wages, materials and plant is collected on site and fed to the contractor's contracts manager, who then passes this raw data to the estimator. The estimator uses it, together with data from other projects, to forecast the next tender price. The consultant quantity surveying firm then analyses the contract BQ from the new tender to provide data which it then uses to forecast the cost of the next similar building and to control its cost during the development of design. When this new design is being built, the feedback cycle starts again. The above procedures, although they purport to represent the usual practice and sound convincing enough, are fraught with problems.

11.5   Occupation costs

In Figure 11.1, there is no feedback shown into the general system from the area of building performance cost (i.e. maintenance, repair and operational costs). Although some public authorities have attempted to obtain data on these costs and pass this information to the design team for consideration in their choice of detailing and specification, this is far from common practice. In the private sector the reluctance to collect and divulge occupation costs has been contributed to by the comparatively small number of clients involved in large continuing programmes of building, and unwillingness to spend the amount of time involved in such an exercise. However, some larger organisations have taken a lead in this area, including the following.

11.5.1   The National Health Service

Operational and maintenance statistics on the acute care NHS estate are collected annually through the Estate Returns Information Collection (ERIC), which is the primary mechanism to record all costs associated with running and operating an NHS trust's estate. It records cost data on two levels: trust level and site level. All costs, including energy, facilities management, maintenance and financing costs, are covered. ERIC enables the analysis of estates and facilities information from NHS trusts and primary care trusts (PCTs) in England, and it is a mandatory requirement that NHS trusts submit an Estates Return. This provides an indication of the status of estates and facilities services in the NHS for the Department of Health. The development of the performance management framework agenda has led to the data now being used to inform a set of performance indicators to measure the performance of a trust's estates and facilities services. Over time these will permit trusts to demonstrate year-on-year improvement in line with the NHS Plan. ERIC also attempts to define the framework for performance measurement and benchmarking of costs in the NHS estate. However, what ERIC does not facilitate is the framework with which to measure the performance of individual buildings within the estate.

11.5.2   IPD occupancy costs service (ITOCC)

ITOCC is the Institute for Professional Development (IPD) Occupiers International Total Occupancy Cost Code, which defines costs, cost ratios and space for most standard types of property such as offices, retail, industrial and warehouses. It also provides a cost framework for other types of real estate. The code comprises four key facets:

- *Transparency*. All occupancy costs are brought together, whether capital or revenue, estate or facilities expenditure.
- *Definitions*. All items are defined in detail to provide a clear and useful document.
- *Usability*. The method is logical and sensible and avoids unnecessary detail. A worked example clarifies and explains the approach.
- *Application*. The code describes the possible uses of more accurate, total cost information in terms of creating key ratios, setting targets, tracking, etc.

IPD Occupiers produce analysis of occupancy costs, space efficiency and property management effectiveness at individual property and portfolio levels, against specific market benchmarks. This is delivered through the UK annual service or bespoke research projects.

11.5.3   BMI Standard Form of Property Occupancy Cost Analysis (POCA)

The aim of the BMI Standard Form of Property Occupancy Cost Analysis (POCA) is to allow standardisation of the system of collection and a single format for presentation. The system has been used for over 20 years by subscribers to the Building Maintenance Information (BMI) service. The database expresses costs per 100 m^2 per annum.

The property occupancy cost analysis presents data in a form that allows comparisons between the cost of achieving various defined functions, or maintaining defined elements, in one building with those in another. The POCA also provides a framework in which the property manager, facilities manager or surveyor may systematically collect occupancy cost data year by year.

An element for occupancy cost analysis purposes is defined as expenditure on an item which fulfils a specific function irrespective of the use or form of the building. The list of elements is, however, a compromise between this definition and what is considered practical.

11.5.4   The EuroLifeForm WLCC model

The EuroLifeForm model, which is discussed in Chapter 7, comprises a design decision support application called the logbook. This allows the building occupier to capture operational cost data during the building life. In line with recent initiatives such as the Chartered Institution of Building Services Engineers (CIBSE) Building Logbooks, the logbook application also facilitates a post-occupancy analysis exercise, the idea being that the building owner will utilise the logbook to record the most up-to-date cost-in-use data on an annual basis, enabling the WLCC model forecasts that were produced at design stage to be matched against actual costs as these accrue.

11.6 Problems with site feedback

Let us consider the contractor's estimating feedback cycle first. The difficulties are as follows (and are further considered in Chapter 16).

Although records of labour and material costs are kept on site, the information is recorded for the payment of wages and checking of invoices, and for monitoring progress on site as an aid to contract management. These purposes dictate a format that is not compatible with the way in which the estimator works. The grouping of items into operations for reference to the contractor's programme creates difficulties in rearranging the information for the pricing of a BQ based on the Standard Method of Measurement requirement to measure items as fixed in place. Consequently, feedback to the estimator tends to be at the very general level of, for example, 'We were a bit low on plaster rates on this contract, Jack', rather than more specific data.

Even if site data could be related back to the estimator, it is unlikely to be of very great benefit because performance of the labour force varies from day to day according to:

- weather;
- supervision;
- industrial and personal relations;
- obstruction by other trades;
- the skill with which the work is planned and organised;
- lack of clear instruction;
- waiting for instructions on design changes;
- waiting for delivery of materials;
- accidents;
- replacement of defective work;
- failure by sub-contractors;
- psychological pressures.

It is unlikely that individual performance on one contract will be repeated on another. It will be explained in Chapter 16 why 'standard costing' methods used in industry do not work on a building site. The estimator can only hope to get close to the total labour content of the project and hope that good site management will avoid wastage of this resource.

No formal record of the performance of labour will be available for work where labour-only sub-contractors are used, and therefore an extremely large area of work is no longer subject to scrutiny and analysis.

Material use and costs will also be variable between sites, and here again no firm data can be expected for the estimator to use. The amounts used will depend on such factors as:

Phase 2

- care in ordering;
- site control and supervision;
- vandalism and site damage;
- replacement of defective work;
- the competence of the workforce.

Plant costs will be even more difficult than the other categories to relate from one site to another, and depend heavily on:

- the quality of site management;
- the amount and type of plant available (especially where the building firm uses its own plant);
- the extent to which the work lends itself to efficient plant use;
- the amount of disruption to efficient use caused by the work programme, and particularly by changes or delays in it.

The problems with feedback mean that very little data is kept in the contractor's office. The database for the majority of estimators still consists of a small black book of labour and material constants, a list of addresses and telephone/fax numbers of suppliers and sub-contractors, and a good deal of experienced judgement.

No wonder it has been suggested that a tender represents the socially acceptable price rather than a scientific appraisal of the resource needs of the project. Indeed there is some evidence to show that estimators bid randomly within plus or minus 10% of the actual contract cost excluding profit. This is considered to be the limit of their ability to forecast reliably what the job will cost, and is discussed later in this chapter.

11.7   Problems with the analysis of BQs

The estimator's problems with the retrieval of site data set the scene for the problems that the cost planner will find when analysing the priced tender document that the estimator has prepared. The vast majority of the information used by the cost planner arises from the BQ, and yet, as we are beginning to see, this document may be based on incorrect assumptions that just happen to work in most cases and are convenient for the estimator to use. It has been said that the only reliable figure in the BQ from the client's point of view is the total (i.e. the tender figure).

11.8   Variation in pricing methods

If we start to break down the fairly similar totals of the different contractors' tenders for the same project into smaller units, we are liable to find substantial variation in each individual subsection when considered in isolation from other sections. The greater the number of categories the greater the degree of variation, for the following reasons:

■ We know that the rates in the BQs are not true costs, but they are not even true prices in the sense of a price on a supermarket shelf. The contractor is not offering to 'sell' brickwork at so many pounds per square metre, but to construct a whole building for a total sum. The rates are simply a notional breakdown of the total price for commercial and administrative purposes. There is thus no reason why any individual rate should be justifiable in relation to either cost or competition.

■ The way in which the contractor prices the preliminary items of plant, scaffolding, etc. and the firm's own profit and overheads, will vary from one firm to another. Some will place these in the preliminaries section while others will include them, in whole or part, in the measured work rates as they see a commercial advantage in not identifying them too explicitly. Consequently the 'cost' of any individual trade, element or BQ item will vary according to the treatment afforded to these factors.

■ It may be in the contractor's best interests to 'load' the prices of those parts of the building which are executed first, such as excavation and earthworks, to improve the project's cash flow (i.e. obtain money earlier in interim valuations to help finance the rest of the work). In a similar fashion the contractor may anticipate variations to the contract and reduce the price of work that is thought likely to be omitted, while increasing the unit rates of any items that are likely to increase in quantity.

■ In addition to the deliberate pricing method of each contractor, there are all the variations in unit rates caused by different assumptions being made by each estimator with regard to the resource requirements. These variations tend to be highest in high-risk trades such as excavator and carpenter, and lowest in those most easily controlled such as glazing and concrete work. The assumptions made will relate to the estimator's view of the firm's expertise and economic structure.

■ In an estimate for a complex product such as a building, it is inevitable that mistakes will be made. These will tend to cancel each other out, but sometimes there is an error of cost significance in a single item, which may not be spotted by the QS.

If other factors, such as the contractor's previous experience of the particular design team or client (e.g. factoring-in a premium for a difficult architect or the QS!), knowledge (or otherwise) of the locality, experience of the type of building proposed or keenness of tendering, are taken into account, it can be seen that data obtained from BQs is likely to be highly variable even for contractors who are tendering for the same job.

Phase 2

11.9   Variation in BQ rates for different jobs

If we now consider the BQ rates of contractors tendering for different projects, we can expect variability due to all the above factors plus a good few more. These additional items will relate to the site and contract conditions for each particular job, and will include the following factors.

11.9.1   Site conditions

There are a number of site conditions that can affect BQ rates:

- problems of access;
- boundary conditions, especially the problems of adjoining buildings;
- soil bearing capacity and consistency;
- topography and orientation of the building, which may affect manoeuvrability on the site and the type of resources (particularly plant) that can be used.

11.9.2   Design variations

The way the building is designed has an enormous effect on the efficient use of resources, and on production method and time. For example:

- If there is a poor repetition of formwork then it is reasonable to expect higher prices if this fact has been communicated to the estimator.
- Wet trades requiring drying and curing time will possibly create more delay than if a dry form of construction is used.

The extra costs, if any, will be represented in the rates or will result in a redistribution of prices, for example more money in preliminaries for the longer supervision required and plant not fully employed, and perhaps less money in the work sections of the BQ.

11.9.3   Contract conditions

The impact of contract conditions is very difficult to anticipate, particularly with regard to their effect on unit rates. In general it can be assumed that the more onerous the terms as far as the contractor is concerned, the more likely it is that a higher tender will be put forward. However, when a contractor is short of work or particularly wants a certain job (say for prestige purposes), then the effect of tough clauses may be less than would normally be expected.

Any contract that requires more working capital for the project, or additional risk, is almost certain to incur a cost penalty, which will be passed on to the client. In addition, conditions relating to the length of the construction period, particularly a shortening of time, and the phasing of the works may result in uneconomic working and will influence the estimator's rates and particularly the preliminary items.

11.9.4   Size of contract

For each contracting firm there is an optimum size of contract that will suit its particular structure and resources. Large firms very often create small-works divisions of the main company to deal with those projects which are small or of a specialist nature, such as restoration or fitting-out work, and which cannot carry the overheads of a giant corporation.

Smaller firms, on the other hand, find it difficult to gear themselves up to a multi-million pound project with its specialist plant, complex labour relations and sophisticated supervision and control requirements. The size of the firm in relation to the contract for which it is bidding will therefore affect its approach to the estimate.

The unit rates for a large project should make allowance for the economies of scale that could be expected. However, the following do not always allow these economies to be made:

- problems of site working;
- lack of advanced mechanical production plant;
- the nature of the industry.

11.9.5   Location

Location will obviously affect the problems of accessibility to the production resources. The transport of materials, workpeople and plant to site, with perhaps accommodation of the workforce on remote sites, makes this an important consideration. On top of these problems are those of local climate, which, even in the UK, may affect the starting on site, the degree of protection required and the interruption of the work programme.

Despite these problems it should be noted that of all the old principal government cost limits, only the housing cost yardstick identified location as a cost variable and divided the country into regions. The other yardsticks made no allowance for regional variations, but did take the problems of a particular site into account by the use of an abnormals allowance.

A contractor's estimator is likely to pass on the problems of a particular area as an extra cost to the client if the firm has had experience of that location. This again will create variation in the rates due to different allowances being made.

Phase 2

11.10   Research into variability

All the problems listed contribute to the wide variability in BQ rates. Derek Beeston of the former Property Services Agency looked at the major items in each work section (or trade) in a large sample of BQs from different government projects and compared the unit rates for each individual item. He found that the variability tended to be different for each trade, and expressed the degree of variability in the form of the coefficient of variation, which enables a comparison to be made between the variability of items of different value and is given by the formula:

Coefficient of variation = (standard deviation/arithmetic mean) × 100%   (1)

So in concrete work, for example, the cost per cubic metre may be £33.00 and the standard deviation £5.00:

Coefficient of variation = (5/33) × 100% = 15.15% say 15%   (2)

This can now be compared with plaster work with a cost of £3.00 per square metre and a standard deviation of £0.30:

Coefficient of variation = (30/300) × 100% = 10.00%   (3)

The standard deviation on its own would not allow the degree of variability in each trade to be compared. The results of Beeston's investigations showed the following:

Excavator	45%	Joiner	28%
Drainlayer	29%	Roofer	24%
Concretor	15%	Plumber	23%
Steelworker	19%	Painter	22%
Bricklayer	26%	Glazier	13%
Carpenter	31%	All trades	22%

It is no surprise that an excavator, with the high-risk problems owing to weather, soil conditions, accessibility, etc. should have the highest variability. The ranking of some of the others is perhaps unexpected and a detailed investigation would need to be undertaken to discover the reasons for their particular performance. However, the table does illustrate very clearly the problems of using BQ data. Once again, success depends on the skill of the user in determining what rate to use against a particular measured quantity. A computer cannot make this judgement.

11.11   The contractor's bid

If we accept the problems associated with obtaining reliable feedback from site and the rather arbitrary value given to unit rates in the BQ, then we might wonder how the contractor arrives at a bid for a particular contract when he is in competition with others. The theory is that the estimator:

- knows how long it will take a team of craft operatives and labourers to execute a job;
- knows the precise amount of material required including wastage;
- can define the exact output of each item of plant to be used and the time it will be on site;
- can add on the measured cost of supervision and overheads;

to arrive at the net cost of the job excluding profit. The process is essentially deterministic; it assumes that the estimator (if clever enough) can determine the exact amount of resources and their cost. However, every contractor knows that this is not possible.

11.11.1 The notion of non-deterministic pricing

Research has suggested that while non-deterministic pricing is theoretically adopted, it is in fact very often used to justify a total that the contractor has in mind from the very beginning. Brian Fine, who used to work for one of the largest UK building companies and later became a management consultant, has suggested that contractors may bid at what they consider to be a socially acceptable price. This is not well defined, but is considered to be the price which society is known to be prepared to pay for a particular type of building. For commercial projects this would probably be related to the income of the office block or output of the industrial plant. For social projects it may be related to the known funds available. In the case of government-funded projects it may be a defined cost yardstick.

It is considered that this value of the building is relatively easily calculated and that generally the building is buildable at around the socially acceptable price. By bidding at this value the estimator overcomes the problem of being unable to predict resource requirements for the project. As the client's anticipated cost is probably based on what has been successfully built before, it is likely that the estimator will have put forward a reasonable estimate, and the labour and material constants are used to justify this.

It is unlikely that many contractors consciously go through the above process, and indeed they often react strongly to the suggestion that they do. However, despite the wide variation in individual BQ rates, the final tender figures are often considerably closer than one would expect. A coefficient of variation for the spread of tenders on a project tends to be in the range of 5–8% and this can be applied to a wide variety of projects.

11.11.2 Support for the notion of non-deterministic pricing

Contractors will occasionally admit that they can write down the cost of the job before they start pricing it, assisted by:

Phase 2

- Working out the yardstick amount before pricing a government job.
- The client naming the approximate value of the contract when enquiring if the contractor is willing to tender.
- The building grapevine of sub-contractors and suppliers, who probably quote several times for the same job but for different contractors, and ensure that the anticipated contract sum is well circulated.

The case is further strengthened by the following:

- *The name of the project appears to influence the cost.* A good example of this phenomenon was the decision by a number of hospital boards some years ago to let nurses' accommodation blocks as a separate contract to the main hospital development. The reason for this decision was that the nurses' accommodation was costing appreciably more than comparable students' accommodation in the education sector. Tenderers had apparently applied the same criteria of complexity, difficulty and high cost associated with hospital projects to the far simpler problem of the nurses' accommodation blocks. Hence of course the high prices, and the suggestion that the real cost was not being estimated.
- *Constants are not reliable.* As stated earlier in this chapter, it is not possible to establish accurate constants of labour, plant and material to apply to the job that is being estimated.
- *Chance of contracts being won.* In a competitive market it appears that contractors take their place in the bidding order by chance. As would be expected with chance, in the long run they appear to win contracts in direct proportion to the number of other contractors they are bidding against.
- *Estimates are tailored to suit the market.* If the estimation of cost really was a deterministic exercise, then the constants used in the estimate would be absolute and unchanging unless, of course, they were amended as the result of ascertained cost feedback. However, what happens when an estimator or firm loses a number of bids in succession? The estimator is told to 'sharpen your pencil' and the firm reduces their rates! In order to survive they must get work.

This practice suggests that the estimate does not mean very much in terms of an accurate prediction. If the firm does get the job and they think it will be 'tight' they put their best management team on the project and very often get a better return than on a job where they foresaw a large profit and did not maintain such tight control! The lower price that they are bidding could be the socially acceptable price.

It would not be fair to suggest that the concept of socially acceptable price has been thoroughly substantiated on the basis of the few arguments considered. However, doubt has been cast on the deterministic estimating process and this in turn should influence the way in which we look at and use cost data.

Obviously the factors of supply and demand related to the capacity of the industry, and its workload, will influence the level of the contractor's bid, and the relationship between these factors and the socially acceptable price has not yet been well defined. However, it is important to realise that market conditions reflect the economic standing of the country, and this aspect may well have more impact on historic cost data, and the price society is prepared to pay for its buildings, than any other.

11.12   The structuring of cost data

In addition to the difficulty of acquiring reliable cost data there is the problem of what to do with it once acquired. The need for a system of cost planning to have access to cost data at different levels of complexity postulates a hierarchy of cost groupings, or even a series of interlocking hierarchies, because some of the criteria to which cost data will be applied are of different types and not merely different scales. One such hierarchy, which we encountered in Chapter 10, is based on the concept of design cost elements. This is a highly pragmatic breakdown of a project for cost planning purposes which, apart from its inherent faults, suffers from the disadvantage that it does not meet the requirements of any other functions in the design/build process. Another hierarchy is the Standard Method of Measurement format used for BQs, which is only of limited use for other purposes. However, its format does tend to pervade documentation produced for process purposes rather more than might be expected, because the information provided in the BQ is:

- generally of a high quality;
- tied back into the contract;
- available to the builder without further cost or trouble;
- structured in accordance with the Common Arrangement of Work Sections for Building Works produced by the Co-ordinating Committee for Project Information (CPI).

Operational cost centres, which are useful for actual control on site, are difficult to relate to the estimating and commercial systems of the industry, except at the highest levels of generality. Because of the obstruction to communication between builder and designer caused by the price mechanism and the associated contractual standpoints, there has been little attempt to link site operation costs to design criteria.

11.13   An integrated system of groupings?

A number of efforts have been made, since the late 1970s in particular, to devise an integrated system of groupings of building work which would meet

Phase 2

the needs of all participants in the project, from the point of view of design process, costing, communication, organisation and feedback. It is argued that such a system would reduce duplication of effort (it is alleged that the same piece of building work may be physically measured up to 15 times for various purposes), improve communication between the parties and permit a far higher quality of costing and cost feedback. The CI/SfB classification system, whose use was encouraged by the RIBA, was one such effort but it was too oriented towards its original purpose of filing design information to have found much favour elsewhere (it is now incorporated in the wider Uniclass system). A much more complex version of the same basic system, CBC, was developed in Scandinavia and relied heavily on computer use at a time (the 1970s) when this imposed considerable constraints on a system.

In Britain, the then Department of the Environment set up a data coordination study with substantial resources and a large measure of contribution from industry and the professions, it being their very reasonable view that this was the only way of obtaining results which would be useful and acceptable to all. Unfortunately the work ceased, for party political reasons, without producing much more than a number of very interesting reports and the construction industry thesaurus, which is used mainly for library, information and reference purposes. Little further work has been undertaken in this field in more recent years and little progress has been made, but there is plenty of (rather dated) background reading available.

11.14   Sources of data

So far this chapter has dealt with some of the more general problems of cost information associated with its use, retrieval and structure. Although these problems should be identified, it has to be recognised that the quantity surveying profession has been using data obtained from BQs for many years with reasonable success. Contractors, too, have been able to prepare realistic estimates from the very generalised feedback that they get from site. The problems associated with data should not therefore be allowed to obscure the fact that present data sources available to the industry have allowed the development of a simple and reasonably effective system of cost forecasting and control with which the various parties are familiar.

From the point of view of the client's cost adviser, information can be obtained from two main sources:

- data passing through the cost adviser's office;
- published material found in the technical press, price books and information systems.

There is no doubt that most cost advisers would normally prefer to use their own data. It will be from a project whose background they know, and they will therefore be aware of all the problems associated with the project – its location, market conditions, complexity, etc. – which influenced its price, and they will also know about the project's outcome. We have already said that traditional cost modelling requires sound professional judgement. This is better served by using information known to the cost planning organisation through involvement in the project rather than received second-hand from a published source, in an abbreviated form and with little background information.

The further detailed breakdown of any structured information is available should it be required. To take the example of an elemental cost analysis, the firm will have available:

- full priced BQs;
- the original measurements;
- working drawings;
- the contract.

These can be referred to should they wish to find further information on why an element cost a certain amount or what were the prevailing market conditions at the time and whether the overall price was considered high or low. It is very unlikely that this level of detail would be available in any published data, however well prepared. To take another example, many QSs/cost planners prepare their own cost indices because they know the weightings and costs included, and these can be manipulated with greater confidence in assessing the needs of a particular project.

The data will refer to the geographical area in which the firm carries out most of its work. Compared to published data, there is a shorter time lag between receiving raw data, processing it and being able to use the structured information in the office. This is particularly relevant to cost indices, where the published information may be anything up to six months behind the times. The choice of classification and structure of the processed data is under the control of the firm. It is therefore possible to ensure that the details most relevant to that particular practice are emphasised and identified, rather than having to accept a standardised published format. If errors and ambiguities occur it is also easier to spot them in your own system rather than in somebody else's.

Having established the preference of most cost planners for their own information, it would be unrealistic to suppose that any but the largest firms or organisations could exist solely on the data passing through their office. The range of projects undertaken by most practices is considerable, varying from government and local authority buildings to commercial and leisure buildings for individuals and corporate clients. The variety of specification, location, size and shape of this range is also large, and it is therefore most unlikely that the office will have up-to-date information which it can apply to all its new projects.

Phase 2

11.15   Published cost data

The common unit rates, which each QS will gather from handling project information, will help in preparing an approximate quantities estimate, but at the coarser level of elements and single price estimates, a wide enough variety of data is unlikely to be readily available. While the cost planner can probably make intuitive assessments, particularly of cost per square metre, it is reassuring to have a check available in the form of published information. Indeed, it is probably the primary role of published information to provide data which enables practitioners to check on their own knowledge and to provide a context for their own decision-making. The following are the major sources of this kind of data, which apply almost entirely to conditions in the UK. Such information as exists overseas is often of a more condensed statistical nature and makes the cost planner's task in most other countries more difficult.

11.15.1   The technical press

There has been a steady improvement in the quantity and quality of cost information appearing in the journals and magazines concerned with design and building management. Until some 30 years ago, almost no information was disseminated in the press regarding costs of buildings. This was always jealously guarded as confidential until *The Architects' Journal* took the lead in refusing to publish accounts of new buildings unless cost information was included. Since then, however, largely due to demand from clients for better cost control, editors have found it necessary to devote more space to cost analyses and forecasts. The type of information that can now be found in such journals as *The Architects' Journal* and (particularly) *Building* magazine includes: elemental analyses; resource costs and measured rates; cost indices with future cost projections; analysis of the economic performance of the industry; and cost studies of different types of buildings. Indeed, with the rapid changes in the level of resource costs and measured rates over very short time spans, the technical journals, published at weekly intervals, have largely replaced the traditional annual standard price books as a source for quick reference material.

11.15.2   Builders' price books

A traditional source of useful information has been the price books, which are normally published annually by several long-established organisations. These too have expanded the scope of their contents to keep pace with current demands and to try and maintain an edge over the technical journals. However, in periods of rapid inflation or market changes they have suffered the drawback of having to prepare their information well in advance of the year of publication. This time lag of several months, and the 12-month period for which the information is meant to be current, means that the rates have tended to be forecasts of what is likely to happen at some unknown point in the year ahead. This problem has more recently been overcome by the publishers offering online computer access to updated information.

Despite the above problems, it is usual to find on the shelves of all the building professions, an assortment of *Spon's Architects' and Builders' Price Book*, *Laxton's Price Book* and *Wessex Price Book* (in 2005 Wessex Electronic Publishing was purchased by the BCIS). Their popularity stems from the very comprehensive nature of the data now included, and subsections normally cover:

- professional fees for building work;
- wage rates in the industry;
- market prices of materials;
- constants of labour and materials for unit rates;
- unit prices for measured work in accordance with the Standard Method of Measurement;
- elemental rates for a variety of work;
- building cost index;
- approximate estimate rates and comparisons;
- cost limits and allowances for public sector buildings;
- European prices and information.

Spon also publish separately a very comprehensive *Mechanical and Electrical Services Price Book*. They also publish a *European Construction Costs Handbook* and an *Asia–Pacific Costs Handbook* for overseas work. As with data from any source, considerable care needs to be exercised in the use of the information in all these books. They assume that a reasonable quantity of work for all items is required. The subletting of all work that is normally sublet (a builder's price for this type of work is likely to be higher than normal market rates) is also assumed. No allowance is included for overhead charges and profit, nor for preliminary items or VAT. The prices are based on a tender price index of 325 in relation to a base of 100 in 1976 (which may not be the same as the index normally used by the cost planner). The user is expected to make all allowances for changes in the location, size of contract, small quantities of particular items and market changes. It will be seen from this how dangerous it is to start picking out prices at random from various publications, all of whose rules may be different, without reading the detailed conditions in each case. However, price books have a very useful role to play in enabling practitioners to check on their own knowledge and assumptions.

Phase 2

11.15.3   Information services

With the advent of cost planning techniques and the need for a wide range of information on different types of project, it was realised that most offices would require information additional to that normally found in one particular practice. It was suggested that if firms could be persuaded to supply their own elemental cost analyses to a central body, then a large pool of information could be gathered which could then be disseminated to the subscribing members. If these analyses could be supplemented with other more general information on trends and economic indicators, then a useful service would be provided to the profession. Therefore, in 1962 the RICS set up the BCIS to undertake this task. For its first ten years the service was only available to QS members of the RICS, but since 1972 it has been available outside the quantity surveying profession. Today there are other cost information services, but the breadth of information and independence of the BCIS, and its sister service BMI, makes these RICS services the leaders in their respective fields. They are dealt with here in some detail.

11.15.4   Building Cost Information Service (BCIS)

The BCIS is a subscriber-based service, which collates and analyses data submitted by its subscribers and incorporates material from other sources. The information is interpreted by the BCIS professional staff of chartered surveyors and is presented in two formats: BCIS Bulletin Service (a hard-copy service) and BCIS Online (an electronic data service). Naturally, all information provided by subscribers is treated in confidence. Approval to publish project-specific data, such as elemental analyses, is sought from subscribers and any other parties to the information. A summary of the BCIS Bulletin Service, which principally deals with capital cost data, is given here.

BCIS Indices and Forecasts is published quarterly. This is a bound report including:

■ the complete series of BCIS Indices;
■ a 24-month forecast of the main tender and cost index series;
■ an executive summary;
■ a thoroughly researched commentary on market conditions with trends projected over the coming 24-month period;
■ a range of individual indices for tender prices, regional prices, input costs and output prices.

A quarterly publication, *Surveys of Tender Prices*, includes a range of current pricing studies incorporating updates of:

■ the BCIS Tender Price Index;
■ the BCIS Building Cost Index.

This publication provides:

- information on average building prices for different building types, new and refurbishment, expressed in £/m² of gross internal floor area and adjusted to current prices using the BCIS Indices;
- functional prices, such as £/pupil place for schools, for around 200 building types;
- studies of price differentials by size of contract, location and type of work;
- results of a survey of the percentage added for preliminaries, dayworks and sub-contract work.

It should be noted that the BCIS Indices are based on an index of 100 in 1985, and so are quite different to those used by Spon's, for example. Previous warnings about care when combining information from different sources should be remembered.

The BCIS Five Year Forecast (produced each summer) provides information on economic factors affecting the construction industry. The forecast includes the following features:

- Building Cost Trends and Summary of Forecasts
- Executive Summary
- Latest Trends
- Economic Background – covering inflation, growth and interest rates
- Materials
- Labour
- Earnings, Wages and Rates
- Market Conditions
- Output
- Housing
- Output Forecasts
- Tender Levels
- Tender Price Forecast
- Assumptions for inflation, demand, labour and materials.

Elemental analyses are a key source of price information from accepted tenders and are published quarterly as loose-leaf data sheets classified for filing by building type. This was covered in the previous chapter.

The information in the *BCIS Quarterly Review* is a selection of BCIS data which gives guidelines on the general level of building prices. It contains average £/m² building prices, the BCIS tender price index, the BCIS building cost index and location factors. There is also a brief commentary on market conditions and tender prices.

The *BCIS Guide to Daywork Rates* is published annually, with an updating service to provide information on the latest changes to the daywork rates. The guide contains all current and historic daywork rates calculated by BCIS, for the wage awards for the major trades in England, Scotland and Wales, including:

Phase 2

- Builders – 10 grades;
- Plumbers – main grades – 7 grades;
- Plumbers – apprentices – 8 grades;
- Heating and ventilating – 22 grades;
- Electricians – main grades – 12 grades;
- Electricians' apprentices (Scotland only) – 4 grades.

Normally published at the end of each calendar year, each rate is calculated in accordance with BCIS's interpretation of the appropriate Definition of Prime Cost of Daywork carried out under a building contract agreed between RICS and the contracting organisations. The example calculations show where differential payments, merit money, etc. should be included.

The RIBA/BCIS Cost Calculator provides a range of costs from the BCIS building project database. These costs reflect the tender costs of actual building projects in the UK, and the calculator is aimed at architects who require more accurate guidance on what typical projects cost to build.

The RICS also provide comprehensive information on occupancy and maintenance costs through the BMI service. The subscription service includes the following data sources.

The *BMI Bulletin* is a regular update for subscribers with the comprehensive BMI Occupancy Cost Information Service data. These paper reports are published throughout the year and compiled in folder format. The *BMI Quarterly Cost Briefing* is available separately on subscription as well as forming a part of the BMI Bulletin Service. The briefing provides guidance on changes in prices and costs, and information on current and forecast trends in maintenance and occupancy costs. Maintenance price indices monitor the movement of contractors' pricing on actual contracts, providing useful information on the market. The price indices contained within the briefing monitor market trends in private housing, public housing and private, industrial, commercial and public non-housing. The briefing also contains information on cost indices, tracking the general movement of costs through different industry sectors.

The latest edition of the *BMI Building Maintenance Price Book* has been thoroughly revised and updated, containing information and data on:

- budgeting;
- estimating;
- cost control;
- schedules of rates;
- letting maintenance work;
- measured term contracts.

The book is in three sections covering:

- Basic costs – labour rates, scaffolding and plant, and materials;
- Labour constants;
- Measured rates.

Finally, the service also provides access to the BMI Special Reports[1], which cover reviews of occupancy, maintenance and rehabilitation costs.

11.15.5   BCIS Online

Since 1984 the BCIS has been operating a computer-based service which gives subscribers unlimited access to the entire BCIS databank from their PC. It can be accessed from anywhere in the world. With a link to the BCIS host computer, a simple menu-driven structure enables the user to select the type of data and the specific information required in the shortest time. The system is straightforward to use. Simple but powerful selection tools cut out time-consuming browsing to locate, for example, the cost analyses needed for a project. This ensures that users have access to the widest possible range of examples as well as the most up-to-date information – particularly important when the market is volatile. Once the required information has been located, users can copy it to their own machine for further use. The system allows analyses to be updated to current or projected pricing levels and adjusted for location. Data can then be printed, displayed on screen, or used in conjunction with approximate estimating and other software such as spreadsheets.

11.15.6   Government literature

The Department of Trade and Industry (DTI) publish the *Quarterly Building Price and Cost Indices* and these can be accessed via the UK government statistics portal National Statistics. Used primarily for those involved in estimating, cost checking and fee negotiation within public sector construction projects, it includes data on tender and output price indices, resource cost indices and location and function studies. This information is produced by the DTI Construction Statistics and Economics Division which has responsibility for the collection, analysis and publication of statistics for the construction sector. The division provides regular statistical analysis of building materials, overseas trade, overseas construction by British firms, price and cost indices and key performance indicators of construction activity. As well as statistical advice, the division provides economic analysis and advice to assist in the assessment of the construction market and in the formulation of efficient policies. The division also produces the *Construction Statistics Annual*. This publication gives a broad perspective of statistical trends in the construction industry in Great Britain through the previous decade, together with some international comparisons and features on leading initiatives that may influence the future.

Finally, the National Statistics portal also provides information on price adjustment formulae for construction contracts through the monthly bulletin of indices. Used in conjunction with the formulae price adjustment method of adjusting building and civil and specialist engineering contracts (to allow for the effects of inflation on the cost of labour, plant and materials), these are also referred to as NEDO Indices, Baxter Indices or Osborne Indices. These are widely used on variation of price contracts.

11.15.7   Technical information systems

Most construction professionals will have access to services such as the Construction Information Service (CIS) and Specify-IT, both providing a wealth of information on product specifications. Prior to the widespread use of these electronic resources, office library systems contained a collection of current trade literature relating to building products. Their major purpose is to provide a ready reference for the practitioner on specification and performance of a wide variety of building products. At one time it was not uncommon to find current price lists included, but prices change more frequently than specifications, and the problem of updating this secondary information has been solved by leaving it out. The use of these services for cost planning purposes is, however, negligible.

11.15.8   Other sources of cost data

While the two main sources of data for the cost planner are the firm's own data and published data, there is a further source available for some types of work that in some ways is a combination of the two. This is the obtaining of information from specialist sub-contractors and specialist consultants, with an ever larger proportion of building cost represented by specialist firms undertaking:

- roofing;
- flooring;
- fenestration;
- doors;
- cladding;
- finishes;
- structural systems;
- landscaping and engineering services.

Like published information it suffers from remoteness, but because there is usually an element of personal contact involved it is often possible to explain and discuss exactly what the circumstances and requirements are. Personal knowledge and contact are also helpful because there is little other guarantee of the soundness of such advice. If prices are obtained from specialist sub-contractors the cost planner must remember to allow in addition for:

- the builder's profit;
- the builder's contractual discount allowance if the sub-contractor's price is net;
- facilities which the builder must provide including:
  - scaffolding;
  - unloading and moving materials;
  - storage;
  - office accommodation;
- use of builder's plant and equipment;
- incidental builder's work in cutting holes, etc.;
- assistance with site fixing. Some firms send only a specialist fixer who does little more than supervise the builder's own site workers.

It is important to find out exactly what the sub-contractor requires in these matters. Some firms are almost entirely self-sufficient, while others lean on the builder as much as possible.

11.16   Future development

Reliable cost data continues to be a vital primary need of the cost planner, and this has become even more apparent as the profession moves towards providing clients with whole life-cycle cost advice. A great deal of research has been carried out on the use of advanced analytical tools, such as data mining and artificial intelligence, in order to best utilise the data that is available. The prediction of trends in construction cost indices is also a rich vein of activity and many larger consultancies will employ economists with the expertise to help advise cost planners on likely changes in materials and labour constants. As these developments filter through to the industry, it should be expected that change will be rapid and fundamental.

11.17   Key points

- An attempt has been made in this chapter to outline the sources of cost data available to the cost planner, and their use and problems.
- The primary source of such data is still the priced BQ, but the reliability of this data is rather questionable and the extent to which it even tries to represent actual costs is open to doubt.
- Price data contained within the BQs is usually confidential and the con-tractor's permission must be sought if any analysis etc. is to be published.

Phase 2

Builders' own cost systems are not usually as suitable for design cost planning purposes as might be expected. Published cost data of various kinds is a useful backup. BCIS and other information services are also very useful.

Note

(1) BCIS Special Report Serial 355, *Occupancy Costs of Industrial Buildings* (2006), provides a useful example of occupancy cost plans for this type of building.

Chapter 12
Construction Cost Indices

12.1 The cost index

Cost indices are fundamental to cost planning since they provide a valuable insight into changes in the cost of an item or group of items from one point in time to another (time series). The Consumer Price Index (CPI) is perhaps a well-known example; it represents the ratio of exchange between money and a basket of consumer goods. In compiling an index, a base date is chosen and the cost at that date is usually given the value of 100 (index number), all future increases or decreases being related to this figure. An index number of 100 is ordinarily used since it avoids the confusion of negative numbers where values fall below the base index number.

Example

Suppose the cost of employing a labourer on site at base date was £160.00 per week and the current figure is £240.00 per week. £160.00 would be given the value 100 and £240.00 would be represented by the figure 150, derived in the following manner:

$$\{(240 - 160) \times 100\}/160 + 100 = \text{cost index in relation to base} \qquad (1)$$

As the base is 100, the number of points of any subsequent index above 100 (in this case 50) also represents the percentage increase since the base date.

However, if we are comparing costs over time in which the data we wish to update is not at base year cost, but has an index of say 120, then to arrive at the percentage increase in cost where the current index is 150 the following calculation is used.

$$(150 - 120)/120 \times 100 = 25\% \qquad (2)$$

Notice that the answer is not the difference in the number of points on the index scale, i.e. 30.

12.2   Use of index numbers

There are a number of uses to which index numbers can be put in the construction industry.

12.2.1   Updating elemental cost analyses

Updating elemental cost analyses is perhaps the most common use for the QS and is an essential part of the elemental cost planning process. Tender information of past projects can be brought up to current costs for budgeting purposes. However, great care is required when updating information beyond a period of, say, two years.

12.2.2   Updating for research

It is extremely difficult to obtain large quantities of cost data relating to the same point in time in order to analyse trends and patterns for cost research. By bringing cost information, obtained at a number of different points in time, to a common date by the use of an index, a much larger sample of data can be examined.

12.2.3   Extrapolation of existing trends

By plotting the pattern of costs measured by an index, it may be possible to extrapolate a trend into the future. However, there are very great dangers in extrapolation. For example, during the 1960s there was a steady increase in cost of about 5% per year. Extrapolations were undertaken using this figure for projects one or two years ahead. This worked quite well for a number of years until 1970, when the cost of building to the client suddenly rose by about 10% in that year. Many QSs (and other professions) had not foreseen this rise and were heavily undervalued in their budgets. It may be thought that inflation is fairly constant in the UK at present – but that is what people thought in the 1960s. Extrapolation based purely on present-day trends can be fraught with risk.

12.2.4   Calculation of price fluctuations

During periods of rapid inflation it was the custom for building contracts to be entered into on a basis whereby the contract amount was adjusted during the progress of the work to take account of inflation since the date of tender.

By applying an index to the cost of work undertaken during a specified period, it was possible to evaluate the increase in costs of resources to the contractor more speedily and with less ambiguity than by using the previous laborious method of comparing every invoice price with the base date cost of the same item. The National Economic Development Organisation (NEDO) index (see later in this chapter) was most often used for this purpose.

12.2.5   Identification of changes in cost relationships

If a cost index is prepared for the different components of a building, or for alternative possible solutions to a design problem (for example steel versus reinforced concrete frame), then it is possible to see the changes in the relationship between one component and another over time. It may then be possible to identify when one solution appears to be a better proposition than another.

12.2.6   Assessment of market conditions

QSs are particularly interested in the price their clients have to pay for a building. If the index will measure the market price, as opposed to the change in the cost of resources, then a measure of current market conditions can be obtained which is of enormous benefit in updating and forecasting cost.

12.3   Approaches to constructing an index

Since the late 1980s there have been a number of attempts at providing a reliable cost index. They include the following.

12.3.1   The use of a notional bill

In this method a typical (or synthesised) BQ is chosen and repriced at regular intervals by a competent QS based on experience of current rates. Unfortunately, rates vary so much that it is extremely difficult to assess the current tender situation in such a way that a reliable index can be obtained from the resultant totals. This method is now to all intents and purposes obsolete.

12.3.2   Semi-intuitive assessment

This method is based on trends in resource costs combined with tender reports by QSs. By receiving reports on current tenders from each of the

Phase 2

project QSs in their organisations, experienced practitioners can adjust their knowledge of the changes in material and labour costs to prepare an index for the current market situation. This works quite well where the market is stable or steadily changing, but these judgements are extremely difficult to make at a time when the economy is subject to stop–go conditions. This method has been superseded by the tender-based index described later.

12.3.3   Analysis of unit price for buildings of similar function

If data could be found giving, say, the cost per square metre of all schools of a certain type being built in the country at one point in time, then the average of these costs could be used to provide the basis of an index. If the exercise were repeated at regular intervals for that type of building, then a regular index could be established. Problems arise, however, owing to such large design variables as specification and shape. Such an index would also be only applicable to buildings that are homogeneous in function and standard and for which there is a regular building programme to provide the data.

12.3.4   Factor cost index

If a typical building is analysed into its constituent resources and the cost of each resource is monitored over time, then a combined average index can be prepared which measures the change in the total cost of the building over the same period. Each resource (labour, plant and material) would need to be given due weighting in the index according to its value in the total building. The construction of, and problems associated with, this type of index are described in more detail below.

12.3.5   Tender-based index

There is a great need for QSs to have a measure of the market price their clients have to pay for their buildings. The accepted tender figure based on the pricing of a BQ is a record of the market price for a particular building at a specific point in time. If the measured items in the BQ are repriced using a standard schedule of base year prices (to give a base year tender), then an index can be constructed by comparing the current tender figure with the new derived base year total. Efficient methods are available to avoid repricing the whole BQ. Details of this method are provided later in the chapter.

This and the previous approach now need to be considered in more detail.

12.4   The factor cost index

It has already been explained that this uses changes in the cost of resources to build up a composite index. To illustrate the method, consider the construction of a simple index for the cost of a brick wall. Assume that the following table represents the cost of the constituents at base year level in the first column, and then at today's date in the second column. The final column represents the index resulting from these figures (calculated as described previously).

	Base year (£)	Current (£)	Index (£)
Bricks	20	28	140
Mortar	2	3	150
Labour	8	10	125
			415

Average index = 415/3 = 138.34   (3)

However, in the above example no account has been taken of the fact that bricks as a proportion of total costs are ten times more important than mortar, all the resources having been given equal status. Consequently, a very rapid rise in the cost of mortar would have a disproportionate effect on the composite index, and this would not be representative of the increase in the cost of a brick wall as experienced by the contractor. To overcome this problem the resources need to be weighted in accordance with their importance, as follows:

	Index	Base year weighting	Extension
Bricks	140	20	2,800
Mortar	150	2	300
Labour	125	8	1,000
		30	4,100

Average weighted index = 4,100/30 = 136.67   (4)

The effect of the weighting is to reduce the index value in this example because mortar (which has increased most in price) does not now share equal status with bricks and labour. A further problem arises at this point. Suppose for some reason the labour force becomes more productive over the time period, by say 25%. Then at base year costs the value of the labour would have been £6.00 instead of £8.00. This will affect the amount of labour involved in building a brick wall and consequently its cost. This change, although affecting the cost to the contractor and possibly the client, will not be represented in the index unless the weightings are changed.

Phase 2

	Index	Current year weighting	Extension
Bricks	140	20	2,800
Mortar	150	2	300
Labour	125	6	750
		28	3,850

Average revised weighted index = 3,850/28 = 137.5   (5)

This is a very real difficulty with the factor cost index because it is not easy to judge productivity over time, particularly with regard to the construction of whole buildings. In fact in the majority of cases, weightings in building factor cost indices are assumed to remain constant until a revision of the index is undertaken and a further analysis of the importance of each resource is made.

Where an index uses the base year weightings for each calculation, it is known as a Laspeyres index. Where the index uses weightings obtained from the current year or point in time that is under consideration, it is known as a Paasche index. Both names derive from their original authors.

Another factor not considered in the above examples is the contractor's profit and overheads. Profit, particularly, can be a function of market conditions and it would be difficult to devise a reliable quantitative measure for this variable. This is why a number of factor cost indices have incorporated an additional component, called a market conditions allowance, very often based on the professional judgement of its author.

12.4.1   Constructing a factor cost index for a complete building

The brick wall was chosen to illustrate the essential ingredients and problems of an index of this type. When constructing an index for a complete building, the task becomes more complex although the same principles apply. A typical procedure can be summarised as follows:

- A typical building (or group of buildings) is selected for analysis into its constituent proportions of labour, plant and material.
- Analysis takes place and the building resources are allocated under various headings to suit the representative cost factors for which information is available. For example, if use was being made of published information for material indices (such as those prepared by the Department of the Environment, Transport and the Regions (DETR)) then the different materials in the building would be analysed according to the structure of that published data.
- The different types of labour (corresponding to different wage-fixing bodies) are identified and the basis for evaluation of a unit of labour cost identified. Usually this would include the following:

  □ Changes in the hourly or weekly wage rate as determined by agreement of the parties to the wage-fixing body.
  □ Changes in employer's 'on costs' and contributions such as holidays with pay, National Insurance contributions.
  □ Changes in the agreed standard working week.
  □ Changes in the average hours worked per week, as recorded in government statistics. This item and the preceding one will affect the degree to which overtime rates are paid for each unit of labour output.
  □ Changes in productivity. One method of obtaining a gauge of productivity is to take the total quantity of materials used by the industry, priced at standard rates (i.e. constant figures) and divide by the total building labour force. If the output per operative increases over a particular time period, then it is assumed that this is the result of productivity and the labour weighting should be reduced accordingly.
  □ Change of location. It may be necessary to assume a particular geographical position for the typical building in order that regional differentials can be ignored. The labour unit cost for each type of craftsman and labourer would be computed and then weighted according to the importance of each in the total labour force. The resultant weighted average for labour would be carried forward for weighting with other resources.

■ Material rates for comparison with base year prices are selected. Although it is possible to identify a long list of cost-significant materials and calculate a separate index for each, it is more usual to rely on published statistics for this type of information. The Department of Trade and Industry regularly produces indices of materials for a range of material classifications. It is then merely a question of analysing the typical building according to the published material types to obtain the weighting to be applied to each material index.

■ Plant types, together with their weighting, are identified and a standard method of evaluation adopted. This may be based on an average of hire rates for the particular item or it may be calculated by considering purchase price, depreciation, maintenance, standing time, etc., to establish a standard resource unit cost which can form the basis of the index. In many published factor cost indices the change in the cost of plant is ignored.

When the weighted index for each of the three basic resource types has been calculated, they are then brought together and weighted again, this time in accordance with the importance of each within the total building cost. Typical values for weighting are, say:

Labour	35–45%
Material	50–60%
Plant	5–10%

The final weighted index represents the change in cost between one point in time and another for the typical building chosen. Any subjective judgement to take into account market conditions, productivity, etc. can then be made if it is considered necessary.

Phase 2

12.4.2   Limitations and uses of factor cost indices

It is important to realise that the factor cost index is actually measuring the change in the cost of resources to a contractor for a 'typical' building. Its main shortcoming from the QS's/cost adviser's point of view is that it takes little or no account of the tendering market. It does not directly measure the change in the price that the client must pay, and although the market conditions allowance attempts to rectify this situation, the source may not necessarily be reliable, having usually been made on a subjective basis as suggested above. Neither does it measure the change in cost of a specific building under consideration by a QS.

It is unlikely that the 'typical building' will have the same mix of resources as the client's project; however on many occasions the difference will not be cost significant. To try and minimise this difference, the typical building may be more of an economic model than a likely real-life building, having, say, a mixture of light concrete block and clay block partitions, so a change in the cost of only one of the alternatives will not unduly distort the index on the one hand or be ignored on the other.

In spite of these drawbacks the method is very suitable for identifying trends in resource costs and relationships. It was particularly useful for evaluating cost fluctuations in contracts that allowed for reimbursement of any changes in cost that occurred during the contract period to the contractor, as is common practice at times of high inflation.

12.5   The tender-based index

A much more attractive proposition than the factor cost index from the private QS's point of view appears to be an index which takes as its source of information the tender document itself. This should record what is happening in the market place and therefore should be much more useful in updating prices for a design budget.

In essence, the tender figure is compared with a figure produced by pricing the same tender document using standard rates at the base year prices. From the two resultant figures an index can be calculated showing the increase or decrease in cost to the client within the current tendering market.

12.5.1   Limitations of the tender-based index

There are drawbacks with the method, including:

■ the questionable validity of BQ rates;

- the difficulty of obtaining priced BQs for jobs which are comparable except for date of tender.

If, however, a large enough sample of projects is taken, then the difficulties can be dealt with by 'trampling the problem to death'. If an appreciable quantity of projects is analysed, it is hoped that all other differences except current levels of cost will cancel themselves out.

An obvious difficulty is the work involved in analysing all the rates in scores of BQs. Fortunately it has been found that much time can be saved by taking only the few largest items in each work section, and that this results in a negligible degree of error.

12.5.2   Constructing a tender-based index

The procedure for preparing and using a typical tender-based index can be summarised as follows:

- Prepare a priced list of typical BQ items at the prevailing base year pricing levels (possibly with an allowance for preliminaries included). This is of course a time-consuming task and the current published forms of this index tend to use the PSA *Schedule of Rates for Building Works*[1], which is available from The Stationery Office. This schedule is very comprehensive and is more than adequate for the majority of buildings.
- Take the priced document for the lowest tender received and note the format (e.g. work section, elemental).
- For each section of the format (e.g. excavator, concrete work) pick up the largest value item in the section, then the second, and so on, until a total of 25% of the section value is achieved. (It is of course the value of the item as extended in the cash column, and not its unit price, with which we are concerned.) This procedure is repeated for all sections with the exception of preliminary items, prime cost and provisional sums, profit and attendance on prime cost items, and daywork percentage additions.
- For each of the items selected, find the corresponding unit price in the base schedule of prices. The difference as an addition or subtraction from the base year rate is then obtained by comparing the BQ unit price of each sample item with its equivalent base year price.
- The extended base year value is calculated for each item (i.e. weighted by its quantity). The total of all the extended items at base year prices is then compared with the total of all the extended items for each section, to obtain an index.
- The index obtained for each section is then weighted according to the value of that section as a proportion of the whole tender BQ for the project being considered (less preliminaries, PC and provisional sums, contingencies, etc.) and a combined index for all sections obtained.

Phase 2

- The preliminaries are usually dealt with by considering them as a percentage addition to the other items in the BQ, excluding dayworks and contingencies. If the rates in the base schedule include an allowance for the preliminary items, then the index figure arrived at so far is not a true one, since net BQ rates have been compared with gross base rates. So, if the 'false' index of BQ items is 130 and the value of the preliminaries in the current tender is 10% (as a percentage of the remainder of the BQ), then the final index would be:

$$130 + (130 \times 10/100) = 143 \tag{6}$$

- To obtain an average index for publication rather than for one particular job, the index numbers for all the tenders which have been sampled in a specified time period are averaged by taking either the geometric or arithmetic mean. The difference between the mean of the current and preceding quarters is used to gauge the movement in tender prices. The geometric mean is usually taken when relative changes in some variable quantity are averaged, because the arithmetic mean has a tendency to exaggerate the average annual rate of increase.

A number of simplifications have been made in the above procedure for the purpose of clarity. For example, in practice, problems arise with items that cannot be matched (these are generally overlooked in the item selection process), and where a sample amounting to 25% of a section's total cannot be obtained. The rules governing the index being used should always be carefully consulted in such situations to discover the method to adopt.

12.5.3   Uses of the tender-based index

The tender-based index has several advantages over a simple factor cost index from the point of view of the client's QS:

- It measures the change in the cost to the client of a particular project over time, taking full account of market conditions in addition to the change in cost to the contractor.
- It is relatively simple to operate once a base schedule of prices has been obtained.
- It allows comparison of the price obtained by tender for a specific project with the national or regional building price trend. This could assist in deciding whether to call for fresh tenders from another group of contractors, or not.
- It allows the relationship between the market for buildings of different function and locality to be plotted.
- It is not based on other indices and therefore any inherent inaccuracies are not compounded.

There are, however, several problems, which have to be acknowledged:

- To overcome the problems due to the variability of price rates in BQs, a large number of projects are required for each index. It has been suggested that at least 80 projects are required for a suitable sample, but this condition is seldom met. Unless access to this number of BQs is available, the trend being plotted may be erratic and not typical. Very few organisations have access to this number of projects and therefore most firms cannot prepare their own index by this method.
- The index relies heavily on the base year schedule, which will have to be regularly revised to take into account new products and new methods of measurement. This is a time-consuming and costly task.
- Because of a lack of suitable projects at any one point in time, the average index may have to rely on an unbalanced sample containing more jobs of one particular function and location than is considered desirable. This may lead to errors in the trend plotted.

12.6   Published forms

There are a large number of published indices available to the construction industry and the main sources are listed below, classified under the two major index types.

12.6.1   Factor cost indices

Of these perhaps the following are the most familiar.

12.6.1.1   DTI Price Adjustment Formulae for Construction Contracts (NEDO)

The one-time National Economic Development Organisation (NEDO) produced a series of indices for building trades based on the resource costs of each trade. These are now maintained by the DTI and are published by TSO and through the Office for National Statistics. They are still popularly known as the NEDO formulae, although the correct name is the Price Adjustment Formulae for Construction Contracts. The indices are based on market prices, nationally agreed wages, etc. and therefore do not take into account the problems of a contractor on a particular site. Although primarily provided for the assessment of price fluctuations on contracts where there is a fluctuations clause, they can be adapted for use in a composite building cost index, as in the BCIS index dealt with next.

12.6.1.2   BCIS building cost indices

These measure the change in basic input costs – labour, materials and plant – and are based on cost models of average buildings. There are nine indices:

- BCIS General Building Cost (excluding M&E) Index;
- BCIS General Building Cost Index;
- BCIS Steel Framed Construction Cost Index;
- BCIS Concrete Framed Construction Cost Index;
- BCIS Brick Construction Cost Index;
- BCIS Mechanical and Electrical Engineering Cost Index;
- BCIS Basic Labour Cost Index;
- BCIS Basic Materials Cost Index;
- BCIS Plant Cost Index.

The model for each BCIS Index is based on the Price Adjustment Formulae for Construction Contracts. The BCIS indices are calculated by applying the work category indices to the models. Each work category index is a compound of resources of labour, materials and plant, and changes in the cost of each of these components are reflected in the work category indices.

Generally, the labour indices are based on the national labour agreements, and the materials indices are based on the Producer Price Indices prepared by the Office for National Statistics. The models were prepared by analysing 54 new work BQs into the building work and specialist engineering work categories.

The BCIS General Building Cost Index is a weighted combination of the Steel Framed Construction Cost Index, the Concrete Framed Construction Cost Index and the Brick Construction Cost Index.

The BCIS General Building Cost (excluding M&E) Index is a similar index to the BCIS General Building Cost Index, but excludes the following installations:

- electrical;
- heating, ventilating and air conditioning;
- sprinkler;
- lift;
- catering equipment;
- other specialised M&E items such as fire alarms.

The cost indices for steel-framed construction, concrete-framed construction and brick construction are all based on cost models containing buildings with each particular form of construction only.

12.6.1.3   The BCIS housing cost index

Originally published in *Building* magazine (and hence often known as the *Building* Housing Cost Index), the BCIS Housing Cost Index (base 1970) is based on fixed weightings of specified labour, materials and overheads found in typical domestic property construction.

Table 12.1   Trends in housing cost indices.

Index	Total % change 1985– 2002	Average annual % change 1985– 2002	Average annual % change 1985– 1992	Average annual % change 1992– 1997	Average annual % change 1997– 2002
BCIS General Building Cost Index	6.1	0.3	−0.5	0.9	1.0
BCIS Steel Framed Construction	5.0	0.3	−0.4	0.9	0.5
BCIS Concrete Framed Construction	3.6	0.2	−0.6	0.8	0.8
BCIS Brick Construction	7.4	0.4	−0.4	0.8	1.3
BCIS General Building Cost Index (excl M&E)	5.4	0.3	−0.7	0.8	1.2
BCIS M&E Cost Index	7.7	0.4	0.2	1.1	0.0
BCIS Labour Cost Index	28.0	1.5	0.7	0.3	3.8
BCIS Materials Cost Index	−10.3	−0.6	−1.4	1.3	−1.5
BCIS Plant Cost Index	4.4	0.3	−1.1	0.5	1.9
BCIS Housing Cost Index	12.8	0.7	−0.3	0.6	2.3

As part of the Scottish Executive 'analysis of historical construction cost movements in social housing report', published in June 2004, trends in the various indices were reported on, as shown in Table 12.1.

12.6.2   Tender-based indices

As tender-based indices are a comparatively recent innovation, there is not the long history of results that could be found with factor cost indices. However, the number of institutions and practices producing an index of this type is steadily growing.

12.6.2.1   DTI's Construction Statistics and Economics Unit building tender price index

The construction of the index is similar to that already described in this chapter, where a sample of items is taken for each work section up to 25% of the value of the particular trade. This index was the original tender price index and has formed the basis for the others that have followed.

Phase 2

12.6.2.2   BCIS tender-based index

The method of construction is similar to the DTI form just described. Two indices are prepared, one relating to fluctuating contracts and the other to fixed price. This goes some way towards overcoming the problem of the different allowances made in tender figures for future variations in price. The sample of jobs from which the index is derived is not restricted to government projects, and therefore it can be expected to cover a wider range of building types than the preceding index.

12.6.2.3   The Sense Construction Tender Price Index

Sense Cost Consultancy (part of Mace) produces an index derived from:

- tender returns from live projects;
- out-turn cost data from the Mace group of companies;
- results of a survey sent to the Mace group of companies' supply chain;
- a review of third party forecasts of general price inflation;
- an internal review of construction cost inflation and general economic activity both globally and in the UK.

Sense have also produced a forecast for tender price inflation during the period 2007–2010; these predictions show a tender price increase over four years, as follows:

- 4.50% in 2007;
- 4.75% in 2008;
- 5.50% in 2009;
- 5.75% in 2010.

Published research by Sense (carried out through their supply chain) suggests that average tender prices will rise faster than building costs, reflecting the fact that the supply chain expects a continued healthy growth in demand for new build and refurbishment work. The survey carried out to collect this data asks a series of questions of the supply chain in an attempt to ascertain whether they are facing cost escalations for their labour, materials and plant, and whether they are passing those increases on to clients by way of increased tender price inflation over the next four years[2].

12.6.2.4   Other indices

Other indices are in general use but do not strictly fall under the previous three headings. One example is the Halifax House Price Index. Since 1984, this index has been used extensively by government departments, the media and businesses as an indicator of house price movements in the UK. A similar approach is taken by the Nationwide Building Society (NBS). The methodology involves tracking a representative house price over time rather than the simple average price approach (used by the Land Registry). Both the NBS and Land Registry methods are based on a large sample of housing data; the latter provides the longest unbroken series of any similar UK index.

The Department of Trade and Industry compiles the Tender Price Index of Social Housing (TPISH). This index is calculated from analyses of accepted tenders, forwarded by housing associations and local authorities. TPISH is an indicator of trends in accepted tender prices for public sector housing building contracts in England and Wales. It is a smoothed quarterly index and includes adjustment factors for location.

12.7   Problems in constructing and using cost indices

It has already been stated that there are a number of problems associated with the construction of cost indices. Such problems can also create difficulties in the use of the index and these, together with some more general points, are summarised here:

- The index will usually be measuring a trend for a typical or model building type, and may not necessarily be measuring the change over time for the particular project that is being developed.
- Where the base date is several years old, the question should be asked whether the index is being based on outdated criteria. This particularly applies to a factor cost index using base year weightings. It may be that the balance of resources has changed considerably in the intervening period and that this is not represented in the index. Over quite a short period the mix of resources on a typical building may alter as a result of the very changes in cost that the index is trying to record; for instance, a rise in metal prices may cause a substitution of plastic for lead and copper in plumbing and roofing work.
- In the longer term, changes in technology may not only make the base mix untypical but may also distort its prices; as new techniques and materials come into greater use their price tends to decrease proportionately, whilst the cost of obsolescent technology tends to rise faster than the general rate of increase. However, every index must have some history in order that trends can be identified.
- Regular publication of the index at monthly or quarterly intervals is required, otherwise the user is put in the position of having to forecast the immediate past (i.e. the time between the last publication and the present) as well as the future.
- The basis of construction of the index must be known in order for it to be used intelligently. The choice of a tender-based or factor cost index will be dependent on what the user wishes to measure, but in addition the weightings of any index are important in judging whether the results can be applied to the project under consideration.

Phase 2

■ Short-term changes in an index must be treated with caution since the inherent errors in the system of compiling the index may well be the equivalent of several points on the scale. In the tender-based indices, a good sample of BQs must be used if bias due to regional variation and building function is not to distort the results.

■ When plotting a trend in an index it is sometimes advisable to use a logarithmic scale for the index against a natural scale for time. The advantage of this method is that if costs are rising at a regular percentage every year, then the index values will be shown as a straight line. Any deviation above or below this line will show immediately as a change in this regular pattern and will assist in detecting a new trend.

■ Cost planners obviously have a duty to keep themselves aware of past, current and future economic trends, and should use this awareness in coming to a conclusion on any extrapolated index value which they wish to use in estimating for future projects. Of course no-one has yet proved that they can accurately forecast the future, and this particularly applies to future index numbers. Anybody who could actually do this would be in the City earning a multi-million pound salary, not cost planning the average building project.

■ Providing reasonable care has been exercised, cost planners should not be blamed if their forecasts turn out to be wrong owing to changed economic or political factors that were unforeseen at the time of making the estimate. For this reason many cost planners tend to use a reputable index such as the BCIS All-in Tender Price Index, on which any responsibility for getting it wrong can be placed. This might be more difficult if an in-house or less well-known index were used.

■ In recent years it has become common practice and far more sensible to give a range of anticipated future costs based on the possibility of certain events occurring. In any case, it is most important that everything should be made explicit when the original estimate is given. The details should include:
  □ the index that has been used;
  □ the point or range that has been chosen together with the reasons for this choice;
  □ the timescale for the project that has been assumed when making the forecast.

12.8   Which type of index to use

By their very existence it is obvious that all cost indices have a role to play in assisting the cost adviser to determine the market forces affecting a project. The choice of index will, however, be determined by the use to which the index is put:

- For short-term forecasting, allowances for fluctuations, etc. the factor cost index can be used.
- For general estimating and cost planning purposes a tender-based index will more reliably measure the change in the cost to the client. It also has the advantage of being able to measure the performance of a tender, region or building type against the national average.
- For very long-term forecasting and budgeting the use of a unit price index based on cost per square metre or unit accommodation may be more applicable.

12.9 Key points

The object of a cost index is to measure changes in the cost of an item or group of items from one point in time to another. The CPI produced by the government every month is a well-known example. In compiling an index a base date is chosen and the cost at that date is usually given the value of 100, all future increases or decreases being related to this figure. Published cost indices are available for:

- building in general;
- various types of building;
- various components of building cost.

    Building cost indices may index either:

- construction costs based on the cost of resources; or
- tenders.

These differ because tenders are more affected by market conditions. The use of a reputable published index gives the cost planner some protection if unforeseen events occur.

Phase 2

Notes

(1) The PSA *Schedule of Rates* (2005) 9th edition (Carillion Services) is one of the most common rate guides in the construction industry and is the standard document for public sector construction work. It contains over 20,000 rates spanning a range of building works and materials. It can be used in measured term contracts (see Chapter 8) but has a five-year shelf life.

(2) This report can be obtained from http://www.sense-limited.co.uk/survey/

Chapter 13
Cost Planning the Brief

13.1   The brief

After the initial budget has been prepared and accepted (Chapters 5, 6 and 7), the cost planning process can begin in earnest, with the brief being agreed and outline drawings being prepared.

13.2   An iterative process

It might be thought that the title of this chapter is misleading; since the cost plan itself is not going to be prepared at this stage, a better title would surely be 'Costing the Brief'. This might be so if that was all that the cost planner was going to do – passively estimate the cost of carrying out the client's ideas. But this will not be good enough. We have seen that this is the stage where the big cost decisions are made, consciously or unconsciously, and design work should not start until the brief itself is right. There is little point in spending large sums of money meticulously – and successfully – cost planning and controlling a design which is not in fact a very good answer to the client's needs.

In the case of profit projects it is often possible to charge higher rents or get a higher price for leases if a better standard of amenity and finish is provided, so that the budget is flexible to a limited extent. In such circumstances the cost disadvantages of larger and more luxurious public spaces, more and faster lifts, better fittings, and so on, will have to be evaluated in relation to income. The acceptable final cost may be quite different to the original rough estimate owing to such decisions.

Clients for private residences often become more ambitious as the drawings progress, and again, provided that they are kept informed of the consequences of their decisions, they may prefer to have the house they now want rather than the cost they first thought of. It is obviously preferable that these matters should be decided at brief stage rather than during design development or, even worse, during construction.

Many authorities consider that the traditional practices of the British building industry make it too easy for the client to make changes during the construction stage and that this is partly, or even largely, to blame for the comparatively high cost of building in the UK compared to some other countries. The cost planner therefore must be helping to develop the brief and examine its cost implications critically (unless specific instructions to the contrary have been given – but clients rarely do that these days!).

Nevertheless, the first task is indeed to cost the brief as presented. With sketchy information and without any sort of design it will be impossible to use any kind of resource-based cost model, so even where it is proposed that the final cost plan will be based on resources, there is little alternative to a product-based estimate at this first stage.

13.3   Preliminary estimate based on floor area

This will be a very approximate estimate; it could hardly be otherwise in view of the lack of information at the cost planner's disposal. There are exceptions to this, for example:

- Where the project is one of a series (such as a chain of service stations or superstores) and information has been accumulated on a number of almost identical buildings.
- Where the first estimate has to be calculated according to a government formula and the building will be tailored to fit the estimate.
- In the preparation of the preliminary estimate based on floor area, previous experience of the designer is most valuable as there are some whose designs always come out above average costs just as there are those who can be relied on to achieve an economical design.

At an early stage the cost planner is quite likely to be told 'This will be an economy job', and only a knowledge of the people or practices concerned will indicate how far this statement can be relied on in framing the estimate. The cost planner should never allow the amount of the estimate to be influenced by opinions expressed by interested parties, or worse still by the figure that the client would like the job to cost, unless these also coincide with the cost planner's own judgement. When an estimate based on early designs has worked out at £3,500,000, it is all too easy to remember that the budget is £2,250,000. The danger then is that the cost planner may assume that the estimate is too high and reduce it a little in order to make the difference less breathtaking. Cost planners may also find themselves in front of a committee or board, one of whose members 'knows a job just like this one which was built for much less than the cost planner's figure'.

Remember that it is the person who has prepared and possibly signed the estimate who will be held responsible for it, not the designer, client or the assertive committee member. This is as it should be. Cost planners are paid for their skilled evaluation of real probabilities, not for telling people what they want to hear.

The initial estimate, like all others, must make quite clear what is and is not included. It is good practice to use a form for preparing the estimate, with a definite section to show inclusions and exclusions. A computer spreadsheet program should similarly present a menu of possible inclusions and exclusions to jog the cost planner's memory and print them out when preparing the estimate for the client. Such items as the following should be specifically dealt with, not just forgotten about or covered by some vague overall clause:

- site preparation, such as demolition or dealing with contaminated land;
- archaeology;
- complete internal decoration;
- joinery fittings;
- venetian blinds and curtaining;
- furnishings;
- office partitioning;
- lighting fittings;
- carpeting in offices and flats;
- site works;
- specialist plant and services;
- shopfitting and security systems.

In fact it is even more important to define these things properly in a preliminary estimate than it is later on, when the detail of the estimate itself may be sufficient to answer many of the questions.

Make sure that professional fees, VAT and project manager's and construction manager's fees (where appropriate) are clearly excluded or included; clients quite rightly think that the estimate shows their total commitment unless they are told otherwise. It is especially important to make the position about VAT clear; it is the usual current practice for cost planners and estimators to ignore this, but the client must not be left in any doubt.

If fees are shown it is important to get them right. Until the mid-1980s this was comparatively easy, as everybody charged according to the fee scales of their various professional bodies. Today, when fees are subject to negotiation or competition, this is more difficult but unless something has already been agreed to the contrary, it will probably be best to assume fees according to scale at this preliminary stage. Reductions can always be made later, if necessary. Professional fees should be estimated to project completion, not just design completion, and should include the various specialists as well as expenses and document reproduction. The latter can be a considerable sum, despite the supposed advent of the paperless office.

The estimate must also be specific about inflation. If it is based on current building costs it should say so, and also make clear that it is subject to revision in the event of increases or decreases in those costs. This is critical, particularly given the current economic climate. In 2007, the *Building* magazine cost update reported significant rises in construction materials prices (particularly steel) and mechanical and electrical services costs (the catalyst being prices for non-ferrous metals). Taking into account the conventional measures of construction cost inflation – building, mechanical and electrical costs – 2007 continues to witness these prices climb away from inflation (based on the Consumer Prices Index). The cost update reports that the Building and Electrical Indices have increased over the past six years at similar rates (average 5.5–5.7% a year).

If, on the other hand, the estimate is based on anticipated cost levels in the future, the basis of calculation should be set out in the estimate so that revision can be made if the prediction proves to be false. But in both cases there are advantages in tying the estimate to a well-known and reputable cost index such as the BCIS, so that the cost planner's personal judgement cannot be blamed if things suddenly change. As already pointed out, the person who could accurately forecast economic conditions two or more years ahead would be in demand in other and more lucrative fields than construction cost planning!

A final problem relates to whether the tender or the final cost is being forecast. Clearly the latter is the preferable alternative, since it is this sum of money which the client is going to have to find in the end, and in any case there is little alternative in the case of projects which are to be built under a contractual arrangement that does not involve a lump sum tender as such. However, in the public sector in particular the custom is often to estimate the tender amount, and it is important in both public and private sectors to be specific about which alternative has been used. At times of high inflation there can be a considerable difference between the two figures, especially if the cost planner makes some allowance for the claims which often arise during the course of the work and which will be reflected in the final cost but not in the tender amount.

As the preliminary estimate cannot be related to the subsequent cost plan (except in total), it is not worth preparing it in any greater detail than is necessary for its primary purpose, that is to give an approximate estimate of cost. Also, although cost planning is very much more than the preparation of an estimate, it gets off to a bad start if the first estimate is badly wrong. This stage is therefore extremely important.

Finding the right rate at which to price floor area is not easy, and the most reliable starting point is a recent similar job from office experience. Failing this, or as a check on it, a wide selection of costs from a published source such as the BCIS system should be used, resisting the temptation to choose only from the lower part of the cost range. The BCIS Online system is a particularly good source, not least because of the way in which it facilitates the adjustments described here.

To adapt the floor area rate from one project to another when shape is not known requires consideration of seven factors:

- market conditions;
- size;
- number of storeys, etc.;
- specification level;
- inclusions and exclusions;
- services;
- site and foundation conditions.

A most important point is to check what is actually meant by floor area. This might be the total area at each floor level measured over the external walls including all staircases, lift wells and similar voids, and including all internal and external walls; or the gross internal floor area (GIFA) measured according to SFCA/BCIS rules, which is the same as the first but excludes the thickness of external walls; or something rather less than the last, perhaps excluding voids, circulation spaces or plant rooms; or the net commercially lettable floor area (often still given in square feet to complicate the issue). Obviously for the same building these would each yield quite different cost factors, and it is important to check that the same option is used for all the data source examples and the new estimate, or that conversion factors are duly applied.

Further points for care include factors which may not be properly reflected in traditional floor area formulae, and which may occur either in the project being estimated or in the cost data examples. The most troublesome of these is the atrium, which was not a common design feature when these floor area formulae were devised and which, if large, cannot be treated as an ordinary void if comparison is being made with non-atrium buildings. A somewhat similar problem arises with buildings that have an unenclosed ground floor used as part of an open car-parking area.

13.4 An example of a preliminary estimate

This section will work through an example in which the above factors are taken into account where applicable.

The proposed project is a new 2,400 m² six-storey office block in a provincial city, with shops on the ground floor and offices on the other five, with a budget of £2,000,000. The previous project on which the estimate will be based is a project of similar quality in outer London, but of four storeys only and also with shops on the ground floor. The total floor area was 1,235 m^2 and the tender price 12 months previously was £680/m^2. The differences between the two schemes are rather more than one would ideally like to see when choosing a suitable cost plan to adjust, but will enable all the issues to be considered and demonstrated.

13.4.1   Market conditions

This adjustment deals with the changes in building prices and in tendering conditions between the time when the previous job was priced (or built) and the anticipated date of tender (or of building) for the new project. Two quite different issues are being considered together here:

- official changes in the cost of resources;
- changes in the market itself, taking into account the capability of the industry and the amount of work available to it, in the regions and at the periods in question.

In doing this it is most important to compare like with like. It may be, for instance, that the project being used as a basis for comparison was tendered for conventionally, so that the prices are those ruling before construction started, whereas the new job might be built using construction management where prices ruling at the time of carrying out the works are what matters.

One of the published cost indices previously described would probably be used, but where the estimates being compared are not too far apart in time (up to a year or so) it is quite possible to make the adjustment for resource costs by simply adding a percentage based on known labour and material fluctuations. It may indeed be necessary to do this for short-term estimates because of the time lag in publishing historical indices.

For this purpose the proportions of labour, material and plant content given in Chapter 12 could be used, with a heavier bias towards materials if the contract is for an expensive building with good finishes and high-class materials. In making this calculation the figures should be taken as fractions of the gross contract amount without deducting overheads and profit, since these should rise more or less in proportion to increases in prime cost. Where do-it-yourself indexing of this type is being used, or where a published index is being used which only reflects changes in the prices of resources, it is most important to remember to make the adjustment for changes in market conditions as well. This can be done intuitively if both projects are local, but if one or both of them is in an area unknown to the cost planner, it may be necessary to make discreet enquiries.

Local changes in the market often tend to be relatively short term (because an overheated local construction sector has to cool down somehow), and thus are not always easy to pick up from published building cost indices.

Phase 2

Local enquiry may be worthwhile in a case such as the present one if the new project is in an unfamiliar region. In this particular example, allowance would have to be made for the difference between outer London and the provincial city, remembering whether or not differences in resource costs (e.g. labour rates) have already been adjusted.

Let us assume an overall upward adjustment of 10% for inflation and region on this contract. This can either be shown at the beginning of the estimate, thus:

$$\text{£680/m}^2 \text{ plus } 10\% = \text{£748/m}^2 \tag{1}$$

or added on at the end.

The first method is preferable, as all the adjustments that follow can then be done at current prices, which is usually easier.

13.4.2   Size, number of storeys, etc.

At this stage no actual drawings or dimensions are likely to be available, so adjustments have to be done by proportioning, that is the use of ratios in comparing the project used as an example and the new project. In the early days of cost planning it was fashionable to use proportioning methods extensively, to show that the new approach was different to old-fashioned approximate estimating. But this point no longer has to be demonstrated, and proportioning has two disadvantages:

■ It is not the QS's usual method of working and thus mistakes are more easily made.
■ Sooner or later approximate quantities have to be used, and these are difficult to reconcile with proportioning.

It is therefore recommended that proportioning only be used when there is no alternative.

The new provincial block will have a larger plan area than the London example; this will tend to reduce the proportion of external wall area to floor area (assuming a similar plan shape) and hence should reduce a little the cost per m² of floor area. However, the new block will have six storeys instead of four, and this will have the effect of increasing the cost as it is a general rule that building higher costs money. But the difference between four and six storeys is not critical, and the likelihood is that this and the previous factor will cancel each other out.

The less that is known about the new building, the less point in adjusting for hypothetical trifles. We are working with a broad brush at this stage, trying to get a ballpark figure as a starting point for the cost planning process. However, a more important point to consider is that on the outer London block one-quarter of the floor area was set aside for shops, whereas on the six-storey building the figure will be one-sixth. As shop areas are normally left in an unfinished state for the shopkeepers to fit out according to their needs and tastes (shell and core), this part of the building will be relatively inexpensive. By reducing the proportion of the building allocated to shops we shall be increasing the overall price/m^2.

Assuming that the saving in floor, ceiling and wall finishes and services between offices and shops will be £72/m^2 of total floor area:

Outer London block
Shops (25%)

£748/m^2 minus 3/4 of £72	=	£694/m^2	
		309 m^2 at £694/m^2	= £214,446
£748/m^2 plus 1/4 of £72	=	£766/m^2	
		926 m^2 at £766/m^2	= £709,316
			£923,762
Check – total 1,235 m^2 at £748/m^2			£923,780

Provincial block

Shops	16.67%	400 m^2 at £694/m^2	= £277,600
Remainder	83.33%	2000 m^2 at £766/m^2	= £1,532,000
			£1,809,600
£1,809,600/2,400 m^2			= £754/m^2

Similar adjustments may be required wherever the buildings are divided into areas with different cost profiles and where the proportions are different on the two buildings. Car-parking areas within office buildings frequently lead to adjustments of this sort.

13.4.3   Specification level

So far the changes have been dealt with by adjusting the elemental rate, as there has not really been any alternative and this was in any case the preferred approach in classical elemental cost planning. But from now on it is possible to think in terms of lump sum changes and adjust the elemental rate at the end, and many cost planners prefer to do this. Both methods are shown here. Whichever method is used, it is important to remember whether or not the building cost has already been adjusted for inflation. If it has been (as is recommended), then any specification adjustments must be done at prices ruling at the date of the new scheme, and not those ruling when the previous or example project was carried out. Alternatively, of course, all specification adjustments could be made at the earlier pricing level, and the inflation adjustment for the total scheme done at the end.

It is assumed that the specifications of the two buildings will be of the same general standard, but:

Phase 2

■ The designer has decided that the cheap floor covering used in the offices in outer London was a mistake, and wants to allow an extra £12/m² for this.

■ The client wants natural stone dressings in lieu of the cast stone used on the previous project.

The adjustment for the flooring is easy. Allow an extra £12/m² over the area of the offices, say two-thirds of the total floor area.

$$1,600 \text{ m}^2 \text{ at } £12/\text{m}^2 = £19,200 \tag{2}$$

This is equal to £8/m² of gross floor area when divided by 2,400.

Calculations like this can be done by simple proportion if preferred (1600/2400 of £12.00 is £8.00), but it is quite easy to make a mistake doing this – inverting the ratio for example! – and the longer calculation is more easily checked.

The stone will be a little more difficult to allow for (in the absence of any drawings, remember). The area of external walling in a medium-rise office building of this sort should be about 10% more than the floor area (again, check this with several analyses if possible). Not much of the wall area will actually be stone; most of it will probably be windows. We will assume that 25% of the area is stone and that natural stone facing will cost £130/m² more than cast stone. So the extra allowance for natural stone is:

$$(110/100) \times 2,400 \times 1/4 \times £130.00 = £85,800$$

which is £35.75/m² of gross floor area.

Note that a number of assumptions have been made about the London example which could have been avoided if a full SFCA/BCIS type of cost analysis had been available, even though it is not intended to produce this first estimate for the new building in elemental form.

13.4.4   Inclusions and exclusions

The client has found that office tenants prefer to do their own partitioning and decoration, whereas on the outer London project these were provided by the developer. This adjustment will most easily be carried out by extracting the relevant lump sum figures from the BQ for the London project, if these are available.

	£
PC sum for office partitioning	26,250
Add for profit and attendance	1,313
Decoration in offices	4,300
	31,863

Phase 2

As it comes from the original scheme this figure must be adjusted for inflation, also for five-sixths of the building being offices in lieu of three-quarters:

£31,863 plus 10% inflation allowance (from above) = £35,049
Division by floor area of offices on London building (926 m²) gives a cost of £37.85/m² of office floor area
The saving on the provincial block by omitting these items is therefore £37.85 × 2,000 m² = £75,700 in total
£75,700/2,400 m² gross floor area = £31.54/m²

By the (alternative) proportion method the calculation is:

£31,863 divided by 1,235 m² = £25.80
£25.80 plus 10% inflation allowance (from above) = £28.38/m²

The omission of partitions in the new building per square metre of gross floor area is therefore £28.38 6 4/3 × 5/6 = £31.53/m². This all assumes that the cost of partitions was available from the BQ or cost analysis for the outer London block. If this was not the case then a telephone call to a firm of partition installers might be the only way of finding out a cost/m². There is no need to tell them that you want to know this so that the partitions can be omitted! This would be a current cost, of course, so does not need to be adjusted for inflation or market changes.

13.4.5   Services

Services form a major part of the cost of any modern building. On a project of any size there will be consulting engineers for the heating, air-conditioning, plumbing and electrical work, and any specialist service installations. The terms of their appointment should include for providing an estimate of the cost of the services for which they are going to be responsible.
  The building cost estimate would be adjusted as follows:

Provincial office block
Estimated figures received from consultants:

	£
Air-conditioning and space heating	165,000
Plumbing	52,000
Electrical installation	108,000
Lifts	118,000
	443,000
Builder's work, attendance, profit 7.5%	33,225
	476,225

£476,225/2,400 m² = £198.42/m²

Outer London block   £
Comparable figures extracted from cost analysis or BQ   186,000
Add 10% for inflation as main estimate   18,600
                                                            204,600

£204,600 divided by 1,235 m² = £165.66/m²
Increase in cost of services compared to outer London block
£198.42/m² − £165.66/m² = £32.76/m²

In view of the early stage at which this estimate is being prepared, the engineering consultants may not feel able to give an estimate for services, but it is important that they should do so. This will avoid a situation where the cost planner assessed the cost independently and the amount required by the engineering consultants when they prepared their detailed scheme proved to be very different. The resulting arguments would do little to improve either the harmony within the professional team or the client's confidence in the team as a whole. Nevertheless, it is quite possible that the consultants, having been asked to establish a preliminary cost target before they have fully established what is required, will play safe and allow themselves a sum of money sufficient to cover almost any eventuality. (It is, after all, in the nature of engineering design to allow for the likely worst case.)

The situation where engineering services appear to be taking rather too large a slice of the whole cake is common enough to be met by most cost planners from time to time, and how it is dealt with depends very much on the cost planner's terms of appointment. If these include a measure of executive authority then the cost planner might well call a meeting to thrash matters out; otherwise it may be only a matter of pointing out the problem to the architect, project manager or whoever is directing the project.

The cost planner should bear in mind, however, that prices per square metre for services are extremely difficult to estimate (there is still a lack of cost data in this important area), and should try to co-operate with the various consultants rather than raise difficulties. It may, for instance, be very reasonable for the engineer to insist on client requirements being defined with more precision before even a rough estimate can be given. If there are no consultants appointed, or if at this stage they are not prepared to commit themselves, the cost planner would be better not making any adjustment to the estimate for the services, unless there are glaring differences between the outline requirements for the two buildings. It is, however, becoming increasingly common for professional cost planners, particularly in the largest firms, to be closely involved in the cost planning of the engineering services which, after all, represent a considerable (and increasing) proportion of the budget for which they are responsible.

Special services, especially transportation services such as lifts, should always be adjusted, as their cost bears little relation to the area of the building and they are much more easily priced on a per unit or per installation basis. The adjustment should be done as a comparison of lump sums, as shown above.

13.4.6   Site and foundation conditions

These are best adjusted on the basis of the so-called footprint area or site area of the building. On most buildings this should be more or less the same as the roof area.

A major factor today is the possibility of having to take measures to deal with a contaminated site. The cost planner must make enquiries of the client, the structural engineer or the local authority to find out if there are any such requirements. This is assumed not to be the case with both the sites under consideration in our example, which are in developed urban areas, but the new provincial project will require piled foundations that were avoidable in outer London. Either the cost of piling per m² of footprint area can be obtained from a suitable analysis, or a certain number of piles can be allowed for, as follows:

Say 90 Nr piles average 4 m long at £75/m   = £27,000
£27,000 divided by 2,400 m²   = £11.25/m²

Note that the complex of pile caps and ground beams in connection with the piles will cost little less than conventional foundations, so there is no saving to be set against the cost of the piles.

13.4.7   Other factors

The client is anxious to have the building erected as quickly as possible and the contract period will consequently be very short. This will tend to increase the cost, both because the site labour force will need to be larger than is economical and because the number of local contractors and sub-contractors who can be relied on to work to such a tight timetable is restricted and competition will therefore be limited. The contract price may therefore have to include a great deal of overtime. This is expected to add about 5% to the cost.

Figure 13.1 shows how the estimate would be set out in an appropriate format, showing the changes in elemental rates. Figure 13.2 shows the same estimate using the usually preferred method of working out the elemental rate at the end. The estimated cost of £1,997,500 is slightly below the budget of £2,000,000, so is likely to be acceptable. It is assumed that the question of professional fees is set out in a covering letter.

13.5   An example using BCIS data

The following example (p. 258) is worked out manually, although in practice the whole thing could be done using the BCIS Online service.

Phase 2

Project	Office Block, Midtown	
Date of estimate	Feb. 06	
Assumed date of tender	Sept. 06	
Gross internal floor area	2,400 m^2	

Basis of estimate		£/m^2
Office block, outer London (Sept. 2005)		680

Adjustment

1 Market conditions

BCIS Index. Add 10%	+	68
		749

2 Size, number of storeys, etc.

6 storeys in lieu of 4 – see notes	+	6
		754

3 Specification level

See notes. Wood block flooring £8

Portland stone £35.75	+	43.75
		797.75

4 Inclusions and exclusions

See notes. Office partitions and

decorations excluded	−	31.53
		766.22

5 Services

See consultant's notes	+	32.76
		798.98

6 Site and foundation conditions

Piling (90 nr 4 m long)	+	16.87
		815.85

7 Other factors

Tendering (see notes) add 2%	+	16.32
		832.17
Professional fees		excluded
		832.17

Total estimated cost   2,400 m^2 at £832.17/m^2 say £1,997,500

Included (and as above)	Excluded (and as above)
	Floor, ceiling and wall finishes
	All furnishings
	Electric light fittings
	VAT
	Professional fees

Figure 13.1 Preliminary estimate of cost, showing the changes in elemental rates.

Phase 2

Project	Office Block, Midtown
Date of estimate	Feb. 06
Assumed date of tender	Sept. 06
Gross internal floor area	2,400 m²

<table>
<tr><td></td><td></td><td>£</td></tr>
</table>

New estimate: 2,400 m² at £680

Adjustment

1 **Market conditions**
   BCIS Index. Add 10%   +   163,200
                                                                    1,795,200

2 **Size, number of storeys, etc.**
   6 storeys in lieu of 4 – see notes (2,400 m² at £6)   +   14,400
                                                                    1,809,600

3 **Specification level**
   See notes. Wood block flooring £19,200
   Portland stone £85,800   +   105,000
                                                                    1,914,600

4 **Inclusions and exclusions**
   See notes. Office partitions and
   decorations excluded   –   75,700
                                                                    1,838,900

5 **Services**
   See consultant's notes   +   78,624
                                                                    1,917,524

6 **Site and foundation conditions**
   Piling (90 nr 4 m long)   +   40,500
                                                                    1,958,024

7 **Other factors**
   Tendering (see notes) add 2%   +   39,160
                                                                    1,997,184
   Professional fees   excluded

Total estimated cost   say £1,997,500 ÷ 2,400 m = £832.29/m²

Included (and as above)	**Excluded (and as above)**
	Floor, ceiling and wall finishes
	All furnishings
	Electric light fittings
	VAT
	Professional fees

Figure 13.2   Preliminary estimate of cost, calculation of elemental rate at the end.

Example

A budget or early brief figure is required for a steel-framed factory of 2,100 m² gross floor area located in Leicestershire. The tender date is expected to be in the fourth quarter of 2006.

Mean cost/m² gross internal floor area of a steel-framed factory adjusted to UK mean location at 1st quarter 2005 prices	341/m²
Adjust for location Leicestershire × 0.915	312/m²
Mean building price of factory in Leicestershire 2,100 m² at £312/m²	£655,200
Allow for local pricing adjustment based on QSs knowledge of local market (say) × 0.95	£622,469
Allowance for inflation to fourth quarter 2005 BCIS All-in Tender Price Index 1st quarter 2005 = 221 4th quarter 2006 = 239 (forecast)	
Adjustment (£622,469 × 239)/221	£673,168
Approximate estimate at fourth quarter 2006	£675,000
Excludes external works, professional fees, and VAT	

13.6 Cost reductions

It may be that the estimated cost is higher than the client can afford, or more than the client wants to pay. It would be very dangerous to reduce the amount of the estimate for this reason, or on some vague grounds such as a general reduction in standard, although of course adjustments could be made for specific items such as (on the basis of the first example) cheaper lifts or a return to the cheaper floor finish.

However, if the total of the estimate is much too high, it is best to face facts and either reduce the size of the building or get the brief or the budget substantially modified. It is the easiest thing in the world to cut an estimate by 10% (or by any other percentage) under the influence of the client's pleadings and the designer's optimism.

It must be emphasised that an overall 10% cut in costs means much more than a 10% cut in standards, as much of the structural work, for example, cannot be reduced in cost. It must also be remembered, as stated earlier, that the estimate will carry the signature of the cost planner, not of those who want to modify it. It can be assumed therefore that if the cost planner amends the estimate from its original figure, there must be some tangible reason(s). These reasons need to be spelt out if a modified figure is presented.

Finally, mention must be made of the threat, 'The job won't go ahead on this figure, do you want it to go ahead or not?' This may be the response of a profit-oriented client, or an architect who does not want to lose a commission. Alternatively, it may be the attitude of a public body that knows a major project will never get the go-ahead if its true cost is known beforehand, but will be very difficult to stop at a later stage when the enormity of its cost becomes apparent. Nobody ever seriously believed, for example, that the Sydney Opera House would only cost the figure of one or two million dollars which was first given; and closer to home, the British Library would have been unlikely to have received the go-ahead if an honest estimate had been given! We are now well into the realms of professional ethics, about which people must make up their own minds in the light of the case before them, but whatever the ethical position the responsibility of the cost planner for whatever figure goes forward must be remembered. To repeat what has been said before, the other parties will adopt a very low profile if things go wrong!

13.7   Data sources

The example of the two office blocks assumed that there was data available on a comparable scheme. In the later stages of estimating, when a cost plan is prepared on an elemental basis, an exact similarity of use between data source and proposed project is not vitally important; costs/m² of external walls, windows, doors, floors, etc. need not vary much between a library and an office block. The costs/m² of floor area of the buildings themselves, however, will vary considerably because of the differing proportion of these items in them.

If there are no truly comparable schemes in the office it may be possible to get information from a professional colleague, or from the BCIS, or as a last resort to use published information. When working from one's own information it is best to use one building as an example because the variables in design and tendering are known, but it is often helpful to supplement this with published information for similar buildings. As many examples as possible should be obtained to try and average out all the variable factors. Published information is always useful for checking and comparing an estimate prepared from one's own sources.

13.8   Mode of working

It may be objected that the above methods of working are far too crude. For instance, instead of guessing a percentage for an accelerated programme for the office block it would surely be better to compare the consequences of a normal programme and an accelerated one by preparing network diagrams and costing out the two sets of resources. Well, of course it would, except that no detailed design exists for the building at this stage, and although it might be possible to study the effects in detail for a known building of fairly similar characteristics, the answer would not be sufficiently valid to justify the amount of work involved (and in any case there was a market forces component to the figure based on restricted tendering which such an exercise would not disclose). Also, the workload to achieve this would be quite heavy; even if computer simulations are used they require a fairly considerable data input.

To repeat what was said at the beginning, what is being looked for is simply a ballpark figure for a hypothetical development ('How many noughts?'), and often there may not be enough firm information even to carry out the very modest range of adjustments set out above. There will be plenty of opportunity to use more sophisticated techniques as soon as there are some tangible proposals to examine, and this process begins with the preparation of tentative sketch designs.

13.9   Key points

- Costing the brief is the stage where the big cost decisions are made, and design work should not start until the brief itself is right. There is little point in spending large sums of money meticulously – and successfully – cost planning and controlling a design which is not in fact a very good answer to the client's needs.
- The preliminary estimates will almost certainly be based on the cost per square metre of floor area, preferably using a computer spreadsheet to jog the memory and make sure that nothing has been accidentally left out. It is usual practice to exclude VAT from the figures. The reports to the client must make clear what is included and what is excluded. Cost planners must be prepared to back their judgement if their estimates come out to a higher figure than the client is expecting.

Chapter 14
Cost Planning at the Scheme Design Stage

14.1 Elemental estimates

Once the initial sketch drawings are available, an outline elemental estimate should be prepared in order to assess the architect's initial solution against the cost parameters of the project. This will be particularly important if the shape or design is rather unusual, or appears inherently expensive. Not unlike the preliminary estimate, it is sometimes called a preliminary cost plan, but this is misleading since it is really only an approximate estimate, and it is probably the elemental format that leads to this terminology. The real cost plan will be prepared when this elemental estimate has been finalised, and it will be in a fresh format.

There are some advantages in using a short list of consolidated elements for preparing the preliminary plan, as the only measurable items will be the main floor and wall areas and there is unlikely to be a specification as yet. For the purpose of this example a short list is assumed, and four main areas need to be measured:

- ground floor area;
- total floor area;
- area of external walls, including windows and doors;
- area of internal walls, including windows and doors.

The following example is based on the architect's first sketch design for the six-storey office block developed from the brief and estimate considered in Chapter 13. This is shown in Figure 14.1.

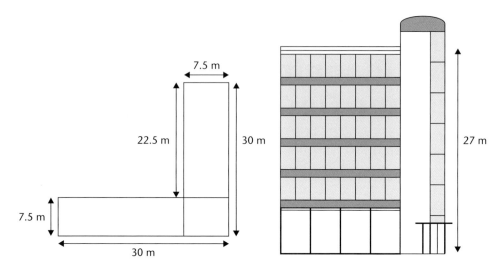

Figure 14.1 Architect's first sketch for provincial office block (not to scale).

Gross internal floor area

Ground floor area	$30.00 \times 7.50 = 225.00$		
	$22.50 \times 7.50 = \underline{168.75}$		
		$\overline{393.75} \times 6$ floors	2,362.50
Tank room	8.00×6.00		$\underline{48.00}$
			$\overline{2,410.50}$ m^2

External wall area

Main walls	$4 \times 30.00 \times 27.00$	3,240
Tank room	28.00×2.50	$\underline{70}$
		$\overline{3,310}$ m^2
Internal wall area (say)		$\overline{1,200}$ m^2

Already it can be seen that the ratio of external wall area to total internal floor area looks rather high (3,310:2,410 = 1.37 to 1), and we may feel that this is going to prove an expensive building. However, the elemental estimate will confirm or reject this theory better than any amount of conjecture. The best rates for pricing would be those obtained from an analysis of the outer London block, since we based the first estimate on a project from that area.

Remember to add in the 10% price increase and the 2% tendering differential that were included in the approximate estimate in Chapter 13, so as to keep the two estimates on the same basis; unless circumstances have changed in the interim, in which case, of course, the total of the estimate will have to be revised. Again, a standard form should be used. The form shown in Figure 14.2 is an example of the sort of thing required. Some of its features would vary according to the preferences of the individual office; for instance some

Phase 2

people might prefer to show the preliminaries as a separate item, or to make a separate element of the frame. On the other hand, it would be possible to combine the finishes with the relevant structural items. However, the selected arrangement should be standard within the office so that the recorded information can be kept in a suitable manner.

The overall rates for elemental estimates can be obtained when doing the ordinary analysis and filed separately, and being large omnibus items there is perhaps less risk than usual in poaching prices from more than one job. With the rise of integrated computer systems this information can be stored and manipulated with far greater ease than was previously the case. While the original practitioners of elemental cost planning believed in showing the elemental costs for each element (as in Figure 14.2), it is more usual today to omit them, and just show the total costs for each.

To return to the particular example: the establishment of rates for the various elements will have been complicated in this instance by the different proportion of unfinished shops compared to the outer London job.

14.2   A typical elemental rate calculation

The rate for floor finishes, for example, will have been obtained as follows:

	£
Project used for comparison (outer London)	
Elemental rate for floor finishes on outer London project	27.90/m²
Add 10% increased costs and 2% tendering	3.35
	31.25/m²

Calculation of floor finish elemental rate for areas excluding shell and core shops:

	£
1,235 m² at £31.25 m²	38,594
Shops 25% = 309 m² at say £10.00	3,090
Balance 926 m²	35,504

which divided by 926 gives £38.34/m² of gross internal floor area.

	£
New provincial block – 2,410 m²	
16.67% shops = 402 m² at £10.00	4,020
83.33% remainder = 2,008 m² at £38.34	76,987
Total 2,410 m²	81,007

which divided by 2,410 gives £33.61/m² of gross internal floor area

	£33.61/m²
Add say 5/6 of £12.00 for more expensive floor finish (an upgrade) in offices (as approximate estimate)	10.00
	£43.61/m²
Say	£44.00/m²

Elemental estimate No. 1				
Project	Office Block, Midtown			
Date of estimate	March 2006		**Total cost (£)**	**Cost (£/m²) of floor area (2,411 m²)**
Assumed date of tender	September 2006			
Ground floor area	(A)	394 m²		
Total floor area	(B)	2,411 m²		
External wall area	(C)	3,310 m²		
Internal wall area	(D)	1,200 m²		
Work below lowest floor finish	(A)	394 × 321.00	126,474	52.45
Upper floors including frame and stairs	(B – A)	2,017 × 123.50	249,100	103.32
Roof including frame	(A)	394 × 164.50	64,813	26.88
External walls	(C)	3,310 × 212.00	701,720	291.04
Internal walls	(D)	1,200 × 50.00	60,000	24.89
Floor finish	(B)	2,411 × 44.00	106,084	44.00
Ceiling finish	(B)	2,411 × 27.50	66,303	27.50
Wall finish	(C + 2D)	5,710 × 13.00	74,230	30.79
Decoration	(B)	2,411 × 4.50	10,850	4.50
Fittings			2,750	1.14
Services			476,225	197.52
Drainage			19,000	7.88
Site works			27,500	11.41
Contingencies			30,000	12.44
			2,015,049	**835.76**
Professional fees			—	—
			2,015,049	**835.76**
(Inclusions and exclusions as approximate estimate unless otherwise stated, VAT not included)				

Phase 2

Figure 14.2   Elemental estimate.

This complicated calculation would have been much easier if the unit costs for the floor finishes had been available for the outer London job, and it would not have been necessary to use the risky proportion method.

Alternative calculation using unit costs
Unit rates taken from detailed analysis of outer London project, plus 10% inflation allowance and 2% tendering allowance.
Quantities measured from architect's drawing of provincial job.

			£
Office areas	1,688 m² at 34.25	=	57,814
Shops	338 m² at 8.95	=	3,025
Tank room	48 m² at 8.05	=	386
Entrance, staircase, WCs, etc.	337 m² at 45.20	=	15,232
			76,457
Add 10% increased costs and 2% tendering			9,175
			85,632

Add extra cost of better floor finish (an upgrade) in offices

1,688 m² at 12.00/m²	=	20,256
Total cost		£105,888
Elemental cost (divide by 2,411)		£43.92/m²
Say		£44.00/m²

14.3   Examination of alternatives

From an examination of the elemental estimate shown in Figure 14.2 we can see that the scheme at £2,015,049 is likely to prove more expensive than the amount of the first approximate estimate (£1,997,500). Much of this increase is due to the slightly increased size of the building as designed (2,410 m² instead of 2,400 m²), but the cost/m² of the floor area is also increased from £832.17 to £835.76, and the estimated cost is now slightly over, instead of under, the budget of £2,000,000. While the increase is not very large (it could be removed by a small reduction in the level of specification), it does make us wonder whether this is the most economical solution to the design problem. At this point the figures in the right-hand column (cost per square metre of floor area) come in very useful in enabling comparisons to be made with other contracts.

It certainly seems that the external walls are costing a lot, which confirms our first thoughts on looking at the plan, but we now have figures to prove the matter. Although it would be practicable to build this scheme within the budget, it would be worth the architect seeing whether an alternative shape would be possible. The rather unsatisfactory appearance of the first elemental estimate should lead to the consideration of alternative schemes. It may not always be possible to improve wall/floor ratios very much, due to site

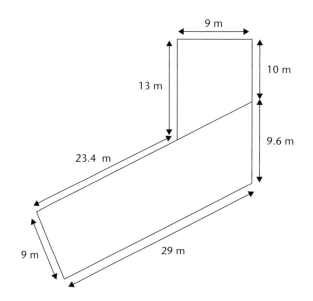

Figure 14.3   Revised sketch plan for provincial office block.

restrictions or other unalterable factors. But suppose that in the case of the provincial office block the architect has been able to produce an acceptable alternative design (Figure 14.3):

Mean length of building = (33 m + 29 m)/2 + (13 m + 10 m)/2 = 43 m
Ground floor area = 43 m × 9 m   = 387 m²
Total floor area 6 × 387 m²   = 2,322 m²
plus tank room   48
                                                          2,370 m²
Length of external walls = 23.4 m + 13 m + 9 m + 19.6 m + 29 m + 9 m = 103 m
Area of external walls = 103 m × 27 m = 2,781 m²
Add tank room walls as before   48
                                                          2,829 m²
Internal wall area (say) 1,200 m² as before.

   The approximate cost for this scheme is shown in the second elemental estimate (Figure 14.4). This is a much more hopeful design with an estimated cost of £1,918,550, slightly below the approximate estimate and the budget. The allowance for site works has been increased, as additional treatment to the front areas would be required owing to the front elevation being set back on the splay.
   It should be realised that outline elemental estimates will not be completely accurate, although the comparisons obtained from them will be valid enough. Each of these breakdowns would only take a few hours of a senior

Elemental estimate No. 2					
Project	Office Block, Midtown				
Date of estimate	March 2006			**Total cost (£)**	**Cost (£/m²) of floor area (2,370 m²)**
Assumed date of tender	September 2006				
Ground floor area	(A)	387 m²			
Total floor area	(B)	2,370 m²			
External wall area	(C)	2,829 m²			
Internal wall area	(D)	1,200 m²			
Work below lowest floor finish	(A)	387 × 321.00		124,227	52.42
Upper floors including frame and stairs	(B – A)	1,983 × 123.50		244,901	103.33
Roof including frame	(A)	387 × 164.50		63,662	26.86
External walls	(C)	2,829 × 212.00		599,748	255.02
Internal walls	(D)	1,200 × 50.00		60,000	25.31
Floor finish	(B)	2,370 × 44.00		104,280	44.00
Ceiling finish	(B)	2,370 × 27.50		65,175	27.50
Wall finish	(C + 2D)	5,251 × 13.00		67,977	28.68
Decoration	(B)	2,370 × 4.50		10,655	4.50
Fittings				2,750	1.16
Services				476,225	200.94
Drainage				19,000	8.02
Site works				45,000	18.99
Contingencies				30,000	12.66
				1,913,600	**807.43**
Professional fees				—	—
				1,913,600	**807.43**
(Inclusions and exclusions as approximate estimate unless otherwise stated, VAT not included)					

Phase 2

Figure 14.4   Revised elemental estimate.

assistant's time, and each hour spent at this stage is worth days spent on cost checks of minor elements later on in a frantic attempt to keep within budget. This is where the process of cost planning really pays for itself.

14.4   Need for care

This is the place for a word of warning. Because we are dealing in such large figures at elemental estimate stage, it is vital that all arithmetic, and if possible all measuring, should be checked. A silly mistake in calculating the principal areas would have serious consequences later on, and just because they seem (and are) simple, the elemental estimates must not be taken lightly. We should remember that everyone makes mistakes, particularly people working under pressure; checking will ensure that there are none. The same of course applies to the preliminary estimate, dealt with in the previous chapter. It is also important to remember that there are many other considerations in designing a building besides cost, and that the cost planner must not be tempted to try and usurp the architect's function when the shape and type of building is under consideration. The cost planner must be prepared to give the architect all possible assistance at this time, but should avoid making detailed suggestions about matters of planning and architecture which are entirely the architect's responsibility.

14.5   The cost plan

When the sketch drawings have been finalised and the budget (modified if necessary) has been accepted by the client, it is time to prepare the cost plan itself.

Although the establishment of the brief and the investigation of a satisfactory solution are being dealt with as two separate and consecutive functions, there may be a certain amount of iteration. Design investigation may suggest modifications to the brief, which in turn will need to be investigated. This is all to the good and will probably result in improved performance. Even if it involves, as it will, a good deal of abortive cost-planning work, the cost planner's work at this stage is relatively cheap compared to the potential benefits.

The cost control of design development, on the other hand, demands considerable resources, and should not be carried out until a satisfactory solution has been defined and agreed. As already pointed out, substantial iteration between the second stage of the cost planning process (investigation of a solution) and the third stage (cost control of the development of design) brings nothing but disadvantages.

The basic principle is to move from the ballpark estimating of outline proposals to the detailed costing of production drawings in a series of steps. The cost plan will probably be based on the most recent elemental cost estimate for the project. It will, however, be developed with a full list of elements instead of the condensed list used so far, complete with an outline specification (agreed with the designer) for each of the elements. This expensive document will form the basis for the system of design cost control. It should not be produced prior to approval of sketch design, since it might have to be done again a second (or third) time if the design changes. In any case, the designer would find it difficult to make decisions on detailed specification matters while the design itself is still undecided. But it should not be left until the design is further developed; that might make it easier to do, but it would then become a record of what has been decided rather than a plan for controlling design – just a rather fancy cost estimate in fact.

The cost plan may take many forms, but the different systems have much common ground.

14.6   Specification information in the cost plan

Opinions differ on how detailed the specification information should be when the cost plan is prepared. The first approach involves giving a fully detailed specification in the cost plan:

- The cost planner/QS will try and get as much detailed information as possible from the architect before preparing the cost plan, and will fill in any blanks with typical specifications within the required cost range.
- When the cost plan is complete, the architect will be supplied with a detailed list showing the type of construction and finish on which the cost is based.
- The architect knows that although anything in the list may be altered at will, any such deviation may have an effect on the planned cost.
- The cost planner will feel that this specification list gives some protection in the possible event of an architect failing to design within the agreed limits, since these limits are written into the cost plan and are not just assumed.
- However, this procedure is open to the objection that the cost planner appears to be taking the responsibility for design out of the architect's hands.

The alternative way of dealing with this problem is to leave the responsibility for design entirely where it belongs, with the architect:

- When the cost plan is prepared, the architect's ideas on specification are incorporated into it, but if these are not forthcoming the cost planner does not provide them.
- Instead, a target cost for the element is devised, based on standards of cost performance achieved elsewhere, and it is left to the architect to design within that cost.

Phase 2

- If the element, when designed, costs more than the sum of money allocated to it, then it must either be redesigned or the elemental cost increased at the expense of other elements.

Few will deny that the theory behind this second approach is sound, but the first alternative is the one which was always preferred by most practising cost planners, and today is almost universally adopted.

Apart from the factor already mentioned of protection against an uncooperative architect, it is said that most architects prefer the discipline of a specification rather than the provision of a cost figure for the element which is unrelated to any specification and may necessitate several attempts at design before it is achieved. The detailed specification method also has the big advantage that if the architect works to the agreed specification, then subsequent cost checks will be unnecessary.

14.7   Elemental cost studies

Cost research or cost studies may be required when the architect is considering design alternatives. It is at this stage that cost models such as those described earlier will be most effective. However, at present these studies would normally be done by means of approximate quantities, about which sufficient has already been said.

Cost studies, if done properly, are expensive in professional time and therefore cannot usually be undertaken in respect of the majority of items in the project. They should therefore be reserved for comparisons of important components, or for projects which are themselves unusual and for which ordinary elemental cost data is of little use. Large-scale cost studies are inappropriate where the component is not a significant part of the cost structure, or where the difference in cost between the alternatives is obviously going to have little effect on overall costs. Where there is little cost difference between one solution and another, the decision should be taken on grounds other than cost and not on the possibility of a marginal saving.

If the contract is to be awarded in the traditional manner with competitive tendering on BQs, no QS can possibly forecast which of two alternatives, with a marginal cost difference, would actually be priced cheaper by the (unknown) successful contractor. In connection with this the cost planner must remember the natural tendency of contractors to play safe when pricing work involving unknown or experimental materials and techniques.

Another point to bear in mind in connection with cost studies is that all constructional systems have certain optimum conditions (of loading, span, etc.) in which they are most economical. There is usually a reasonable spread of conditions on either side of the optimum in which the system remains reasonably economical, but once the conditions get outside this range costs will start to rise steeply. Even though the particular system may still be perfectly feasible in these unsuitable conditions, it is likely that there will be a cheaper and better solution. It is also worth noting that a high standard of fire resistance, thermal insulation or sound insulation normally costs money, and that it is usually wasteful to employ methods of construction or materials which offer these advantages if they are not required.

The next sections look at some of the principal elements in a building, and the particular aspects of design costs that affect each of them.

14.8 Foundations

The type of structure will normally influence the type of foundations required; a structure that imposes heavy point loads cannot be carried on conventional strip foundations, for example. Piling is almost always a very expensive solution and is normally employed only where conventional foundations are impossible because of the depth at which a bearing formation exists, combined sometimes with the waterlogged nature of the ground. Different systems of piling have particular advantages and disadvantages, but the specialist piling firms (most of whom offer several systems) are usually more than ready to offer their services in connection with early cost investigations. As previously noted, the cost of the piling itself will usually be a complete extra, because the complex of pile caps and ground beams associated with this form of construction is often as expensive as conventional foundations.

14.9 Frame

Cost studies may be necessary to determine the type of frame to be used or indeed whether there should be a frame at all.

External walls of habitable buildings need certain qualities of weather resistance and heat insulation, and such a wall capable of bearing quite substantial loads costs no more than a non-load-bearing wall with similar weathering and insulation performance.

In conditions where normal external walling would be capable of carrying the weight of floors and roof the cost of a frame is a complete extra, and it follows that for small buildings of not more than three storeys a frame is likely to be an expensive solution, even though it may be desirable for other reasons. A frame for small buildings becomes economically justified only where it is necessary to have a very high proportion of the external walls glazed, or in big sheds such as warehouses and the like where the walls do not need good insulation or aesthetic qualities, and cheap non-structural materials such as profiled cladding can be used.

A frame, if required, is likely to be of steel or reinforced concrete, with light alloys and timber as recent and rather specialised competitors. As with other

Phase 2

comparisons between materials, if either steel or reinforced concrete held all the cost advantages, the other would have disappeared from the market. For multi-storey construction there is a general tendency for reinforced concrete to be cheaper in normal conditions of loading, although the difference may often be small. The structural steelwork industry is continually improving its techniques of design and construction, particularly when the cost competition from reinforced concrete becomes severe. Although steel is inherently cheaper than reinforced concrete for most normal applications, the saving can sometimes be wiped out by the measures necessary to give the structure an adequate fire resistance. This is almost certain to be the case if in-situ concrete cladding is required to beams and columns, so the type and standard of fire resistance is an important factor in any such cost comparisons.

Steel therefore gives its best comparative cost performance as a frame to single-storey buildings or in roof framings where fire resistance is not normally required by the Building Regulations and where the steel may be used without any protection other than paint. For such applications reinforced concrete is not usually competitive, except in some proprietary systems for constructing factories or sheds out of standard pre-cast units. Steel frames have advantages over in-situ frames in site programming, and some contractors (particularly small or medium-sized firms) prefer working with a steel frame rather than erecting a reinforced concrete structure of their own, and this preference would be reflected both in tendering enthusiasm and in completion time for the project.

In designing cost-effective steel frames, attempting to save weight of steel can be counterproductive if the result is increased complexity – by using compound welded members, for instance, in lieu of simple sections. A steel fabrication firm's advice should be sought as to the point where savings are swallowed up by fabrication costs. Most constructional steel firms and specialist reinforced concrete contractors are very helpful in giving cost information for a new project.

If a consulting structural engineer is appointed for the project, the cost planner would obviously work with the consultant in providing cost information. As alternative frame designs may affect other elements, the structural engineer may not be able to ascertain the full cost implications without assistance. As well as a choice of materials, frame design will involve a choice of frame shapes and spacings. The most uneconomical solution will occur where heavily loaded beams have to span long distances, especially if they are also restricted in depth. The spacing of columns and beams will have an important effect on cost, particularly if the columns are expressed on the elevations and covered with expensive cladding.

If the frames are spaced too closely, the savings in sections will not pay for the additional frames, since although frames spaced at 2.5 m centres will each be carrying two-thirds of the weight of frames spaced at 3.75 m centres, they will cost much more than two-thirds of the latter. This is because the choice

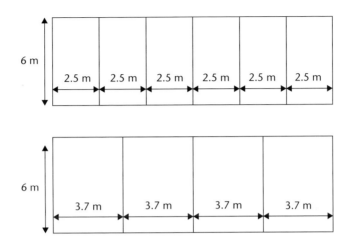

Figure 14.5 Alternative grid layouts.

of section for a beam is affected by its own weight and span as well as by the load it carries, and the span of the beams will not have been changed (see Figure 14.5). Since the most economical spacing will depend on the span and the loading, it is necessary to consider each case individually. For normal floor loadings, however, it is unlikely that spacing as close as 2.5 m will give the most economical solution, while spacings much in excess of 5 m will begin to produce additional costs on the floors and roof owing to their increased span, even though the frame design itself may still be economical.

14.10   Staircases

Since the structure of a concrete staircase represents less than a third of its total cost (the remainder made up of finishes and balustrades) and since the whole elemental cost is only a minor one, it is doubtful whether the staircases of a building are the most fruitful field for cost studies. However, as well as the staircase and its finishing, balustrades, etc., the true total cost of a staircase would be quite considerable, including:

■ the surrounding walls of the staircase together with foundations;
■ windows;
■ wall finishes;
■ doors;
■ roof, etc.

The most rewarding approach is therefore likely to be a reduction in the number of staircases rather than in the details of their design.

Phase 2

14.11   Upper floors

In contrast to staircases, upper floors are an important element and require comparatively little work in the evaluation of alternative designs, which may therefore be well worth doing. As with some other elements, it is difficult to arrive at true cost comparisons without considering the frame element as well.

14.12   Roofs

The architect may require cost studies of alternative roofs to which most of the preceding remarks apply. In addition, the comparative costs of flat roofs and pitched roofs may be involved. These will depend more on the level of the two specifications than on the basic difference in roof type. However, for medium to large spans a satisfactory pitched roof is likely to be rather cheaper than a flat roof of comparable quality, partly because of the simplicity of spanning large areas with roof trusses rather than deep beams. It is also often possible to use the resulting roof space as part of the accommodation area.

Pitched roofs also tend to be more durable than flat roofs, which are sensitive to poor workmanship and poor design detailing unless very well specified. Heavy maintenance issues with cheap flat roofs have led to the avoidance of these wherever possible.

14.13   Rooflights

Rooflighting is not normally the most economical method of lighting rooms where windows are a possible alternative, and individual domelights or small lantern lights are probably the least satisfactory from a cost point of view. However, there are many reasons why the architect may choose this means of lighting, for instance, openable lights for natural ventilation or trickle vents, and the element is not important enough for this to affect the total building cost very considerably. On some current buildings steeply pitched roofs effectively form the walls of upper storeys, and these require the same level of fenestration as a wall. This is one of the areas where the theory of elemental cost planning breaks down, and it will almost certainly be necessary to consider the walls and the roof together for cost planning purposes.

14.14   External walls

External walls may be the most important structural element, particularly in a multi-storey building, and as a tremendous range of constructional methods and finishes may be used, this is a suitable element for cost studies.

In the type of building where the external walls comprise a series of repeating panels, it will be sufficient to study a single panel in detail. It will often be difficult to consider this element in isolation from windows, internal finishes and frame. Note that ordinary brick or concrete block walls finished with facing bricks externally are likely to be far cheaper than any other construction of comparable performance and durability. Therefore a considerable area of plain walling on a building may enable smaller areas of luxury construction or finish to be used while still keeping the overall elemental cost quite reasonable.

14.15 Internal walls and partitions

The comparative costs of traditional partitioning methods are well enough known for cost studies to be unnecessary, but if it is proposed to use a special type of partition on a large scale, then a cost investigation would be worthwhile. Comparative costs of small areas of special partition or glazed screens, on the other hand, would not have much effect on the total building cost and would not usually be worth investigating.

If the BCIS or similar list of elements is being used, it is important to remember that the cost of finishings to traditional partitions is included in other elements, but the complete cost of self-finished proprietary partitioning is included in the partitions element, so in this instance yet again the elemental costs will not be strictly comparable.

14.16 Windows

Windows are often a major element of modern buildings and have a very large cost range. Cost differences will normally be due to performance requirements rather than the actual materials used in manufacture; standard metal, uPVC or timber windows for housing schemes are competitive in cost, as are high-class metal or hardwood windows for prestige buildings, but the latter category may be four or five times dearer than the former.

Apart from the actual material and section of framing, the cost of windows is substantially affected by performance, particularly as regards double glazing and weather stripping, but there are also factors affecting individual windows or groups of windows rather than the fenestration of the building as a whole:

- *Size of window*. Small windows tend to be high in cost per square metre, because of both the greater intricacy of the window itself and the cost of forming the opening in the wall; these two costs vary according to perimeter rather than area.
- *Size of panes*. Very small panes increase the cost per square metre, as do panes that are so large it is necessary to glaze them with stout float glass

instead of sheet glass. It is unlikely that any saving on window frame or glazing bars would counterbalance the cost of plate glass unless an extremely expensive type of window was being used.

- *Opening lights.* This is probably one of the most important factors in window cost, particularly where a high standard of weather resistance is required. The heavy comparative cost of opening portions of windows (as against areas of fixed glazing) makes it very difficult to compare window costs on an overall square metre basis unless the pattern of opening and fixed areas is very similar. Unfortunately, it is not even possible to compare the costs of opening portions on a square metre basis as so much of the cost (hinges or pivots, fasteners, framing to angles) varies according to the number of sashes rather than their size.
- *Decoration and glazing.* Some types of window come to the site ready glazed and self-finished, while others require to be glazed on site and painted.
- *Special types of glass.* For example, solar reflective glass or the laminated glass required by the Building Regulations in some windows or screens.

Because of the above factors it is often necessary for any cost studies of windows to require individual consideration of all the window types in the building, rather than being confined to a typical window and the results applied to the remainder on a square metre basis.

14.17   Doors

Except on buildings that are divided into a large number of very small rooms, such as hostels or flats, or where non-standard sizes or fancy ironmongery are used, the doors are not usually a significant element. Because of this, because they are one of the components most subject to heavy mechanical wear and because they are also one of the most noticeable features of the building, it is probably a mistake to be excessively concerned with cost when designing this element.

14.18   Floor, wall and ceiling finishes and decorations

Cost studies of these elements are likely to be comparatively simple, and where large areas are involved are certainly worth doing.

14.19   Engineering services

On most buildings engineering services are one of the most important elements. It is essential that cost studies be carried out, and these will be fairly meaningless unless whole life investigations into running costs, performance and updating/replacement are taken into account. Most cost planners are unlikely to have access to the data on which such studies should be based and will need to work closely with specialist engineers, preferably with those who are going to be responsible for the design.

14.20   Joinery fittings

Of all minor building components, single joinery fittings involve the most work in arriving at a cost, and so cost studies of such fittings should not usually be attempted. Where a fitting has been designed for repetition (such as a bedroom fitting for a large hostel or a typical bench to be used as a basis for furnishing a set of laboratories), the position is of course different.

14.21   Cost studies generally

We have seen many ways in which cost studies can contribute to the design of a building. However, it is important to realise that it would not be economically possible to employ more than a few of them on any particular building, as a complete cost study of every element, major and minor, would not produce savings commensurate with the amount of time expended. The only occasion where such a course might be practicable is where a large number of similar buildings are to be erected, as for instance a standard house design for a large housing authority.

14.22   Preparation of the cost plan

We now come back to considering the preparation of the cost plan itself. This may take many forms, but all the systems have much common ground. The plan will almost certainly be prepared by elements and the cost will be expressed per square metre of floor area even if it is calculated in a different manner, in order to be able to make comparisons between other schemes of different size on a common basis. If there is a cost target it will by now have been set up with some degree of finality and the architect will want to go ahead with the working drawings.

Since cost planning involves cost control, the architect must design in accordance with the cost plan, which must therefore be available before the working drawings have proceeded very far. The following are needed in order to prepare such a plan:

- a drawing and a standard of specification to which the cost plan can be related;
- a cost analysis of a comparable project.

Phase 2

It is not necessary that the comparable project should be for a similar use, as long as there is some reasonable compatibility as buildings. For instance, a police station and a health clinic are both public buildings with fairly similar storey heights and are divided into both small and large rooms. The elemental analysis of a police station could therefore be used to prepare the cost plan for a health clinic if nothing better was available, whereas between say a church and a multi-storey block of flats there is no resemblance whatever.

The cost plan must be prepared to a standard format, for reasons that have already been emphasised in connection with analyses and preliminary estimates. The form should show:

- the chosen list of elements;
- preliminaries and insurances (if separate);
- contingency sum;
- professional fees (if to be included);
- the cost index value which is being used.

It is common practice in some offices to allow a design margin of between 1% and 5% as a design contingency in addition to the contract contingency, and this margin may come in useful during the cost check when it can be absorbed into any element that needs it.

Although specialist computer software is available for cost planning work, many cost planners prefer to use an ordinary spreadsheet package adapted to the practices of their own particular office. Whichever method is used, the software should guide the user through the process of preparing the cost plan, making sure that all necessary adjustments are made and that nothing is forgotten. Prior to the large-scale adoption of computers for the purpose, much the same control was obtained by the use of standard manual forms and procedures. The first page of a typical manual form is shown in Figure 14.6 and the accompanying specification in Figure 14.7.

14.23 Method of relating elemental costs in proposed project to analysed example

There are two ways of doing this, either by using unit costs of the elements or by proportion. In spite of what may be thought, these are really two ways of doing much the same thing. In both methods the unit quantity of the element (e.g. area of external walls) has to be measured from the sketch drawings of the proposed project.

Using unit cost method:

- This quantity is priced out at the elemental unit rate per square metre for the same element obtained from the comparable analysis, where this is applicable, but if specifications are different then approximate quantities are more likely to be used.

COST PLAN

				Note	This cost plan is based on the
Project	Office Block, Midtown				attached outline specification,
Date of cost plan	Apr. 06				and both documents should
Assumed date of tender	Sept. 06				be read simultaneously
Total internal floor area	2,390 m²				

	Unit quantity	Unit cost (£)	£	Total cost (£)	Elemental cost/m² (£)
1 Work below lowest floor finish					
Ground floor area	390 m²	321		125,190	52.38
2 Structural frame	2,390 m²			125,600	52.55
3 Upper floors					
225 mm Hollow Pot	386 m²	60	23,160		
150 mm in-situ RC	1,585 m²	41	64,985		
			88,145	88,145	36.88
4 Staircases					
RC Staircases					
1nr 25 m rise	25 m	1,225	30,625		
1nr secondary 21.5 m rise	21.5 m	900	19,350		
			49,975	49,975	20.91
			continued	**388,910**	**162.72**
			1		

Figure 14.6 Front sheet of typical cost plans.

■ The approximate quantities will give the total cost of the element, and this total is then divided by the floor area of the proposed project to give the elemental price per square metre.

When the proportion method is used:

■ The ratio of unit area to floor area in the proposed project is compared to the corresponding ratio in the analysed example.
■ The elemental cost per square metre is then adjusted in proportion to the difference in the two ratios.
■ With this method one is not concerned with the actual unit cost (either in total or per square metre) at all, although the total cost for the element will be required for comparison purposes at cost check stage.

What are the advantages and disadvantages of the two methods? The proportion method is the logical choice for use where the elemental cost per square metre is regarded as an index of performance rather than the result of a specification, and it tended to be preferred by central government departments in the early days of cost planning. However, the unit cost method is almost exclusively used today. It is more in line with the traditional quantity surveying approach and most cost planners would feel less likely to make mistakes when using it, particularly if the workings became complicated. The

Phase 2

OUTLINE SPECIFICATION for COST PLAN

Project	Office Block, Midtown
Date of cost plan	Apr. 06
Assumed date of tender	Sept. 06
Total internal floor area	2,390 m²

Work below lowest floor finish

Foundation to walls and/or columns
In-situ concrete piles average 4 m long (approx. 90 nr) reinforced concrete pile caps and ground beams

Basements, walkway ducts, etc.
Semi-basement 1.5 m deep approx. 8 m × 4 m for boilers

Also, lift pits approx. 0.75 m deep in waterproof concrete

Rising walls
Clay common bricks in cement mortar built off ground beams (approx 0.75 m high generally to damp-proof course).
Reinforced concrete walls (0.3 m thick) around stairwell and lifts

Ground floor slab
Generally, 150 mm slab reinforced with fabric on building paper and hardcore. Note thickness of hardcore to make up levels (0.65 m average)

Damp-proof courses and membranes
Lead-cored bituminous dpc in walls, three-coat Synthaprufe cold-applied bituminous emulsion, containing synthetic rubber latex, two-coat asphalt tanking to semi-basement

Structural Frame

Generally
Reinforced concrete beams (at 3.50 m c/c) and columns

Figure 14.7   Specification notes to accompany the cost plan.

unit cost method is more easily related to approximate quantities either at cost plan or cost check stage, and remember that at one or other of these stages fairly detailed quantities have to be introduced.

In order to illustrate these systems in use, two examples are worked out below using both methods. It will be seen that either method will allow differences in quality as well as quantity to be adjusted.

Example 1

Total floor area	5,000 m²
External wall area	4,000 m²
Wall:floor ratio	0.80
Cost of external wall per m² of floor area (elemental cost)	= £75.00
Cost of external wall per m² of wall (unit cost)	= £93.75

Proposed project

Total floor area	4,800 m^2
External wall area	3,600 m^2

Elemental cost of external walls on proposed project

Unit cost method

3,600 m^2 of wall at £93.75 = £337,500

£337,500/4,800 = £70,312

Total cost 4,800 × £70,312 = £337,500

Proportion method

Wall:floor ratio = 3,600/4,800

£75.00 × (0.75/0.80) = £70,312

Example 2

Total floor area 5,000 m^2
Internal doors 35 nr
Cost of doors per m^2 of floor area (elemental cost) = £2.10
Cost of doors each (unit cost) £300

Proposed project
Total floor area 4,800 m^2
Internal doors 50 nr
Cost of doors each £330 (better quality)

Elemental cost of internal doors on proposed project

Unit cost method

50 doors at £330 = £16,500

£16,500/4,800 = £3.44

Total cost 4,800 m^2 × £3.44 = £16,500

Proportion method

£2.10 × (5,000/35) × (50/4,800) × 33/30 = £3.27

14.23.1   Sources of data

If the analysed example will not yield all the necessary information for preparing the cost plan, this information must be obtained elsewhere. One approach is to take a price from another analysis. This is potentially dangerous; we have seen how a priced BQ will probably contain compensating errors and that these will be incorporated in the analysis. If we take items from more than one analysis we may get cumulative instead of compensating errors. It would be possible (indeed tempting) to take a number of elements from different analyses, all of which are priced too low, but this would make the total of the cost plan quite unrealistic. This cannot happen where a single analysis is used throughout.

Phase 2

If prices have to be obtained from other analyses, it is wise to collect as many examples of the same item as possible so that any errors are likely to cancel out in the average. The alternative is to use approximate quantities. These are likely to be very approximate indeed (remember that there are no working drawings) and herein lies the danger; it is much easier to leave things out than to put extraneous items in. Therefore, when using approximate quantities the cost planner must be very careful to include all the items included in the description of the element (e.g. skirtings and their screeds with 'floor finishes', or rainwater pipes with 'roofs'). It is also important to include the percentages for preliminaries, insurances or contingencies if these are included in the elemental rates.

Needless to say, the quantities must be priced at the same level of market rates as the cost plan itself, although the rates will be higher than ordinary BQ prices because of the need to include the cost of secondary work which would be measured separately in a full BQ. The approximate quantities themselves will be similar to those normally employed by QSs for estimating purposes, except that the elements must be kept separate. So, for example, the plastering on the underside of a concrete roof slab could not be included in the overall rate for the roof but would be worked out separately as part of ceiling finishes.

In some offices it is the custom to keep a complete schedule of prices for use when pricing approximate quantities. This certainly gives a consistent level of pricing, but it is doubtful whether any but the largest organisations would find the preparation and maintenance of such a list to be an economic proposition.

Examples of use of information from analysed building

Analysis
Total floor area 2,000 m²
Element: internal partitions
Elemental cost: £27.00 m²

	£
1,400 m² of 75 mm breeze partition at £13.00	18,200
900 m² of 225 mm and 300 mm brick walls at £33.00	29,700
120 m² of glazed partitions at £50.80	6,096
2,420 m²	53,996

Average unit cost = £53,996 ÷ 2,420 = £22.32 m²

Proposed project
Total floor area 3,100 m²
Total area of partitions (measured roughly from drawings)   3,500 m²

If we do not know the type of partition required, or if the proportions of various types are likely to be similar to those of the analysed example, we can use the average unit rate or work by the proportion method:

3,500 m² at £22.32 = £78,120 total cost divided by 3,100 = £25.20/m² elemental cost

(1)

But if even at this stage we can see that the proportions of the various sorts of partition are quite different to the analysed example, we must use the individual unit cost:

		£
20% 75 mm breeze partition	= 700 m² at £13.00	9,100
10% 225 mm brick wall	= 350 m² at £33.00	11,550
60% glazed partition (cheaper type)	= 2,100 m² at £45.00	94,500
10% timber stud partition (no rate for this: price built up by approximate quantities)	= 350 m² at £23.00	8,050
		123,200

£123,200/3,100 m² = £39.74 elemental cost

When severely altering the proportion of items like this it is necessary to watch out for any freak rates. For instance, it is possible that the rates for glazed partitions in the analysed example may have been much too low, and being a very small part of the whole job this will not have had much effect on the total price. On the proposed project, however, the proportion of glazed partition has been increased from 10% to 60% and such an error would be much more serious. This emphasises the point made before, that the smaller items in the analysis may well have the least reliable prices.

14.23.2   Elemental costs

By whatever means the total cost of the element has been obtained, it should always be expressed per square metre of floor area, for the following two reasons:

- For comparison of reasonableness with other buildings. If the partitions, for instance, are costing far more per square metre of floor area than on other buildings of the same type, it is worthwhile investigating the reasons. These may be perfectly sound; the cost may include built-in cupboards or there may be a greater degree of insulation required between rooms, but it is also possible that there is inefficient planning of the accommodation and this would also be reflected in other elements.
- The elemental price per square metre enables the importance of any extravagance or economy to be judged in a way that is not possible when considering unit prices only.

14.23.3   Comparison of elemental costs

Consider the unit costs for three different specifications of internal doors:

Phase 2

- Hardboard faced skeleton framed flush doors hung on steel butts to softwood frames, no architraves – unit cost £45.00/m^2.
- Plywood faced semi-solid core flush doors hung on steel butts to softwood frames with softwood architraves – unit cost £82.50/m^2.
- Mahogany veneered plywood faced solid core flush doors hung on solid drawn brass butts to mahogany frames with mahogany architraves – unit cost £175.00/m^2.

From a comparison of the unit costs it would seem that the third specification is prohibitively expensive for normal work compared with the more usual second specification. Yet to adopt the better specification in a typical school of 3,000 m^2 with 70 internal doors would only increase the elemental cost per square metre by about £2.16/m^2, and to use the completely inadequate first specification would only save £0.88/m^2. Considering that the total cost per square metre of the school would be around £600, these increases and decreases are quite insignificant.

On the other hand, a marginal difference in some other elements, such as the particular choice of hardwood for wood block flooring, which may only alter the unit price by 10%, will have as great an effect on the cost of the school per square metre as a 100% difference in door prices. It will soon become obvious to the cost planner which items need the most time spent on them at cost plan stage – in other words, which elements have the greatest cost significance.

It would be worthwhile to use approximate quantities for a major element such as external walling where a difference of £1–2 in the unit cost per square metre would have an important effect on total cost, and to check the cost plan very closely when drawings subsequently become available. Such elements as internal doors or ironmongery on the other hand can be dealt with much more rapidly.

14.24   Presentation of the cost plan

The cost plan should be presented as neatly as possible; it will almost certainly be going to the client and should reflect well on the firm that has prepared it. To what extent expense should be incurred in order to give the plan an impressive appearance is a matter for individual preference, but if the document has been prepared by computer, a high-quality (preferably colour) printer should be used. Another matter for personal decision is the extent to which the detailed build up of the plan should be made available to the architect. Whether this should be done in all cases will depend on the degree of mutual confidence that exists and also on the available facilities for copying a large quantity of rough working.

With the cost plan should go the specification, if any, on which it is based; this will necessarily be more in an outline form than the type of specification that will eventually be embodied in the BQ. It may either be included as part

Phase 2

of the cost plan itself or prepared as a separate document. A typical example was shown in Figure 14.7. If a separate document is used, it is important that there should be a space for each specific item so that it can be quickly seen whether everything has been dealt with or not. The specification should preferably be filled in by the cost planner in consultation with the architects, rather than being sent to the architects for them to fill in by themselves.

14.25   Key points

- Early estimates of the cost of the scheme will probably be done using a short list of some half-dozen principal elements only. But because such large figures are being dealt with it is even more vital than usual that all arithmetic, and if possible all measuring, should be checked.
- The cost control of design development demands considerable resources and should not be carried out until a satisfactory solution has been defined and agreed. Substantial iteration between the second stage of the cost planning process (investigation of a solution) and the third stage (cost control of the development of design) brings nothing but disadvantages. The basic principle to be adopted is to move from the ballpark estimating of outline proposals to the detailed costing of production drawings in a series of steps.
- The cost plan will probably be based on the most recent outline elemental cost estimate, but developed in as much detail as a full elemental cost analysis complete with outline specification (agreed with the designer) for each of the elements. It is generally preferred that the cost plan should incorporate the specification on which it is based. Although elemental estimates can be based on approximate quantities or proportion methods, the former is usually preferred today. Cost studies of individual elements are generally expensive and should be limited to elements of major cost significance or to elements, such as floor finishes, where not much work is involved.

Phase 2

Cost Planning and Control at Production and Operation

Chapter 15
Planning and Managing Project Resources and Costs

15.1 Introduction

Once the cost plan has been approved by the client and the working drawings are available, the resources required to realise the project need to be considered.

Developers normally need to involve themselves with the building industry in order to get their building work undertaken. The building industry and the civil engineering industry together are often referred to as the construction industry, and many firms (certainly most of the large ones) operate in both sectors, even though their staff and organisations will probably be separate. Smaller contractors often specialise in either civil engineering or building work (such as groundworks in the case of the former). It is, however, very difficult to separate the two halves of the construction industry statistically, because of the overlap which occurs. For instance, the construction of the foundations of very large buildings could almost be classed as civil engineering (and you will find it so referenced in a library, if this operates the Dewey Decimal Classification or UDC system), and there may be quite a lot of building work on some civil engineering projects. A further sub-division of the building industry is into housing and other work, many main contractors having divisions which specialise in one or the other (Taylor Woodrow, for example, has a housing division called Bryant Homes).

The building industry in the UK has changed considerably over the last 40 years, in two ways. First, most general building contractors now undertake only a small proportion of their turnover using their own directly employed operatives, the majority of their work being outsourced to specialist or trade sub-contractors. The general contractor's role has changed from being primarily a provider of resources into being a provider of management and financial services. Very often the supervisory and administrative staff and a handful of labourers will constitute the entire site workforce of the general contractor.

Secondly, between the end of World War II and the late 1970s, the building industry was geared largely to serving central and local government departments as its direct clients. Because of changes in the country's economic structure, this is no longer the case. In much of the country, and particularly in the south-east, the industry has had to accustom itself to the different norms of private enterprise clients who are more interested in results than in procedures.

15.2   Nature of the construction industry

The UK construction industry accounts for a significant proportion of the economic output of the UK economy. As of 2006, the industry can be summarised in statistics as follows (*Construction News* 2006):

- Construction's output is £102.4 billion at current prices.
- It accounts for 8% of Gross Domestic Product (GDP).
- Output to grow by 0.8 per cent in 2005 and 2.4 per cent in 2006.
- One in 10 people employed in the UK work in construction.
- Contractors employ 1.68 million people.
- The 'black' construction economy is valued at £10 billion a year.
- There are 192,404 construction firms in the UK.
- There are about 700 PFI and PPP projects with a capital value of £43 billion.

A summary graph of the industry is shown in Figure 15.8. Because it operates with a relatively small investment in fixed assets in relation to its large turnover, the industry is very flexible and is able to accommodate itself to major changes in workload much more easily than industries which are plant based. Its workforce can be reduced in periods of recession without the archetypical political crisis caused by the closure of, say, a car factory, and perhaps because of this it has often been used as an economic regulator by government. Similarly, it can expand its efforts rapidly in order to meet demand without requiring major investment.

15.3   Problems of changes in demand

Nevertheless, violent changes in demand do harm the industry. Workers who are made redundant in times of economic downturn or recession do not always return to the industry afterwards, and the problem is compounded by the need to continually train skilled workers. Firms argue that they are unable to afford to do this when the order books are empty, and in times of economic upturn they usually argue that there is little time to complete this training. In Liverpool, this was recognised as a major problem by Sir Joe Dwyer FCIOB, Chair of Liverpool Vision (an independent company established to bring together key public and private sector agencies to deliver the regeneration of Liverpool city centre). In a speech to the Liverpool Centre Chartered Institute of Building in 2006, he argued that one of the pivotal problems facing UK construction is the severe underinvestment in training by contractors and the haemorrhaging of the knowledge base. It is also difficult to keep a balanced management team of the right size when demand fluctuates wildly, and, as stated above, management expertise is now probably the general contractor's main stock-in-trade.

A further problem concerns the supply of materials, because material and component manufacturers are plant based and find it more difficult to respond rapidly to changes in demand than the construction industry itself. If a boom is regional in character, this problem can be mitigated by importing materials and components from elsewhere in the UK or Europe, but for bulk materials in particular (where transportation forms a large part of their cost) this can often be an expensive alternative. It is therefore very important for the cost planner to look at the general and local economic situations when forecasting costs. In the early 1980s tender prices remained almost static, although official costs of labour and material inputs continued to show an increase. In the five years between the third quarter of 1980 and the first quarter of 1985 tender prices increased by less than 1%, but the official index of building costs rose by no less than 38% in the corresponding period. The same thing happened even more violently between 1990 and 1996. This imbalance between tender prices and building costs was partly due, of course, to tenderers' willingness to take a cut in profits during difficult times. But it was mainly due to the fact that real costs did not increase, whatever the official figures might have said.

Theoretical inflationary increases were totally counterbalanced because:

- Labour could be easily obtained and no longer had to be bribed to work for one employer rather than another.
- With workers fearing the sack if they did not perform, productivity tended to improve.
- Under the stress of competition, materials arrived when they were supposed to and merchants started to offer discounts instead of requiring extra payments for prompt delivery.

For the above reasons management of projects and the meeting of deadlines became much easier, and it was no longer necessary to make a substantial allowance in tenders for risk.

Although the above situation occurred on a national scale, it is important to look out for similar supply-and-demand imbalances occurring regionally or even locally. Much of the building industry consists of smaller firms operating within a limited radius, and labour is only mobile to a limited degree.

15.4   Costs and prices

There are two ways in which building costs can be estimated or analysed, based on:

- the prices charged for the finished building or parts thereof;
- the cost of the resources required to create them.

Most of the cost planning and cost control procedures traditionally used by QSs on behalf of the client or design team have depended on prices for finished work-in-place, obtained from BQs or elsewhere, because this is what the client will actually have to pay. However, these prices may be little more than notional breakdowns of the contractor's tender offer and may have more of a marketing than a cost basis.

Practical building contractors are often scornful of this approach, and it is often suggested by people from the construction side that the cost planner should be more concerned with costs (or real costs as the proponents of this argument like to call them). This is an important issue, but it is not the simple choice between fiction and reality that is implied in this argument. In fact any figure in this context is simultaneously both a price and a cost; it just depends where you are looking at it from. It can generally be said that the seller's price is the buyer's cost. Thus:

- The contractor's price is the client's cost.
- The sub-contractor's price is the contractor's cost.
- The materials supplier's price is the sub-contractor's cost of materials.

Today, because so much of the work on major projects is being undertaken by sub-contractors on a price basis, and the independent professional QS is increasingly involved in the management of projects, it is doubtful whether many general contractors now know much more about production costs than the QS. But in a market-based economy, the whole notion that there is such a thing as real costs is mistaken anyhow. For cost-planning purposes, both the finished product price approach and the resource cost approach have their strengths and weaknesses, and good cost planners should understand these and know when each should be used rather than simply adopting the one normally used by their profession in the past.

We have already looked at finished product prices in some detail, and now we need to understand how resource costs are in fact incurred.

15.5   The contractor's own costs

Builders like to think they know about costs and they think of them in resource categories, which are rarely thought of separately by traditional QSs as they manipulate their all-in unit rates.

15.5.1   Contractor's direct costs

15.5.1.1   Direct site labour

Direct site labour costs relate to the tradesmen and labourers actually producing the work. At one time these costs would have been described as wages, being almost entirely composed of this single item, but today they will include substantial payments in respect of National Insurance schemes, holiday schemes, training schemes, etc.

In the strictest sense of the word there may be almost no wages paid to production employees, because so much of the work is done by sub-contractors. In addition, for various reasons to do with taxation, and perhaps because of the inherent British dislike of the master/servant relationship, the custom grew up whereby the contractor's few production workers were self-employed and were taken on as labour-only sub-contractors for a fee. However, recent government legislation has inhibited this practice.

Labour costs are of particular concern to the contractor because they have to be paid out weekly in ready cash as they are incurred and cannot be postponed or put on a credit basis as can most other commitments; the importance of this will be shown later in this chapter. It is convenient to include the labour-only sub-contractors under the heading of direct site labour, as they will also usually require weekly cash.

15.5.1.2   Materials

The materials costs will usually be paid by means of monthly credit accounts, payment being due at the end of the month following that in which the materials are delivered (so that materials delivered in January are paid for at the end of February, rather like credit cards). Such settlement by the contractor usually entitles the firm to a cash discount, 2.5% being the most common figure although 5% may occasionally be allowed. It is not unusual for contractors to delay payment beyond the cash settlement date, sometimes for a further month or even two months. Since they may lose their cash discount for the sake of one or two months' credit, this could be an expensive way of borrowing money compared to a bank overdraft, but it has the advantage of being ready and convenient and of not requiring collateral security.

Within reason, and being careful not to let things get out of hand, builders' merchants are used to acting as financiers to the industry in this way (particularly in hard times). This has advantages for the merchant. If a building firm owes a substantial amount it is difficult for it to withdraw its custom from the supplier, or even reduce its level of buying substantially, as immediate settlement of accounts might be called for. The merchant therefore has a captive customer who cannot afford to be too fussy over prices or delivery dates.

In many cases the contractor will have been paid by the client for the materials, or for the work in which they are incorporated, before settling with his merchants. Because of the increasing tendency to buy in fabricated components rather than making things on site, a larger proportion of expenditure now falls into this category and a smaller proportion into labour, than was previously the case.

15.5.1.3   Small plant

Small plant covers hand tools and small mechanical plant of a kind that can be directly associated with specific pieces of finished work.

15.5.2   Sub-contractors and major specialist suppliers

Because of the increasing tendency to employ specialist sub-contractors rather than using the builder's own labour and materials, this category has expanded over the last 20 years and is now usually the largest head of expenditure. In terms of payment methods it very much resembles the materials category, but sub-contractors have always tended to be less accommodating than materials suppliers with regard to extended credit.

If sub-contractors and suppliers are nominated or named by the architect, they are in a position to complain to the architect if payment is delayed, and since such complaints cast doubts on the builder's solvency, the builder will usually make efforts to pay them fairly promptly.

The builder will often have been paid by the client for the work before settling with his sub-contractors, and some large firms of builders have attempted to formalise this in the past by inserting a pay-when-paid clause in their sub-contracts. This practice has now been outlawed by the HGCRA, which, subject to certain exceptions, gives all the parties to a building contract the right to know the amount to be paid and the right to be paid on a determinable date.

15.5.3   Site indirect costs (preliminaries)

The term 'preliminaries' is sometimes used to describe these items because they usually appear at the beginning of a BQ under this heading. They comprise all the items of site expenditure that cannot be attributed to individual items of work but to the project as a whole, or to substantial sections of it. Such costs may include:

- *Salaries and wages of site staff.* The salaries paid to management, supervisory and clerical staff employed on the site. In former times these payments would have been limited to a foreman and a few junior site staff, but today the total costs of on-site management will often be greater than those of the directly employed production workers on the site.

- *Site offices, messrooms and facilities.* At one time these were a comparatively minor item, but today the temporary site office buildings can form a major multi-storey complex.

15.5.4   Major plant

Large items of mechanical plant may be dealt with in one of two ways:

- charged to the job when brought on to the site and credited (less depreciation) when removed;
- an hourly or weekly hire charge made plus a charge for bringing the plant on site and removing it.

Most large contractors find it convenient to set up a subsidiary plant company, which will charge the plant out to the sites on a hire basis, while plant hired in from outside will be similarly dealt with and paid for by credit account. Lorry (truck) transport is usually arranged in a like manner. It is unusual for a builder to own large items of plant except through a subsidiary company. There are substantial advantages in hire purchase through a finance company, since the interest charges (less tax) will be lower than the return which the contractor expects to make on the working capital employed; it would therefore be uneconomic for the contractor to invest any of this capital in plant purchase.

It is difficult to attribute the costs of major plant items to specific pieces of direct work, and they are usually treated as a site indirect cost. However, they can sometimes be allocated to major cost centres, such as excavation or concrete superstructure.

15.5.5   Off-site costs (also called establishment charges or overheads)

Off-site costs represent costs incurred in running the company as a whole and which cannot be attributed to any one particular contract: head office expenses, builder's yard, salaries of central management and directors, insurances and interest on loans.

Off-site costs are usually allocated to projects as a percentage of the direct costs. This can operate unfairly against small simple projects which require little head office input, and some major contractors have separate small-works departments run on more economical lines, to avoid loading their smaller jobs with a large overhead and therefore making it difficult for them to compete with smaller firms.

15.5.6 Profit

Often referred to in the industry as mark-up, profit is strictly speaking not a cost; it is in fact the difference between the builder's cost and the client's price.

15.6 Two typical examples

The following cost breakdowns relating to two large office buildings in the Home Counties of England in the mid-1980s, undertaken under ordinary lump sum contracts by a main contractor, may be of interest. The figures for project B overstate the proportion of builder's direct work, since it was not possible to separate out some of the small sub-contractors undertaking traditional trade work. Profit is included under the various cost heads, since because of commercial secrecy it was not possible to separate them out.

	Project A	Project B
(1) Builder's direct costs	21.0%	37.0%
(2) Sub-contractors and major specialist suppliers	70.0%	49.5%
Total direct costs	91.0%	86.5%
(3) Site indirect costs	6.5%	10.0%
Total site costs	97.5%	96.5%
(4) Off-site costs	2.5%	3.5%
Total costs	100%	100%

15.7 Cash flow and the building contractor

The pattern of cash flow on a project is vital to a contractor. As an example, a simple contract for a £410,000 building, to be erected in 30 weeks, has been chosen. The prime cost sum to the contractor is £400,000, leaving £10,000 profit. This represents 2.5% on turnover, which is fairly usual, but the real profit percentage may be very different to this small figure, which is so often quoted as an example of the poor financial returns of the building industry.

Table 15.1 shows the weekly outlay on labour, plant hire and overheads, the weekly value of material deliveries, and the sub-contractors' accounts which are received monthly. The table also shows the total prime cost at the end of each month, together with the QS's valuation. In this example it is assumed that 10% of the value of the work is withheld by the client each month as 'retention' until the total of the retention fund amounts to £25,000, after which remaining work is paid for in full. This arrangement would be

Table 15.1 Weekly outlays for a 30-week, £410,000 building.

	Week No.	Wages, plant hire and overheads	Materials delivered	Sub-contractors' accounts received	Total prime cost and overheads	QS's valuation	Valuation less retention
March	1	1,000	2,000				
	2	1,500	1,000				
	3	1,500	1,000				
	4	2,000	3,000		13,000	10,000	9,000
April	5	3,000	10,000				
	6	3,000	10,000				
	7	3,000	6,000				
	8	4,000	6,000	10,000	68,000	70,000	63,000
May	9	4,000	10,000				
	10	4,000	3,000				
	11	5,000	20,000				
	12	5,000	10,000				
	13	5,000	10,000	10,000	154,000	170,000	153,000
June	14	5,000	17,000				
	15	6,000	15,000				
	16	6,000	10,000				
	17	5,000	5,000	10,000	233,000	250,000	225,000*
July	18	5,000	5,000				
	19	5,000	10,000				
	20	4,000	5,000				
	21	3,000	3,000	30,000	303,000	325,000	300,000
August	22	3,000	2,000				
	23	3,000					
	24	2,000	5,000				
	25	2,000					
	26	2,000	5,000	30,000	357,000	375,000	350,000
September	27	4,000	10,000				
	28	4,000					
	29	3,000					
	30	2,000		20,000	400,000	410,000	385,000
	TOTAL	106,000	184,000	110,000	400,000	410,000	385,000
						retention	25,000
							410,000

*Maximum retention now withheld

less generous to the contractor than the current requirements of the JCT Standard Building Contract, and if the latter were being used the contractor would have rather more cash in hand than is shown in the examples.

It will be seen from Table 15.1 that, as usual, the project progresses slowly in the early stages (following the classic S-curve of expenditure plotted against time (Figure 15.1)); and because much of the early expenditure is in any case related to setting up rather than producing finished work, the first

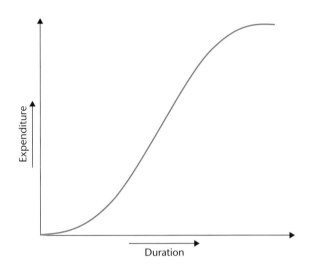

Figure 15.1   Typical project S-curve of expenditure plotted against duration.

valuation by the QS does not meet the full cost. However, by the time the job is halfway through it is showing a handsome profit (perhaps the loading of BQ rates for the early trades may have helped), although the finishing off, again as usual, is not very profitable.

Table 15.2 illustrates the contractor's cash flow, on the assumption that everything progresses as would be expected. The client pays the valuations approximately 14 days after they are issued, and the contractor pays the nominated sub-contractors as soon as he receives the money. Materials are paid for at the end of the month following delivery (in practice the sub-contractors and materials suppliers would allow discounts for such prompt payment, but these have been ignored in the example for the purposes of clarity). Labour etc. has to be paid for weekly as the costs are incurred. At the conclusion of the project the client pays half of the retention sum and pays the balance six months later.

The cash flow in Table 15.2 is also shown graphically in Figure 15.2. What is the first thing we notice about the cash flow as shown in Table 15.2? Although this is a £410,000 contract, and although the contractor is certainly bearing the risk (and undertaking the organisation) of a project of this size, such a figure has nothing to do with the contractor's financing of the job. Except for one or two weeks, the contractor never has more than £25,000 'sunk' in the contract and in fact often has no money invested in it at all; for example, by the middle of June the firm has received £43,000 more than it has paid out. If we therefore say that the firm could carry out the job on a working capital of £25,000, its £10,000 profit is not 2.5% but 40%, a very different figure. In fact the true situation is even better than this, because the contractor's average working capital (setting positive cash flows in some weeks

Table 15.2   Payments and receipts in £s based on Table 15.1.

	Week No.	**Payments** Wages etc.	Materials	Sub-contractors	Total	Amounts received	Cumulative cash flow
March	1	1,000			1,000		−1,000
	2	1,500			1,500		−2,500
	3	1,500			1,500		−4,000
	4	2,000			2,000		−6,000
April	5	3,000			3,000		−9,000
	6	3,000			3,000	9,000	−3,000
	7	3,000			3,000		−6,000
	8	4,000	7,000 (March)		11,000		−17,000
May	9	4,000			4,000		−21,000
	10	4,000		9,000	13,000	54,000	20,000
	11	5,000			5,000		15,000
	12	5,000			5,000		10,000
	13	5,000	32,000 (April)		37,000		−27,000
June	14	5,000			5,000		−32,000
	15	6,000		9,000	15,000	90,000	43,000
	16	6,000			6,000		37,000
	17	5,000	53,000 (May)		58,000		−21,000
July	18	5,000			5,000		−26,000
	19	5,000		9,000	14,000	72,000	32,000
	20	4,000			4,000		28,000
	21	3,000	47,000 (June)		50,000		−22,000
August	22	3,000			3,000		−25,000
	23	3,000		27,000	30,000	75,000	20,000
	24	2,000			2,000		18,000
	25	2,000			2,000		16,000
	26	2,000	23,000 (July)		25,000		−9,000
September	27	4,000			4,000		−13,000
	28	4,000		27,000	31,000	50,000	6,000
	29	3,000			3,000		3,000
	30	2,000	12,000 (August)		14,000		−11,000
October	32			18,000	18,000	35,000	
			Release of retention	5,500	5,500	12,500	13,000
	34		10,000 (Sept.)		10,000		3,000
April	60		Release of retention	5,500	5,500	12,500	10,000
	TOTAL	**106,000**	**184,000**	**110,000**	**400,000**	**410,000**	**(10,000)**

against negative in others) is much less than £25,000, as can be seen from Figure 15.2. This is therefore not a £410,000 job as far as the contractor's budgeting is concerned.

Before becoming too excited at the prospect of making 100% (or more) per annum profit as a matter of course, look at Table 15.3 and the accompanying

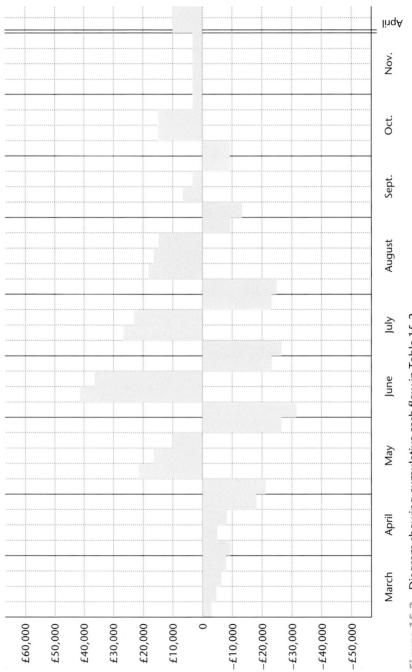

Figure 15.2 Diagram showing cumulative cash flow in Table 15.2.

Table 15.3 As Table 15.2 but with late and underestimated payment in £s by client.

	Week No.	Wages etc.	Materials	Sub-contractors	Total	Amounts received	Cumulative cash flow
			Payments				
March	1	1,000			1,000		−1,000
	2	1,500			1,500		−2,500
	3	1,500			1,500		−4,000
	4	2,000			2,000		−6,000
April	5	3,000			3,000		−9,000
	6	3,000			3,000		−12,000
	7	3,000			3,000		−15,000
	8	4,000	7,000 (March)		11,000	8,000	−18,000
May	9	4,000			4,000		−22,000
	10	4,000			4,000		−26,000
	11	5,000			5,000		−31,000
	12	5,000		9,000	14,000	52,000	7,000
	13	5,000	32,000 (April)		37,000		−30,000
June	14	5,000			5,000		−35,000
	15	6,000			6,000		−41,000
	16	6,000			6,000		−47,000
	17	5,000	53,000 (May)	9,000	58,000	88,000	−26,000
July	18	5,000			5,000		−31,000
	19	5,000			5,000		−36,000
	20	4,000			4,000		−40,000
	21	3,000	47,000 (June)	9,000	50,000	70,000	−29,000
August	22	3,000			3,000		−32,000
	23	3,000			3,000		−35,000
	24	2,000			2,000		−37,000
	25	2,000		27,000	29,000	75,000	9,000
	26	2,000	23,000 (July)		25,000		−16,000
September	27	4,000			4,000		−20,000
	28	4,000			4,000		−24,000
	29	3,000			3,000		−27,000
	30	2,000	12,000 (August)	27,000	14,000	48,000	−20,000
October	34		10,000 (Sept.)	18,000	28,000	34,000	
			Release of retention	5,500	5,500	12,500	−7,000
May	64		Release of retention	5,500	5,500	22,500	10,000
	TOTAL	**106,000**	**184,000**	**110,000**	**400,000**	**410,000**	**(10,000)**

Figure 15.3. This represents the cash flow on the same project where the client is not being quite so helpful. The monthly payments are being delayed for a further two weeks, and the QS is being prudent with the valuations, finally undervaluing to the extent of £10,000 (possibly because of variations which have not been agreed), although the full amount is eventually paid over.

Phase 3

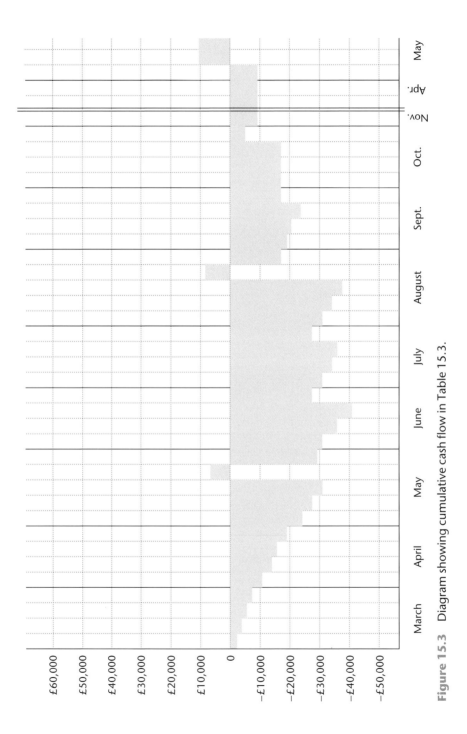

Figure 15.3 Diagram showing cumulative cash flow in Table 15.3.

This is still the same project with the same costs and the same profit, but as far as the contractor is concerned it is on a completely different scale. The contractor now has an almost permanently negative cash flow, often amounting to £30,000–40,000. This project will involve two or three times the capital commitment of the previous one and as far as the contractor's financing is concerned it is a project of more than double the size, although the profit is still the same. On the other hand, if the contractor were to complain, the client might wonder what all the fuss was about. As Figure 15.4 shows, the effect of the different payment pattern on the client's cash flow is proportionately very small.

Many contractors would confirm that this kind of situation is not unusual, nor does it by any means represent the worst that may befall them as regards deferring of payments, although the Construction Act 1996 does afford them some protection. The contracting firm's remedy has often been to defer payment in turn to its own suppliers, which in the above example could more than restore the cash flow figures to their former satisfactory state (Table 15.4 and Figure 15.5). It is possible that delaying payment in this way might cause the contractor to forfeit some of the cash discounts that the materials suppliers give. These are normally of the order of 2.5%. In the event of all the suppliers refusing to give any discount, the contractor would stand to lose a total of £4,600 although in practice it would be unlikely that more than a few suppliers would do so unless the delays became very serious. However, the loss of a few cash discounts might be regarded as a small price to pay for the benefit of a cash flow so much improved that the project could run without any capital commitment at all after the first 16 weeks.

Finally, however, a dreadful warning: suppose there is an exactly similar job, where the client pays punctually and fully (as in the first example) but where the contractor has underpriced the work by 10% and is therefore not going to make £10,000 profit but £20,000 loss. Also, because of financial difficulties, the contractor is paying the suppliers one month late. The resultant cash flow (Table 15.5 and Figure 15.6) looks very satisfactory until the end of the project in September. It is in many ways better than that of the soundly managed contractor in Figures 15.2 and 15.5, and requires a working capital of about £10,000 for a few weeks only. However, instead of a handsome profit this contractor will finish by losing the whole of his working capital twice over. If the firm has several such jobs going on concurrently, so that the negative cash flows on one coincide with the positive flows on another, it will be able to keep trading for some time before the crash comes, and meanwhile its cash flow figures may look fairly healthy.

It will be seen how difficult it is to distinguish the early symptoms of insolvency when looking at a contractor's accounts, and indeed it is quite possible that a badly managed contracting firm may not itself realise what is happening until it is too late.

Phase 3

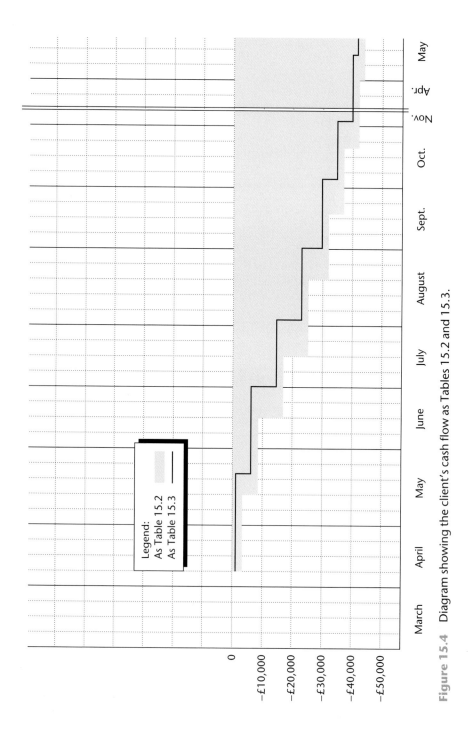

Figure 15.4 Diagram showing the client's cash flow as Tables 15.2 and 15.3.

Table 15.4 As Table 15.3 but with one month's delay in payments (£s) to materials suppliers.

	Week No.	Wages etc.	Materials	Sub-contractors	Total	Amounts received	Cumulative cash flow
			Payments				
March	1	1,000			1,000		−1,000
	2	1,500			1,500		−2,500
	3	1,500			1,500		−4,000
	4	2,000			2,000		−6,000
April	5	3,000			3,000		−9,000
	6	3,000			3,000		−12,000
	7	3,000			3,000		−15,000
	8	4,000			4,000	8,000	−11,000
May	9	4,000			4,000		−15,000
	10	4,000			4,000		−19,000
	11	5,000			5,000		−24,000
	12	5,000		9,000	14,000	52,000	14,000
	13	5,000	7,000 (March)		12,000		2,000
June	14	5,000			5,000		−3,000
	15	6,000			6,000		−9,000
	16	6,000			6,000		−15,000
	17	5,000	32,000 (April)	9,000	46,000	88,000	27,000
July	18	5,000			5,000		22,000
	19	5,000			5,000		17,000
	20	4,000			4,000		13,000
	21	3,000	53,000 (May)	9,000	65,000	70,000	18,000
August	22	3,000			3,000		15,000
	23	3,000			3,000		12,000
	24	2,000			2,000		10,000
	25	2,000		27,000	29,000	75,000	56,000
	26	2,000	47,000 (June)		49,000		7,000
September	27	4,000			4,000		3,000
	28	4,000			4,000		−1,000
	29	3,000			3,000		−4,000
	30	2,000	23,000 (July)	27,000	52,000	48,000	−8,000
October	34		12,000 (August) Release of retention	18,000 5,500	30,000 5,500	34,000 12,500	3,000
November	38		10,000 (Sept.)		10,000		−7,000
May	64		Release of retention	5,500	5,500	22,500	10,000
	TOTAL	**106,000**	**184,000**	**110,000**	**400,000**	**410,000**	**(10,000)**

When times are bad in the construction industry, it is common for contractors to submit very low tenders in order to obtain work and so keep their organisation employed, and they can then easily become short of cash. It is also not unknown for contractors who have already got into this position, either for the above reason or through bad management, to continue to quote

Phase 3

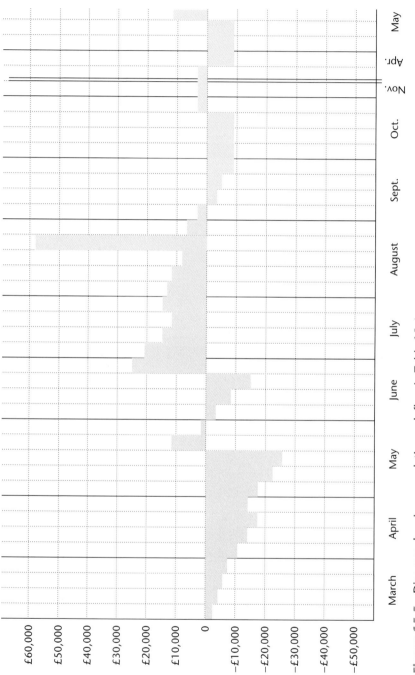

Figure 15.5 Diagram showing cumulative cash flow in Table 15.4.

Table 15.5 As Table 15.4 but builder's work underestimated by 10%, and one month's delay in payments to materials suppliers (£s).

	Week No.	Wages etc.	Materials	Sub-contractors	Total	Amounts received	Cumulative cash flow
			Payments				
March	1	1,000			1,000		−1,000
	2	1,500			1,500		−2,500
	3	1,500			1,500		−4,000
	4	2,000			2,000		−6,000
April	5	3,000			3,000		−9,000
	6	3,000			3,000	8,100	−3,900
	7	3,000			3,000		−6,900
	8	4,000			4,000		−10,900
May	9	4,000			4,000		−14,900
	10	4,000		9,000	13,000	49,500	21,600
	11	5,000			5,000		16,600
	12	5,000			5,000		11,600
	13	5,000	7,000 (March)		12,000		−400
June	14	5,000			5,000		−5,400
	15	6,000		9,000	15,000	81,900	61,500
	16	6,000			6,000		55,500
	17	5,000	32,000 (April)		37,000		18,500
July	18	5,000			5,000		13,500
	19	5,000		9,000	14,000	65,700	65,200
	20	4,000			4,000		61,200
	21	3,000	53,000 (May)		56,000		5,200
August	22	3,000			3,000		2,200
	23	3,000		27,000	3,000	70,500	42,700
	24	2,000			2,000		40,700
	25	2,000			2,000		38,700
	26	2,000	47,000 (June)		49,000		−10,300
September	27	4,000			4,000		14,300
	28	4,000		27,000	31,000	48,000	2,700
	29	3,000			3,000		−300
	30	2,000	23,000 (July)		25,000		−25,300
October	32			18,000	18,000	33,500	
			Release of retention				−3,900
	34		12,000 (August)		12,000		−15,900
November	38		10,000 (Sept.)		10,000		−25,900
April	60		Release of retention	5,500	5,500	11,400	−20,000
TOTAL		**106,000**	**184,000**	**110,000**	**400,000**	**380,000**	**−20,000**

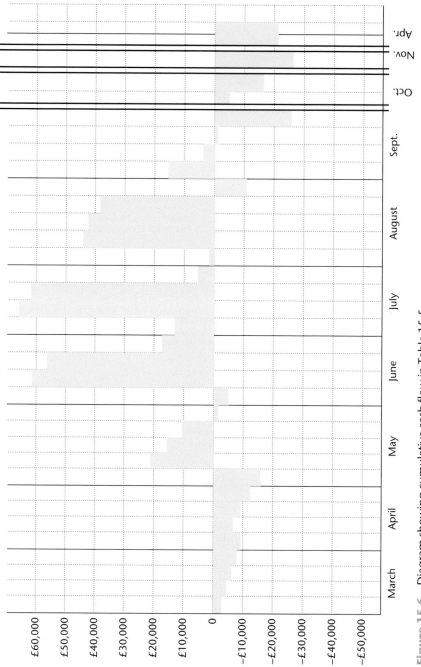

Figure 15.6 Diagram showing cumulative cash flow in Table 15.5.

absurdly low prices. This is because they urgently need the cash flow from new work in order to pay their past debts. If the contracting firm in the above example is receiving payments on another job by September, it will be able to pay its suppliers and keep going (although the new job in turn will be getting into even worse difficulties in due course and will need an even more drastic dose of the same medicine).

QSs do well to be suspicious of a building company that is expanding its operations rapidly at prices which its competitors cannot match. However, we can see from the earlier examples that even a soundly managed contracting firm will find it very tempting to get deeply into debt at the bank if it is able to get returns of more than 50% on money which it is borrowing at less than 15%. So cash flow assessment of accounts must be used carefully, but these examples do demonstrate that quite small percentage differences in estimating can have a dramatic effect on profitability. This is in fact the principal weakness of the construction industry, that the difference between a substantial profit on capital and a substantial loss can lie inside the normal margin of error in estimating.

A greater capital investment in a project on the part of contractors might lead to greater stability, as the required profit on turnover would then have to be much higher than 1% or 2%. If the contractor had to make 10% profit on turnover to get a reasonable return on capital, then estimating errors of 2% or so would have a proportionately smaller effect on the firm's overall profit percentage and would not make the difference between boom or bust. Moves to reduce retentions in recent times have had exactly the opposite effect; however, retention bonds are now coming into vogue.

15.8 Allocation of resource costs to building work

The contractor would appear to be in a much better position than the consulting QS as far as knowledge of actual building costs is concerned, but this is not necessarily the case. Keeping an accurate record of costs in the four categories:

- builder's direct costs (site labour, materials, small plant);
- sub-contractors and major specialist suppliers;
- site indirect costs;
- major plant,

is not very difficult, but translating them into usable data for cost planning future projects is another matter.

The contractor would in fact be ill-advised to use the actual costs from a particular project for cost planning a quite different one, for two reasons. First, many of the factors which affect site costs on an individual project have nothing to do with the design of the building and will not repeat from job to job. These include:

- weather;
- supervision;
- industrial and personal relations;
- obstruction by other trades;
- the skill with which the work is planned and organised;
- alternatively, lack of clear instruction;
- waiting for delivery of materials;
- accidents;
- replacement of defective work;
- failure by sub-contractors;
- psychological pressures.

Secondly, in practice, costs are rarely kept in any greater detail than the activity or operation. An operation has been described by the Building Research Establishment as 'a piece of work which can be completed by one man, or a gang of men, without interruption by others', such as the whole of the brickwork to one floor level, or the whole of the first fixings of joinery. Unfortunately, each operation is unique to the project and the cost information cannot be reused in this form.

15.9   System building, modular assembly, prefabrication and cost

No discussion of building costs would be complete without considering the effect of modular assembly, system building and prefabrication on costs. The history of system building in the UK witnessed accelerated development during the 1950–60s, although it is argued that this had as much to do with politics as with architecture; this building form provided a rapid response to the social housing needs in the UK at that time. It has been argued that this led to the construction of buildings that were not only poor in architectural terms, but also significantly underachieved in terms of performance and client/end user satisfaction. Examples such as the tower block Ronan Point in London (the reports on which could be described as the antithesis of system building) contributed in part to the current stigma attached to prefabrication of buildings in the UK. This resulted in an industry that was dominated by traditional forms of construction, particularly during the period 1970 to mid-1980s.

The 1990s, however, witnessed a renaissance in system building and modular assembly. The Egan Report could be identified as one of the prime movers, introducing the building as a product that could be manufactured more efficiently and to a higher standard away from the site itself. Furthermore, newer forms of procurement strategy such as PFI/PPP have also featured in this re-emergence. This sea change is perhaps all the more remarkable in an industry that is renowned for its insistence on resisting the incorporation of benefits found by other industries, automotive and aviation being just two examples.

Off-site manufacturing techniques have progressively developed within UK construction since a renewed interest in the 1990s, borne from the demands for a more sustainable industry and changes to traditional procurement practices in the public sector. Prefabrication and standardisation have, to some extent, been tainted by the stigma created by the 'nightmare' designs of the 1950s and 1960s. From the accelerated development during this period, the construction technique was all but abandoned in the UK – yet comparatively thrived in Europe and the USA. The UK's insistence on traditional construction methods and the characteristics of the property market made it difficult for the development of off-site manufacturing techniques to be presented as a credible alternative to conventional bricks and mortar construction. The advancements in technological and manufacturing capabilities, mainly in the automotive and aviation industries, have been applied to the fabrication of specific-user packages, mainly dedicated to public welfare and educational facilities. Choice and individuality is not compromised by standardisation needed to exploit the advantages of a dedicated production process. New materials have further advanced the viability and application of this building type, with the manufacturers taking note of the past mistakes regarding inadequate design and specification.

15.9.1   Some background to prefabrication and modular assembly

The history of prefabrication and standardisation of buildings can be traced back to the Roman occupation of the British Isles (Gibb 1999). The advantages of prefabrication were embraced in times of increased demand and immediate need. Yet off-site manufacture has had a rough ride. After the formation of local authority housing in the 1890s, and particularly after 1919, consecutive governments pressed for or insisted on standard designs and persuaded local authorities to experiment with new construction methods. But the systems that were developed fell short of expectation and none produced the anticipated savings in the vital factors of labour, money and improvement. This instigated a return to traditional practices (Morton 2002). Across the Atlantic, the production of prefabricated dwelling comparatively thrived, but it was not until 1945 that the first shipment of 30,000 temporary prefabricated houses was imported from the USA. These homes were well accepted and a subsequent housing drive of the early 1950s led the government to issue substantial advice and directions to local government on efficient house construction techniques (Morton 2002). This is further evidence that the UK accepted prefabrication as a stopgap solution to immediate problems.

The UK indulged in a radical overhaul of the way it built, through experimentation in industrialised building techniques. Through the early 1960s the

Phase 3

designs became high-rise and some gave adequate accommodation, but unfortunately many were found to be more expensive than conventional dwellings, with faults embedded in the design and the utilised materials, therefore making them unpleasant to live in (Morton 2002).

The well-documented disaster of 1968, which is cast as the critical event and is generally claimed to have sealed the fate of the industrialised housing programme, was the partial collapse of the tower block Ronan Point in the London borough of Newham after a gas explosion. This accelerated not only the decreasing use of prefabrication, but also the rejection of high-rise living in the UK (Morton 2002). The revival of system building, as identified by Gibb (1999), to some extent came through the changes in the NHS and government reformation of public infrastructure financing through PFI schemes. This led to a revival of interest in cost-effective design and construction techniques. Also contributing to this was the need for a more sustainable built environment and one that was more accommodating to change. The beginning of this century has seen the continued growth of off-site manufacture where applicable and beneficial. This is a contributing factor to the degradation of the stigma that flowed through both the industry's and its clients' veins throughout the twentieth century.

15.9.2   *Applications and suitability for prefabrication and modular assembly*

CIRIA report 176 (CIRIA 1999) identifies typical applications of modular building: hotels/motels, offices, retail outlets, prisons and residential schemes. Many construction product developers have now sought to position themselves as niche operators within the modular assembly market, two examples in the UK being Kingspan and Yorkon. Kingspan, for example, together with HLM Architects and Cyril Sweett, have developed a range of example designs of off-site modular buildings, which comply with NHS Estates and Commission for Architecture and the Built Environment (CABE) guidelines for district general and community hospitals, primary care centres and general practices. The options given by manufacturers for the designs of modular buildings have been seen in the past as a limiting factor of system building. Now, with the building not specified as a generic building type, the manufacturer can offer several options that fall in the boundaries of the function. The options are centred on the main elements of the building, including the services and fabric of the building, this making it more applicable nationally as it can fit into the majority of areas with aesthetic differences. Modular assembly forms have also offered learning and student accommodation packages that work on the same theme of function centred design, which is tailored to the client's needs through the options list. This is a considerable improvement on the buildings offered through the twentieth century.

15.9.3   Labour and skills availability in the industry

Concerns about the shortage of skilled operatives and the inadequacy of training have recurred throughout the last century (Morton 2002). This is an ever-present problem identified by both the Latham Report in 1994 and the Egan Report in 1998. A product of this shortage has been development of new technologies, as seen with prefabricated timber-framed housing. An article in *Construction News* (2004) pointed out that timber-framed housing manufacturer Pace and architects Cartwright Pickard joined forces to develop an off-site building system for the house builder. The article identified that one of the key drivers behind the development was the innovative response of the construction method to the housing and skills shortage. Reaction to the skills shortage is raising issues, as emphasised in one quote: '. . . the contractor is trying to de-skill the site-process, reducing the number of labour intensive jobs on site . . .' (Thompson 2000). The general consensus is that the Egan Report is the sole factor for the resurgence in off-site construction, but as the following quote shows, the nature of the industry as reactive innovator is still apparent: 'Sir John Egan may have spawned innovation across huge swathes of the industry but necessity is still the mother of invention' (Thompson 2000).

15.9.4   Improvement in health and safety

The main effect on health and safety is the level of control gained by taking construction from the site to the factory. CIRIA (1999) insists that pre-assembly can improve health and safety through moving work to a factory environment where health and safety measures can be implemented more effectively. This was highlighted in a report which stated that 'Champions of prefab stuck to their faith in the undeniable truth that there is much work on a building site that would be more efficiently and safely done in a factory' (*Construction News* 1993).

Whilst pre-assembly implies that larger items are being handled, this should not increase the risk to health and safety. Unit installation on site should be planned in advance, providing the opportunity to identify potential hazards from the outset and thus facilitate a safer installation programme (CIRIA 1999). At Winterton Primary School (Norfolk) in 2004, a prefabricated sports hall was erected on to an existing structure using the Quantum CF system (this was essentially the name given to a team that was established to develop the method of design and installation, CF standing for Composite Fastbuild) (Taylor 2004). On this project, a team of five workers – three contractors, a crane operator and a lorry driver – completed the majority of the erection and installation work. The erectors could carry out the work

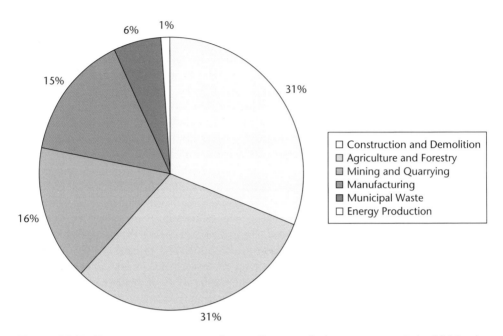

Figure 15.7   European waste output. Source: European industry waste statistics 12 March 2005 (www.defra.gov.uk/environment/statistics/waste/kf/wrkf13.htm).

efficiently and safely from a mobile access platform. This method mitigated the need for scaffolding, although this would have posed a higher safety risk if the building was constructed through traditional methods of brick and block wall construction.

15.9.5   Whole life-cycle costs and the environment

Use of pre-assembly reduces environmental impact from site processes, because on-site work is lessened and there is more control of the process (CIRIA 1999). This is valid in terms of the procurement process. Construction is putting a serious dent in the world's natural building materials like stone; Figure 15.7 highlights the waste produced by the construction industries of Europe. Note the comparison of manufacturing to the building industry, which highlights a vast difference in waste output. This is something the industry needs to work on to sustain the environment, and the government is pressing for this through taxation changes and guides to sustainability. There are fewer components to be brought to site than with traditional building construction, therefore saving in deliveries to site; it radically changes transportation and handling logistics.

Phase 3

Generally, larger units will be installed using mechanical lifting. However, far from this being seen as a disadvantage, it tends to result in an increase in on-site productivity by removing the ad hoc approach towards materials delivery and movement that is commonplace on many traditional construction projects (CIRIA 1999).

As with any building, maintenance or changes by the user will have to take place during the life-cycle of a project. Maintenance interventions to a traditionally built structure, fabric or services may be more intrusive and extensive in getting to the problem than with system buildings, which will be designed to accommodate maintenance. For example, modular plant rooms located on the roof of a building could be removed and replaced in one operation over a weekend shutdown (CIRIA 1999).

In the housing boom of the 1960s, the industry had a concentrated demand through which the industry served the state as a client. Yet the demand in the private sector was, and still is, variable and unfocused (Bosch and Phillips 2003). Since the privatisation programme of the 1970s and 1980s, construction work has focused on private clients; a plethora of demand now exists from small maintenance tasks to large infrastructure programmes (Bosch and Phillips 2003). The industry has benefited from sustained growth following the recessional period of the early 1990s. Figure 15.8 highlights this, taking into account the collective distribution of work across the industry.

With the healthy development of the economy and the shift to flexible outsourcing, the industry has seen fragmentation and inefficient co-ordination of resources and control of production. The result is an increase of constraints on enhancing productivity or adoption of new technologies, resulting from the lack of investment in development owing to cost and the use of legislation to control the industry (Bosch and Phillips 2003). The industry's demand has a cyclical fluctuation with concentrated growth in domestic and non-domestic areas. The industry is witnessing a growth in the deficit of skilled labour, and growing material costs. This presents a problem for the industry in the form of creating new methods of construction that do not rely on natural materials with high production cost and the need for skilled labour in their application (Hillebrandt 2000).

15.9.6 Client benefits

'If you are going to build these things before you get to site, you've got to be sure that the client isn't going to change its mind' (Taylor 2004). This quote is written in the context that the client's position has to change with prefabrication as the design must be finalised prior to the start of the fabrication of the building. This differs from traditional method construction as the building and design can be altered during the procurement period (although this does of course present problems anyway). But with the design stage traditionally

Phase 3

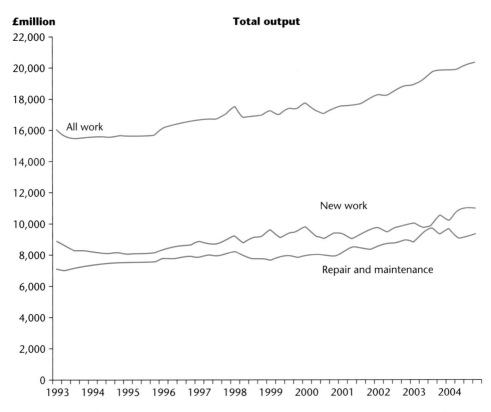

Figure 15.8   UK construction industry: volume of output at constant 2000 seasonally adjusted prices (DTI *Construction Statistics Annual* (2004)).

rushed to get the work started on site, there is more room for conflict. 'Litigation, another hidden cost in construction, can be reduced by greater prefabrication,' claims Mr Neale, a senior lecturer at Loughborough. 'It would certainly reduce on-site conflict and it is more likely that whatever is prefabricated would be done on time' (*Construction News* 1993).

This approach could be applied to the alterations and instructions by clients through architects, and the work of agreeing fees would ultimately be borne by the client in cost and time. With off-site construction 'once you start changing things on site you lose the whole benefit of off-site manufacture' (Taylor 2004). This is true and does pose a considerable disadvantage to the client, but with a pre-assembly project the aim is to get it right first time by taking more care and paying more attention to the client's brief. Therefore the main reason for using standardisation or pre-assembly on a project is that it can significantly improve predictability and efficiency in various ways,

Phase 3

because it builds on the combined knowledge and experience of the participants (CIRIA 1999).

15.9.7   Procurement strategies

The use of alternative approaches to the traditional-style procurement of buildings has been prevalent in system building-led projects. One common example is design and build, a procurement route that allows a building or engineering contractor full control of the design and construction process (Mosey 1998). It has been present in many industries for a considerable time. Hughes and Murdoch (2000) observed that 'when somebody buys something, the usual process seems to be to buy a product that has been designed by its producer'. This was the practice within the construction industry until the emergence of architecture as a profession distinct from construction, which led to the emergence of general contracting (Murdoch and Hughes 2000). Mosey (1998) argues that for the construction client that is not itself a professional in the design and construction business, there is great comfort in knowing that the person to whom they are talking is the person who is responsible for making their project a reality. This of course means that the contractor has overall responsibility in terms of project cost, as the risk of the cost exceeding the price falls entirely on the contractor's shoulders. This is also true for time control in meeting the project programme (Murdoch and Hughes 2000). With this procurement option having an uneven distribution of risk, more importantly drawing the risk away from the client, the price of the contract can be affected.

Single-point responsibility and fixed price contracts mean that the contractor carries more of the risk than a general contractor, and thus, risk attracts inflated costs (Murdoch and Hughes 2000). Offsetting this, the ability to allow the contractor and the client to share savings is an important facet of the contract. This creates an incentive to the contractor, as any savings made by completing the project below the price may sometimes be shared between the client and the contractor (Murdoch and Hughes 2000). An advantage in terms of time for the project inception as a whole is the allowance for the contractor to start early. As the contractor is carrying out the design work, there are opportunities to overlap the design and construction processes and thus to make an early start on site. The reason for this is the design does not have to be fully complete, as with traditional procurement, therefore the contractor can begin work on site if the foundations design is complete yet the superstructure design has not been finalised (Murdoch and Hughes 2000). As discussed previously, design and build has been used by many industries including manufacturing. A system building is a product and the parallels between industries (manufacturing and construction) have led to the use of design and build on many off-site-manufacture building projects.

Phase 3

15.9.8   Cost planning issues and the economics of modular assembly

It is difficult to obtain competitive tenders for this sort of work because, if prices are given by two or more firms, each firm will quote for its own patent system and it will be almost impossible to compare the value of the tenders without a considerable amount of work. In these circumstances cost comparisons based on cost modelling techniques of the type described in this book may provide a truer picture than an attempt to check detailed quantities and prices, even where these are provided.

However, in its more highly developed forms, system building may involve the delivery of the whole building in the form of a kit of parts ready to fit together in a few days (in the case of a house) or in a few weeks in the case of a larger building such as, say, a Travelodge Hotel. The economics of this practice are not easy to evaluate, partly because of the advertising and hard selling which often go with it, and partly because the economics are in fact much more complicated than the more usual site building costs.

The cost advantages that the prefabricating firm possesses are, at first sight, considerable:

- It has the benefit of planned mass production under factory conditions, safe from the weather hold-ups which affect site output and free from the difficulties of supervision and quality control which occur when operatives are working all over a scattered site.
- It can employ expensive but money-saving plant which would be too cumbersome, valuable or delicate for site use, and which in any case could not find full employment for its output on a single site.
- It can take advantage of modern methods of handling and transportation to bring its large fabricated units direct from the factory to the position on site where they are required.
- In fact, we might tend to accept the view that the lack of development of factory-based building techniques simply shows that the construction industry is hopelessly old-fashioned.

However, there are very sound reasons why prefabrication/modular assembly has historically had a much smaller impact on the industry than was at one time expected. One reason was the popular dislike of an environment composed of factory-produced buildings, although this might have been easier to overcome if there had been a really strong economic justification for them. In fact the economic gains have usually proved to be disappointing. What are the reasons for this? In the traditional site production methods of building:

- The site builder has none of the heavy overheads of factory production (the firm will not have to pay rent for its site workshops and will probably not even have to pay rates).

- Many of the cheap techniques (such as bricks and mortar) which the site builder uses are not suitable for off-site production.
- The crude handling methods which can be used with the small and rough components used for site assembly are cheaper than the tackle required for the careful handling of large units.
- The site builder is able to provide a made-to-measure building instead of one 'off the peg'.

There are other economic problems which have emerged with experience of prefabrication, including:

- A broken brick is simply a broken brick and there are plenty of unbroken ones to use, but if the corner of some special component is damaged another one will have to be ordered.
- The cost of replacement is the least part of the difficulty; the disruption to programme caused by the resultant delay can be far more serious.
- On low-rise projects such as housing, factories, health buildings, etc., the site preparation, levelling, foundations, drainage and site services, roads, paths, car parks, fencing, landscaping, etc. involve so much work on site and site organisation that the prefabrication of the basic superstructure is only dealing with part of the problem and not always the most significant part. This particularly applies if the finishings and services are not even included in the package.
- This problem is tending to increase rather than decline in importance, because today there are very few projects where the site is a level open field on which a factory-produced building can easily be placed; in fact the arrangement of the building or buildings is more usually dictated by the shape and configuration of the site.
- The building of the various in-situ connections between standard units (especially if these are at different levels on an undulating site), or the work necessary to accommodate them to irregular site boundaries, is piecemeal work which is difficult to organise efficiently, and it can more than swallow any cost savings generated by factory production of the main units. In such circumstances it might well have been better to build an in-situ building designed from the start to fit its location.

15.10   Key points

The construction industry is exceptionally flexible in dealing with local and national economic variations, but it adjusts its prices accordingly. Contractors' costs and prices therefore do not move in step with each other, except in very settled times – and it is the contractor's price that is the client's cost.

Contractors' costs comprise:

Phase 3

- direct costs (labour and materials);
- sub-contractors and major specialist suppliers;
- site indirect costs (preliminaries);
- major plant;
- off-site costs (establishment charges or overheads).

The building industry works on a very small capital commitment compared to its turnover, and its cash flow pattern is therefore vitally important. Prefabrication of complete buildings (except single-storey warehouses, sheds and the like) has not so far proved to be of economic benefit.

Further reading

Harvey, R. and Ashworth, A. (1997) *The Construction Industry of Great Britain*, 2nd edn. Butterworth Heinemann, Oxford.

References

Bosch, G. and Phillips, P. (2003) *Building Chaos: An International Comparison of Deregulation in the Construction Industry*, pp. 188–207. Routledge, London.

CIRIA (1999) *Standardisation and pre-assembly adding value to construction projects* (Report 176), pp. 24–48. Construction Industry Research and Information Association, London.

Construction News (July 22nd 1993) Emap Construct Ltd.

Construction News (September 23rd 2004) Materials: Cutting edge off-site kits go high-spec. Emap Construct Ltd.

Department of Trade and Industry (2004) *Construction Statistics Annual 2004*, pp. 141–184. The Stationery Office, Norwich.

Gibb, A.G.F. (1999) *Off-site Fabrication*, pp. 8–24. Whittles Publications, Caithness.

Hillebrandt, P.M. (2000) *Economic Theory and the Construction Industry*. Macmillan, Basingstoke.

Hughes, W.H. and Murdoch, J. (2000) *Construction Contracts: Law and Management*, 3rd Edition, pp. 41–69. Spon Press, London.

Morton, R. (2002) *Construction UK: Introduction to the Industry*. Blackwell, Oxford.

Mosey, D. (1998) *Design and Build in Action*, pp. 1–4. Chandos Publishing, Oxford.

Taylor, D. (2004) A new school of thought in prefabrication. *Construction News*, 14 October.

Thompson, R. (2000) The Egan assessment. *Construction News*, 23 November.

Chapter 16
Resource-based Cost Models

16.1 Effect of job organisation on costs

Many major costs on building projects are not directly related to the quantity of work produced but are concerned with time and with occurrences (or non-occurrences) of various kinds. The main quantity-related cost is materials, so that given prudent procurement and effective control of waste, there should be little variation in the cost of this part of the work whichever contractor is appointed and however the work is organised and executed. The real scope for gain (or loss) lies in the non-quantity-related items, and these depend on the way the job is organised and managed. This is where the competition between contractors takes place, especially today when, as we have seen, the contractor's main stock-in-trade is management skill.

16.2 A well-managed construction project

A well-managed project is one where both plant and workers have clear un-interrupted flows of work. Money is not spent on removing or returning workers or plant from the site, or on unproductive time waiting on site because of gaps in the workflow, or on moving workers or plant unnecessarily around the site. Materials are channelled to the spot where they are needed at the right time, and with a minimum of double handling.

It is not merely a question of the actual time spent hanging around; people work much more productively when they can see a clear task in front of them and where they feel they are participating in an efficient operation. It is a constant complaint by contractors that they are usually unable to recover the cost of disruption of this state of affairs when delays or interruptions are caused by the client or the design team.

16.3   Traditional versus resource-based methods of cost planning

Recognising the increasing importance of these management-based costs gives scope for comparing the cost planner's traditional price-oriented approach unfavourably with a resource-based method, which can take account of method and workflow. But the traditionalist's reply must be that conventional estimates are based on successful past tenders, and that few successful tenderers will have worked out their prices on the basis that the job will be badly run, whatever may have happened afterwards. In fact it might be claimed that the traditional price-based method automatically allows for good management without ever having to bother about the details of its implementation, and this facility is especially valuable in the early stages of estimating before the project has been fully designed.

16.4   Value added tax (VAT)

Value added tax is currently payable on building work other than housing, and also on architects', QSs', engineers', and planning supervisors' fees, and other professional fees. The rate set by Revenue and Customs at the time of writing is 17.5%.

It is customary to exclude this amount from estimates and tenders. This practice is well understood within the construction industry, and has been followed in this book. But this must be clearly pointed out in any figures given to clients, who may otherwise think that the estimate represents their total liability.

16.5   Resource-based cost models

The detailed cost models examined so far have been largely based on the measurement of finished work in place, and its valuation from BQs or other market price-oriented data. Quite apart from the general fallibility of BQ rates, these models have embodied two major fallacies:

- that the production cost of a building element is proportional to its finished quantity;
- that the cost of a building element has an independent existence which can be considered separately from the rest of the building.

We have been aware of these drawbacks almost from the start and are able to come to terms with them through the exercise of professional skill. However, it is worth considering whether it would not be better to base our estimates and cost control procedures on production cost criteria, as is done in most (perhaps all) other industries.

Phase 3

One major reason why we do not normally adopt this approach has already been given, which is that under lump sum contracting arrangements, it is the market price rather than production costs which concerns clients and their consultants. This reason, however, is not valid if we are considering a cost-reimbursement or management type of contract, or if we are looking at a situation where cost planning is being executed within a design and build organisation. There is, however, another quite different criticism of resource-based cost planning. As already discussed, a great deal of the resources of modern building are attributable to plant and organisational costs. Both the initial estimating and the subsequent refining of the estimate therefore require the envisaging of technological solutions. Whether the building is masonry, steel-framed, pre-cast concrete frame and panel or in-situ concrete will demand entirely different approaches with probably significant cost differences for any particular configuration and set of user requirements. This may in fact sound like an argument in favour of a resource basis rather than a criticism, but the principal disadvantage is that it moves the design considerations of a building into the production field much too early in the process.

An economical structural system may be postulated by the resource-based cost planner and the configuration of the building developed, for example, to suit the radius of action of the tower crane (or cranes) placed in the most efficient positions, or to suit the repetition of pre-cast units. The architect's traditional approach, as we have already discussed, is at direct variance; the form of the building is evolved primarily as a set of user-oriented spaces, and the best technological solution for that configuration is then investigated. Clearly in both cases some compromises may have to be reached, but the basic issue is whether:

- user needs and environmental considerations come first and construction methodology follows; or
- production efficiency should be the consideration from which design develops.

There are too many existing buildings that remind us that construction optimisation lasts for months but the consequences last for years (although not always as many years as the designers intended). Unfortunately production managers are no less indolent than the rest of us and prefer to postulate easy solutions rather than applying themselves to devising an efficient way of building a user-oriented design. It ought to be possible to take the latter approach provided the details of construction have not been developed too far before the building contractor is called on. Nonetheless, while it is true that the QS's traditional elemental cost planning approach enables the early design process to proceed with some semblance of cost control prior to construction decisions, such decisions do have to be addressed sooner or later.

Once this point is reached, a resource-oriented approach has much to recommend it in cases other than the competitive price-in-advance situation,

Phase 3

particularly where the contractor who will be undertaking the work is a party to the cost planning exercise. If this is not the case, then the method is of doubtful advantage; there are many different ways of organising a construction site and each contractor has preferred methods and equipment, so there is unlikely to be an optimum solution that any builder would automatically accept.

Much of the advantage of a resource-based estimate is that it can be used for production cost control purposes, and this benefit will be lost if the estimate has been prepared by someone else and the chosen contractor throws it out of the window. A resource-based estimate will deal quite separately with the different cost components of:

- labour;
- plant;
- materials;
- sub-contractors,

rather than amalgamating them into a series of all-in rates as is the QS's custom. However, having said this, a very substantial part of the total cost will comprise sub-contractors' work.

Much of the resource-based estimate will therefore have to be based on the specialists' all-in prices in exactly the same way as a product-based estimate. So the claim by the builder's estimator to know more about the so-called real cost of building than the independent professional cost planner becomes somewhat questionable. The resource-based estimator's immediate concern will therefore be with that part of the work usually undertaken with the contractor's own resources: normally the structure of the building including excavation. The resource-based estimator's real expertise (and the area where the product-oriented estimator is weak) is in determining and pricing the site organisation and the project duration, including the management of all the specialists' programmes and the integration of their work with the building structure and with each other's efforts. In a design and build, contracting or construction management organisation, the estimator will almost certainly have the assistance of a planning engineer or project manager in this work. They may look at alternative configurations for the proposed building, as well as alternative means of obtaining the same configuration.

As far as estimating the cost of the contractor's own work is concerned, the materials present few problems since unlike labour and plant, their cost is reasonably related to the quantities of finished work. Anybody preparing a resource-based estimate will, therefore, have to start off by measuring (or assuming) quantities of finished work in order to determine material requirements, and also as a first step in looking at the scale of the project and the distribution and interrelation of its parts. The person doing this will tend to work in bulk quantities (e.g. cubic metres of concrete) split into categories and locations that seem to be organisationally significant.

Before proceeding much further, however, an outline programme for the works will need to be prepared.

16.6   Resource programming techniques

The estimator, in respect of each different estimate, will need to decide the principal operations to be undertaken and their methodology and duration. The operations cannot be considered in isolation since they will be inter-related by two factors:

- *The need to utilise labour and plant effectively*: so that operatives and machines are not unproductive (downtime) for long periods between operations, are not required to be working in two different places at once and do not spend too much time relocating from one part of the site to another.
- *The inescapable sequence of building*: so that, for example, the walls and columns cannot be constructed until the foundations are completed, and the first floor cannot be placed until the ground floor walls and columns have been built, etc.

There are various techniques in common use to assist the estimator in this task, which tend to rely in the first instance on graphic methods. Some key techniques are explained briefly here but a useful reference for more detailed coverage is Cooke and Williams (2004).

16.7   The Gantt chart

The Gantt chart is perhaps recognised as the most traditional method of programming and is often referred to as a bar chart (this is technically incorrect however). The chart takes its name from Henry Gantt (1861–1919), who developed the technique in 1910[1].

The widespread use of software applications such as Microsoft Project has facilitated the creation of sophisticated Gantt charts that previously were often time consuming and expensive to produce. These applications were primarily geared towards the project management market but today the creation of these charts is often performed by a subordinate to the project manager such as a scheduler or programmer.

On a Gantt chart a horizontal chronological scale is used; this is ordinarily divided into days, weeks or months depending on the time horizon, and the various operations comprising the project are listed vertically down the left-hand side. The timing and duration of each operation is then indicated by a horizontal bar spanning the relevant period of days/weeks and shown on the same line as the operation it refers to. Figure 16.1 shows a simplified example, covering the construction of a new workshop. The bars follow the

Phase 3

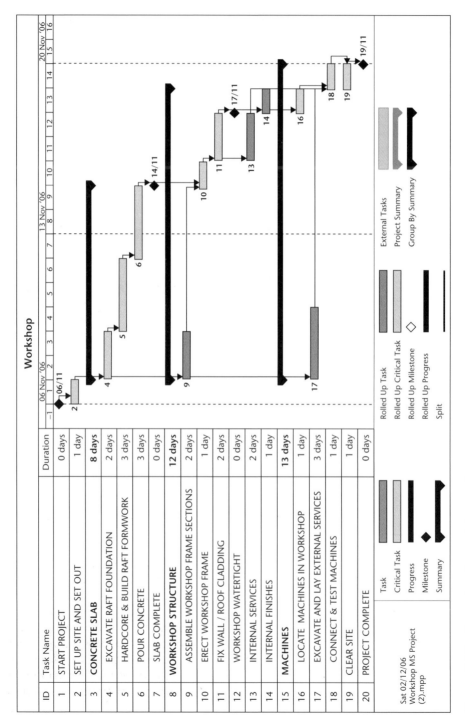

Figure 16.1 Gantt chart for workshop project.

classic pattern of moving diagonally from the top left-hand to the bottom right-hand of the chart as progress is made through the project. The bar chart is simple and easy to follow. It provides a useful, at-a-glance indication of the dependences and interdependences of various operations within the contract period. You will often see such a chart on the wall of construction site offices and it is very popular for the purpose of monitoring progress and logistics management.

For small projects the Gantt chart is easy to interpret, as Figure 16.1 shows, but for long complex projects the Gantt chart can become quite cumbersome. Larger Gantt charts, for example, are difficult to appreciate on the average-sized computer screen. Gantt charts are often criticised for communicating relatively little information on project complexity; and that complex projects cannot be communicated effectively using such a chart. Furthermore, Gantt charts represent only part of the constraints of projects since the focus is primarily on schedule management. While Gantt charts can identify task dependencies, displaying a large number of these can result in an untidy chart.

Because the horizontal bars of a Gantt chart have a fixed height, they can misrepresent the planned workload (resource requirements) of a project. In Figure 16.1, Tasks 5 and 6 appear to be the same size but in reality the magnitude of the tasks may be different. A related criticism is that all activities of a Gantt chart show planned workload as constant. In practice, many activities have front-loaded or back-loaded work plans, so a Gantt chart with percent-complete shading may actually miscommunicate the true project performance status[2].

16.8   The critical path diagram (or the network diagram)

'Network diagram' is perhaps a generic term that accounts for precedence diagrams and critical path diagrams, which are special applications of this approach. The network consists of a series of nodes joined together by lines or arrows, and normally moves from a start at the left-hand side to a finish on the right.

Figure 16.2 shows a network diagram for the project in Figure 16.1; the boxes or nodes represent the activities shown on the left-hand side of the chart in Figure 16.1; and the lines or arrows joining them simply illustrate dependencies. Traditionally, the nodes were usually drawn in the form of sequentially numbered circles which are large enough to contain some numeric information; the length of the lines joining them is of no significance and is chosen arbitrarily to suit a clear layout of the diagram. Modern charts, such as that shown in Figure 16.2, use boxes containing information on each task, such as duration, start time or finish time. Often these boxes will provide additional information such as early/late start time (EST/LST) and early/late finish time (EFT/LFT) and float (slack allowed in that task).

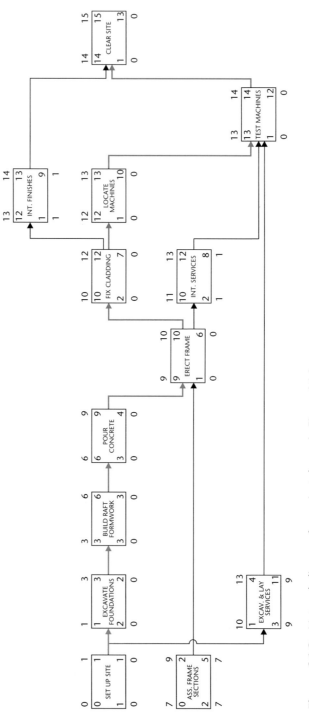

WORKSHOP

Figure 16.2 Network diagram for project shown in Figure 16.1.

A network in an unquantified form is simply a precedence diagram. The diagram can thus be drawn without any idea of the length of time that any operation will take, and in this form reflects the earliest planning stage, in which the planner is identifying those operations which fundamentally depend on each other and those which do not. Once this has been done it will never need to be redrawn (unless the project itself alters); any changes will be made only to the numeric information within the nodes.

Remember, on a building project it is fundamental construction dependence which counts and not convenience or the efficient use of resources, which are considered at a second stage.

Two important points are worth noting:

- The network can be expanded to include not merely the construction but also the whole of the planning and design process and can thus become a tool in the hands of the client's professional representatives. This capability is particularly useful where overall time requirements are important or where design and construction are to run in parallel on a management, design and build or fast-track contract.
- The operations may be shown at a strategic level (build walls) or at a much more detailed level (build ground floor external walls, build in sills and lintels, and so on).

For eventual control purposes a fully detailed network may be used, but this is expensive to prepare and would be inappropriate at planning stage, before the detailed design has been finalised. On the other hand, a simple planning network cannot show the interdependence of meshing operations. For instance, if the erection of pre-cast beams can commence after only some of the columns have been completed, this would require the work of each group (or even pair) of columns to be shown as separate operations. This problem occurs throughout the project, but at planning stage is only likely to be considered in those parts which the planner feels to be of crucial import-ance – probably major structural work rather than finishes.

Once the network has been drawn and the precedences and dependencies settled, it is time to quantify the problem. The procedures for doing this are sufficiently standardised for there to be a number of computer packages, as described previously, and the work then entails identifying the estimated durations on a schedule.

If manual methods are being used, however, a suitable method is to mark durations in the node boxes, which are often divided into quadrants (or sometimes seven parts) for the purpose of quantification (see Figure 16.3). In this example we shall consider the node consisting of quadrants. In the top left-hand quadrant is written the reference number of the operation concerned (full description of the operation probably being written out in a numbered list), and in the top right-hand quadrant the estimated duration in

Phase 3

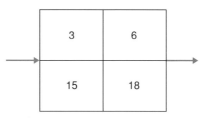

Figure 16.3   Activity node on the critical path diagram.

days, weeks or whatever unit is being used. When all the individual durations have been inserted, the total project time can be calculated.

Beginning at the start and working along the arrows, the shortest possible elapsed time to the completion of each operation is inserted in the bottom left-hand quadrant. This is arrived at by adding the duration of the operation concerned to the highest elapsed time of any of the other operations which precede it and which are therefore arrowed into it. By the time the end is reached the total time for the project will have been calculated.

In respect of each operation, we will now know the earliest time at which it can be completed. Where the operation concerned is crucial, this will also be the latest time at which it can be completed if the project as a whole is to be finished on the calculated date. However, many less important operations may be able to be delayed without delaying the completion of the whole project. In order to identify the crucial operations, the bottom right-hand quadrant of each node may be used to show the latest possible date at which the operation can be completed without causing delay. This is calculated by working backwards from the finish and, in the case of each operation, calculating from the earliest start time of any of the operations that immediately follow it. When this task is completed it will be found that for some operations the earliest and latest dates for finishing are the same. These are the critical operations, and the sequence of arrows which joins them forms the so-called critical path through the project. In the case of a non-critical operation the difference between the earliest and latest finishing date represents the float, or the period for which that operation can be delayed without affecting the completion date for the project.

All these figures can of course be set out in the form of a schedule, instead of on the diagram itself. It should be noted that if part of the float for a particular operation is utilised, this may affect the float available for succeeding operations. If all the float is used up, then the operation and its successors will have become critical and the critical path will have changed. It is therefore necessary for the project controller to keep a close eye throughout the progress of the job on operations with a very small float, as well as those already on the critical path.

16.9   Resource levelling (or smoothing)

There may well be two or more operations in a network which do not depend on each other in the construction sequence – for example, foundation excavation and drainage trenches. It will not matter which is done first, and they could in theory be done simultaneously (whereas the foundation excavations and the concreting of them could not). However, their relative timing may depend on resource utilisation. To dig the foundations and drainage trenches simultaneously would involve bringing an unnecessary number of diggers to the site, whilst to leave several weeks between the two operations would involve either machines standing idle, or taking them away and bringing them back. If operations were on critical or near-critical paths, then the additional expenditure would have to be accepted (or the time for completion altered), but otherwise their floats would be adjusted to give a more economical sequential programme. The adjustment of a programme in this way to make the best use of expensive plant, and to give gangs of operatives a consistent and balanced work pattern, is an important part of the planning exercise, and is often called resource smoothing.

16.10   Resource-based techniques in relation to design cost planning

The use of resource-based methods is not practicable at the earliest budgeting and planning stages as the cost planner who is intending to use these methods needs to start off with some idea of the configuration of the proposed building: its size, shape, height and preferred technology. The cost planner may well suggest a suitable combination of these requirements, and if working for a construction firm these may be based on systems already rationalised in the organisation and which can be built quickly and efficiently. For some types of development (e.g. industrial premises) this could be a valid approach. Alternatively, the cost planner may begin with a solution proposed by the architect, test it in terms of resource use and then investigate alternatives.

At this stage the main concern will be with those factors that can be identified as major time constraints and production cost constraints. By using skills derived from experience, the cost planner will be looking for a solution that gives effective use of labour and equipment by providing smooth and continuous workflows, and which enables the most economical type of plant to be used. For example, if one single heavy component has to be lifted at the extreme reach of the tower crane, this will dictate a much heavier and more expensive type of crane to hire and run, probably for the whole duration of the contract. The crane cycle will also dictate the intervals at which components, or skips of wet concrete, can be lifted into position, which in turn may determine the size of gang and the method of working. Alternatively, an

Phase 3

efficient method of working may require a different cranage configuration to that first thought of.

In turn, formwork usage will have to be considered and the concreting programme planned to allow for optimum reuse, with sufficient time allowed for any necessary alterations. This is where the contractor may look for savings; for example, by keeping column sections constant all the way up the building, because it may be that the cost of the extra material thereby required in the upper storeys will be more than saved on labour, plant and formwork. Decisions of this kind can often only be made in the light of the actual construction programme envisaged, and whether site-mixed or ready-mixed concrete is to be used.

A further matter to be considered is the staffing of the site, and the number and type of supervisory and control staff. The contractor will also be very concerned, even at an early estimating stage, with the integration of the engineering services work with the construction programme. Some of this integration concerns the way in which the contractor and the various sub-contractors will have to work together, because the installation of pipes, conduits and components may be incorporated into the design in a way which means that they will interfere with other contractors' work sequences.

The planner/estimator will also be on the lookout for major items of equipment whose placing may cause difficulties. Equally important will be the question of whether, and if so when, the permanent engineering facilities – lighting, heating, toilets, and passenger and goods lifts in particular – may be used to facilitate the construction work.

The duration of the complete project, and of major stages in it, will have a substantial effect on plant, supervisory and establishment costs, and this duration may in the end be determined as much by service sub-contractor requirements as by the builder's own work. In highly serviced buildings, in fact, they may be the prime determinant.

The criticism is sometimes made that, in assessing and balancing all the above matters, builders' estimators and planners tend to attach too much importance to optimising their own part of the work, whereas in the end it is often the work of the specialists and their co-ordination which proves to have dictated the time for a project. This is perhaps where the independent professional cost planner can take a more balanced view.

Resource-based estimates are certainly no cheaper to prepare than a QS's elemental estimate. Much the same sequence will therefore be employed. First, strategic estimates are prepared, taking account of major factors only, often those which are of importance in a comparison between alternative schemes. A resource-use plan at this level may also be used in the preparation of a competitive tender based on BQs. Then a detailed production plan and estimate will be prepared, probably incorporating a full critical path network. This would be intended for use as a production control document, and would be unlikely to be prepared until the design itself had been finalised and the

Phase 3

contractor was expecting to be chosen to construct it. It would almost certainly involve consultation with major sub-contractors, and the use of an appropriate computer package.

16.11   Obtaining resource cost data for building work

Keeping an accurate record of costs split into:

■ site labour;
■ materials and sub-contractors;
■ plant;
■ establishment charges;

and allocating them to the right contracts is comparatively easy, but the lump sums obtained do not really contribute very much to the understanding of how costs are incurred.

A detailed site-costing system is much more difficult to arrange. In theory it might be possible to cost each contract in terms of finished work, attaching costs to the measured items (or groups of items) in the BQs, but this is rarely done because it is difficult and expensive to keep records of time and material for a multiplicity of items and subsequently process them. In addition, many costs are not directly related to specific quantities of work, such as the following four types:

■ *Quantity related*. These are the costs which bear a straight relationship to quantity of finished work; many material costs fall into this category as do some components of labour cost. With this type of cost, twice the quantity of work costs twice as much.
■ *Occurrence related*. These costs are related to a particular event, or occurrence, such as the bringing of excavation plant to the site, or the moving of a plasterer's equipment from room to room. While the scale of the occurrence is obviously affected by the scale of the work, which is anticipated, its costs will not vary in proportion to actual quantity of work executed.
■ *Time related*. Some costs (such as the hire of a major item of plant) are related to a length of time, and not to the amount of work done in that time.
■ *Value related*. Fire insurance, for instance, will be related to the value of the project. Establishment charges are also often allocated on a basis of project value and can be included under this heading.

In addition to these four types of cost, it must be remembered that some of the work done will relate to temporary items (such as erection of site huts and conveniences) that do not form part of the permanent works. On many projects the shuttering of concrete work is a very large item in this category.

Phase 3

16.12   Identification of differing variable costs

It is all very well to make these rather academic distinctions between different types of cost, but in practice they may be difficult to separate. For instance, the plasterers are unlikely to keep separate the time involved in moving from room to room, and the charge for a major item of plant may well incorporate several cost types, for example:

- *Quantity related*. Fuel, and part of hire charge.
- *Occurrence related*. Bringing to site and removal, moving, assembling and dismantling.
- *Time related*. Part of hire charge.

However, even an arbitrary division into these different types will be better than trying to allocate total cost pro rata to quantity of finished work, and would form a more suitable basis for using cost information for estimating or cost planning purposes. Previous revisions of the Standard Method of Measurement have taken a welcome step towards identifying costs under these various categories.

16.13   Costing by operations

In practice, costs are rarely kept to any lower level than an operation, which has been defined by the Building Research Establishment as a piece of work that can be completed by one man, or a gang of men, without interruption by others. A typical operation might be the whole of the brickwork to one floor level, or the whole of the first fixings of joinery.

It is comparatively simple to record costs on site for such overall parcels of work, and to allocate materials to them, and it is much more meaningful to attach the time-related and occurrence-related costs to total operations than to try to split them up among quantities of measured work to which they bear little relation. If the contractor's estimate can be similarly subdivided, it will be possible to compare cost with estimate at each stage of the project. Comparing actual costs with estimate is one of the main purposes of costing by operations; as already explained, this approach does not lend itself to the provision of cost planning data for future projects.

16.14   Use of resource-based cost information for design cost planning

The contractor would appear to be in a much better position than the consulting QS/cost planner as far as real cost information is concerned, yet would be ill-advised to use the actual costs from a particular project for cost planning another quite different one, for two reasons.

First, many of the factors that affect site costs on a particular project have nothing to do with the design of the building and will not repeat from job to job. These include site conditions, the weather, industrial and personal relations, the skill with which the work was organised, accidents, late delivery of materials, defective work, and failure by sub-contractors. Second, although the operation is an excellent concept for collecting costs, it is very difficult to reuse the information arising from it because each operation is unique. The only way in which, say, first fixing of joinery on one project can be compared with another quite different one is by looking at the quantities of the different types of work involved, and the whole essence of operational costing is that the costs are not broken down in this way. If they were, we would be back to costing work items from the BQ and, as explained, this is impracticable. In some instances, the operation may embrace more than one cost-planning element.

However, the contractor's operationally based estimates have neither of these disadvantages when considered as a database for cost planning:

■ They are based on an assessment of average costs, and as such will have a closer relationship to future tender prices than ascertained costs from a particular project.
■ They will break down each operation into measurable characteristics for pricing purposes (the estimated costs of which may be based either on experience or work study methods) and these can be used for the purposes of design cost analysis.

In so far as the contractor has access to this kind of estimate, and the consulting QS has not, the contractor may be said to have an advantage in cost knowledge. Against this, the average QS is much more experienced than the average builder in translating project cost data into terms which are relevant at early design stage, and this expertise is essential to cost planning and control; the QS's need is for a BQ which reflects more closely the way in which operational estimates are built up. For instance, if the preliminaries section of the BQ were always priced in a consistently itemised way (like the other sections of the BQ), the consulting QS could analyse and manipulate this important part of the cost in detail instead of as a percentage of the measured work. It must also be remembered that the proportion of project cost represented by the execution of the measured builder's work is steadily declining, and the proportion represented by specialist work, fixed charges and management costs is increasing.

16.15 Key points

■ Since all contractors pay much the same prices for labour and materials, the differences in their costs are mainly to do with how well they organise and manage their projects.

Phase 3

- In order to prepare a resource-based estimate they therefore need to plan the construction, and this is almost impossible to do until some sort of drawings exist.
- A resource-based estimate is therefore not appropriate for the earliest stages of budgeting or planning the accommodation to be provided for the client. The contractor will need to plan the construction using a bar chart and/or a network diagram.
- The proportion of total costs represented by the main contractor's direct costs is steadily declining. It is much more difficult to use the costs of a previous job to forecast the cost of a new one than it is to do the same thing with prices.

Notes

(1) *Work, Wages and Profit* by H.L. Gantt, published by The Engineering Magazine, NY, 1910 (Source: Wikipedia).
(2) http://en.wikipedia.org/wiki/Gannt_Chart (accessed 2006).

Further reading

Perera, A.A.D.A.J. and Imriyas, K. (2004) An integrated construction project cost information system using MS Access™ and MS Project™. *Construction Management and Economics*, **22**(2), 203–211.
Pilcher, R. (1994) *Project Cost Control in Construction*, 2nd edn. Blackwell Science, Oxford.
Raftery, J. (1991) *Models for Construction Cost and Price Forecasting*. Royal Institution of Chartered Surveyors, London.

Reference

Cooke, B. and Williams, P. (2004) *Construction Planning, Programming and Control*. Blackwell Publishing, Oxford.

Chapter 17
Cost Control (1): Final Design and Production Drawing Stage

17.1 Cost checks on working drawings

After the cost plan has been prepared and agreed, it will be the cost planner's responsibility to ensure that the work shown on the production drawings and measured in the BQ corresponds with the plan, as otherwise the tender figure is likely to be quite different from the amount of the estimate. Any necessary adjustments must be made before the contract documentation, including the BQ, is finalised, *not afterwards*. For this purpose all drawings should be cost checked as they come into the cost planner's office; this must be a matter of routine, and a system for tracking the cost checking routines performed on the drawings should be devised.

Unfortunately, this stage in the QS's work (the receipt of working drawings and subsequent revisions and details) is usually hectic and there is little time to spare for things like cost checks. It is unfortunate that these checks can best be done by a senior person, who is usually very busy in other directions. Where the cost plan has been based on approximate quantities tied to a fairly detailed specification, and where the design has proceeded in accordance with it, the work involved in cost checking should be minimal. If, however, the design has been radically changed for some reason, it may be better to start the process again and prepare a fresh cost plan at as early a stage as possible, rather than attempting to check the detailed drawings against a cost plan prepared for a somewhat different building.

Cost checking procedures are quite the most time-consuming and error-prone part of the whole process and the more they can be legitimately minimised the better. Should it prove to be necessary to undertake a thorough cost check of the whole design (perhaps because the cost plan did not give the architect sufficient guidance), then the drawings will have to be prepared by elements if the feedback from the cost checks is to be of any real use.

Phase 3

The extent to which cost checking should be carried out will depend on:

- *The amount of extra time that the QS can be allowed.* This will vary from job to job, but it must be realised that if lack of time prevents essential cost checks from being carried out, then the whole attempt at cost planning will largely be a waste of time.
- *The amount of apparent alteration to the scheme since the cost plan was prepared.* If the cost plan was based on a very similar project for the same architects and client, and the details of specification and design were therefore known fairly well, the cost check might entail little more than a quick glance at the drawings to see that nothing is substantially different. Even where the circumstances are not quite as ideal as this, it may still be possible for the QS to be satisfied by a quick inspection that the drawings show what the cost plan envisaged. The drawings should be stamped and initialled, even so.
- *The amount of detail in the cost plan.* If it was possible to take out fairly full approximate quantities, the cost check may simply consist of comparing these quantities with the working drawings and checking that the specification is unaltered.
- *The degree of confidence that exists between cost planner and architect.* If the architect is known to be cost conscious, capable of and enthusiastic about cost design, the cost checks will be much less important than where this confidence is lacking.
- *The familiarity of the type of project.* Most cost planners would feel fairly confident of cost planning a school, whereas a planetarium would be a very different proposition. In the latter case the cost plan would probably incorporate rather a lot of assumptions and it would be necessary to check these in detail.
- *The importance of the element.* It would be a very self-confident cost planner who would not bother to check the external walling, or the roof of a single-storey building, even under the most ideal conditions. On the other hand, if time is pressing some of the smaller elements can often be ignored.

17.2 Carrying out the cost check

Again this is a suitable occasion for using standard forms. Where the architect is designing by elements the procedure will be considerably simplified, but even if this is not the case the check must still be done by elements. The design of the cost check form may be left to the tastes of the firm or other organisation. It should show at least the following information:

- *Number of check.* The checks on a particular project should be numbered consecutively, starting at 01.

- *Total estimated cost of project after completion of previous check.* This amount will be obtained from the previous cost check form, or if this is the first check it will be the unaltered total of the cost plan.
- *Reference number* of drawing(s) or other information being checked.
- *Element being checked.*
- *Amount allowed in cost plan in respect of element* (or as amended by previous checks).
- *Estimated cost of element* as calculated from the drawing(s), or other information, being checked.
- *Difference (plus or minus) between the two last.*
- *Estimated cost of project after this difference has been added to or deducted from the previous estimated project cost.*

A typical example of a cost check form is shown in Figure 17.1. As each cost check is carried out, a copy of the cost check form should be sent to the architects so that they have an up-to-date running total of the project. If any check shows a substantial increase or decrease, it would be as well to discuss

COST CHECK

Contract:
Cost check:
Element:

Date:

Gross internal floor area:

Total cost of project forward from cost check No. 2

			£ total	**£/m²**
Total cost of element from updated cost plan			£125,600.00	52.55
Elemental cost check (see attached dimensions)				
380 m³ reinforced concrete in beams and columns	£90.00	£34,200.00		
2,950 m² formwork to beams and columns	£16.00	£47,200.00		
31 tonnes reinfct (as Mr Smith of Consulting Engineers 28.06.99)				
	£31,200.00	£37,200.00		
Sundries		£10,000.00		
Revised cost of element carried forward		£128,600.00	£128,600.00	53.80
Amount of saving/extra			£3,020.00	£1.26
Revised total cost of project carried forward to next cost check		£1,907,222 + £3,020 = £1,910,240		

Phase 3

Figure 17.1   Example of cost check form.

the matter with the architects rather than merely sending the results of the check to them. Similarly, they should be warned if the general standard of detailing appears to be more lavish than was allowed for.

If a drawing shows changes in more than one element a separate form should be used for each. The rough workings in connection with each check can be done on dimension paper and stapled to the office copy of the form.

A list should be kept of all drawings or other information received, with columns for marking:

- that the cost check has been carried out;
- the reference numbers of the cost check form or forms;
- the elements involved.

The purpose of this is to enable the checker, on receiving a drawing showing, say, part of the roof, to look back through the list to see whether any adjustments have already been made to this element.

If it appears that a detailed cost check of any particular drawing or information is not necessary, the list can be marked accordingly. However, all drawings and information (such as sub-contractors' quotations or replies to queries) must come to the cost checker in the first instance, even if someone else is waiting for them. In connection with cost checking it is as well to remember the total cost of the project and also the inherent margin of error in cost planning. A cost planner whose estimates consistently get within plus or minus 5% of the accepted tender is doing well. In these circumstances it is obviously not worthwhile to spend much time cost checking isolated details of a large project, as the possible differences in cost are so small compared to the probable overall margin of error.

Prices for cost checking may be obtained from the same sources as have been used earlier. But as there will be a tendency for the items to be more detailed than at cost plan or estimate stage, the cost planner is likely to be using either built-up rates, price book rates or actual rates from the BQ from which the example analysis was prepared. Again it must be remembered that the contractor's permission must have been obtained for this latter course.

It will be very tempting to lift prices from BQs for other projects at this stage, but the errors that can arise from this practice have previously been pointed out. If prices are built up or obtained from price books, the level of market prices on which the estimate is based must be remembered. It will also be necessary to include a realistic percentage for preliminaries and insurances; this will not necessarily be the same percentage as the contractor showed in the analysed example.

17.3 Use of an integrated computer package at production drawing stage

Figures 17.2a–d illustrate an extract from the output of an integrated computer cost planning package (CATOpro) relating to an urban site development

URBAN SITE REDEVELOPMENT SITES A, B and C 2.0 SUBSTRUCTURE						HOUSING ASSOCIATION
Description	+ Quantity	Unit	Rate £	Calc	Total £	Notes
1 Clear/strip site	10,587.00	m²	2.00		£21,174.00	Site layout
2 Strip foundations	2,061.00	m²	60.00		£123,660.00	To houses
3 Extra over allowance for stepped foundations	1.00	Item	3,000.00		£3,000.00	
4 Piles, ground beams & pile caps		m²	105.00		£17,325.00	To flats
5 In-situ reinforced concrete ground slab with dpm and screed	2,226.00	m²	60.00		£133,560.00	
6 Allowance for cut and fill earthworks	1.00	Item	25,000.00		£25,000.00	
U/D 4,350 m²					£323,719.00	

Figure 17.2a   Revision to the substructure element of sites A, B and C.

for a housing association. Perhaps the biggest advantages of a computer package such as this is that the figures for the total project are automatically updated whenever a change is made to any element or part element, without the opportunities for error or omission which are always present with a manual system. Figure 17.2a shows a revision to the substructure element of sites A, B and C. The input to Figure 17.2a is the measurement of approximate quantities from the drawings, either manually or entered by digitiser, or measured and calculated using a lower level of the computer package (not shown here). Figures 17.2b and 17.2c (level 2) show the next stage, the revised elemental summary page for sites A, B and C. The various figures can either be inserted directly on an elemental basis per element or automatically derived from a level 3 cost check. In this case the figures for '2.0 Substructure' have been transferred by the computer from the level 3 elemental cost check which was shown in Figure 17.2a. Figure 17.2d (level 1) shows the updated summary of cost for the whole project, automatically updated from the more detailed levels of working of which examples have been given. The figure for sites A, B and C has been transferred by the computer from the level 2 elemental summary (Figure 17.2c). It will be noted that costs are given in £/ft² as well as in £/m², this being the unit of measure preferred by most real estate clients.

17.4   Use of resource-based techniques at production drawing stage

This technique has been introduced in the previous chapter. If this approach is being used, a detailed production plan and estimate would be prepared at

Phase 3

Phase 3

Element Code	Element	%	Cost £/m²	Cost £/ft²	Quantity	Unit	Rate £	Sub Total	Total £	Notes
1	1.1 Demolitions									In enabling works
2										
3	**TOTAL**									
4										
5	1.2 Alterations									In enabling works
6										
7	**TOTAL**									
8										
9	2.0 Substructure	10.84	74.42	6.92				3237190	3237190	
10										
11										
12	**TOTAL**	**10.84**	**74.42**	**6.92**				**3237190**	**3237190**	
13										
14	3.0 Superstructures									
15										
16	3.1 Roof	6.77	46.46	4.32				202,098.00	202,098.00	
17	3.2 External walls	10.51	72.14	6.7				313,805.00	313,805.00	
18	3.3 Upper floors	4.51	30.97	2.88				134,728.00	134,728.00	
19	3.4 Windows and external doors	12.18	83.61	7.77				363,725.00	363,725.00	
20	3.5 Internal walls and partitions	6.42	44.08	4.1				191,741.00	191,741.00	
21	3.6 Internal doors	3.57	24.51	2.28				106,600.00	106,600.00	
22	3.7 Stairs	3.06	21.03	1.95				91,500.00	91,500.00	
23										
24	**TOTAL**	**47.02**	**322.8**	**30**				**1,404,197.00**	**1,404,197.00**	
25										

Figure 17.2b   Revised elemental summary page for sites A, B and C (level 2).

URBAN SITE REDEVELOPMENT SITES A, B and C   **HOUSING ASSOCIATION**

Element Code	Element	%	Cost £/m²	Cost £/ft²	Quantity	Unit	Rate £	Sub Total	Total £	Notes
26	4.0 Internal Finishes									
27										
28	4.1 Wall Finishes	6.34	43.49	4.04				189,177.00	189,177.00	
29	4.2 Floor Finishes	4.33	29.72	2.76				129,268.00	129,268.00	
30	4.3 Ceiling Finishes	2.48	17.03	1.58				74,097.00	74,097.00	
31										
32	TOTAL	13.15	90.24	8.38				392,542.00	392,542.00	
33										
34	5.0 Fittings and Furnishings									
35										
36	5.1 Fittings	5.42	37.17	3.45				161,700.00	161,700.00	
37										
38	TOTAL	5.42	37.17	3.45				161,700.00	161,700.00	
39										
40	6.0 Services Installations									
41										
42	6.1 Mechanical Services	13.11	90	8.36				391,500.00	391,500.00	
43	6.2 Electrical Services	9.78	67.11	6.24				291,910.00	291,910.00	
44	6.3 BWIC with services	0.67	4.6	0.43				20,000.00	20,000.00	
45										
46	TOTAL	23.56	161.71	15.03				703,410.00	703,410.00	
47										
48										
U/D	4,350/m²	100	686.34	63.78				2,985,579	2,985,579	
									2,985,579	

Figure 17.2c   Revised elemental summary page for sites A, B and C (level 2).

Phase 3

URBAN SITE REDEVELOPMENT					HOUSING ASSOCIATION	
Summary	**Cost**	**Calc**	**Total £**	**Area**	**Cost £/m²**	**Cost £/ft²**
1 Enabling works	77,231.00		77,000.00			
2 Sites A, B and C	2,985,579.00		2,986,000.00	4,350	686.00	64.00
3 External works associated with units	288,216.00		288,000.00	4,350	66.00	6.00
4 Preliminaries	327,300.00		327,000.00	4,350	75.00	7.00
5 Infrastructure	809,084.00		809,000.00			
6 Contingencies	226,660.00		227,000.00			
Project	**4,714,070.00**		**4,714,000.00**			

Figure 17.2d   Updated summary of project cost (level 1).

this stage, probably incorporating a full critical path network. This would be intended for use as a production control document, and would be unlikely to be prepared until the design itself had been finalised and until the contractor was expecting to be chosen to construct it. It would almost certainly involve consultation with major sub-contractors, and the use of an appropriate computer package.

17.5   Cost reconciliation

If tenders have been called for and received, the tender which is most likely to be accepted should be reconciled with the cost plan. If, as should occur, there is little difference between the totals, this is still worth doing as there are quite likely to be large compensating discrepancies in the various elements and these may provide information for future use. The reconciliation is in fact a comparison of the final cost plan (as amended by cost checks) with a cost analysis of the tender. This also ensures that the cost analysis will be done at the earliest possible moment instead of being left until someone has time to do it. There may be all sorts of explanations for a considerable difference between the cost plan estimate and the actual cost of a particular element: perhaps the cost planner made a mistake; quite possibly the builder's estimator made a mistake or did a deliberate price adjustment. However, if there is a definite one-way divergence between cost plan and tender right through all the elements, it is likely that either the cost planner misjudged market levels or that the tenders themselves are abnormally high or low. If the cost planner feels that the latter is the case, then this could be pointed out. Note that permission is never required to analyse the contractor's tender for cost reconciliation purposes. It is only if the analysis is to be published, or if the individual prices are to be used for cost checking other projects, that the question arises.

17.6   Completion of working drawings and contract documentation

The completion of the working drawings and the contract documentation will normally mark the end of the cost planning process as such, although the vital task of cost control will have to continue throughout the project to ensure that the planned cost is achieved.

17.7   A critical assessment of elemental cost planning procedures

It has been suggested that elemental cost planning techniques will be justified if they consistently succeed in forecasting cost within a margin of 5% up or down, and indeed more recent research suggests that even this target may be over-optimistic, given the variability which exists in tender levels. It might well be asked whether all this work is worthwhile if, at the end of it all, the tender is liable to differ from the estimate by such an amount. Simple single price rate methods of estimating have often come far closer than this and involve much less work. The answer is threefold:

- Most budgets are flexible by at least 5% or, if they are not, it is simple enough to make the requisite modest alterations in the scheme.
- It must be emphasised that traditional single rate methods cannot be relied on to achieve anything like this standard of accuracy, and that for every spot-on square metre estimate there is another instance of such an estimate being up to 50% out. But where an estimate, however prepared, forecasts the cost to within 1% or 2% under traditional competitive tendering conditions, then luck as well as skill will have played its part in the result.
- Elemental cost planning achieves a balance and economy in the building which cannot be attained by any other method using traditional tendering procedures. A real-life example concerns a technical school which was estimated on a square metre single price basis and where the difference between estimate and tender was much less than 5%. Satisfactory though this was, nobody was able to answer a very simple question from the architect during the design stage as to whether that estimate would cover a certain type of wall cladding. There was nothing in the estimate with which it could be compared.

In view of the alleged unreliability of estimates prepared by single rate square metre methods, it is as well to explain again why these methods can safely be used for the preliminary estimate before the cost plan is prepared. It is because the cost plan will show up any error in the estimate before working drawings or BQs have been prepared, and the necessary adjustments can be made before any of the detail work is started. This is a very different situation

Phase 3

to discovering the error at tender stage when time and money have been spent designing and tendering for an impossible scheme in detail.

While elemental cost planning has made an outstanding contribution to the study, forecasting and control of building costs, it nevertheless suffers from a number of basic limitations which have prevented it from developing much beyond the stage it reached within the first few years. One might wonder whether it is doomed to be like the lead–acid electric battery which, although very useful for over a century, has never been capable of development into the kind of power-storage unit that the world is waiting for.

The limitations of elemental cost planning are as follows.

- Nobody has yet produced a set of elements, each of which performs a single function but which can be easily cost related, nor are they ever likely to. This means, among other things, that it is not really possible to compare the cost performance of the same element on two different buildings, nor, within one element, to compare two different technical solutions concerning one building. In order to attempt this task (which the QS is often asked to do), it is necessary to consider all sorts of extraneous matters, some of which are difficult to quantify. It might be thought possible to overcome this limitation by the use of sophisticated computer programs that would enable the interaction of the elements to be worked out exhaustively. This might well be possible, but it would be expensive and at the present time is not worth doing because of the next limitation: the costs that are being manipulated may bear no relation to fact; they are not really data in any scientific sense.

It is a waste of time and money to use sophisticated methods to manipulate inaccurate data, as is recognised in the computer world where the maxim 'garbage in, garbage out' is well known. In fact, such processing is worse than useless, because the fact that sophisticated analysis methods have been employed leads people to attach undue importance to the results. They would immediately recognise the original data as unreliable if it were presented to them in an unprocessed form.

It must be remembered that there is no pressure on contractors to insert rates in the BQ which represent carefully estimated production costs for each of the items, and there are many reasons (as set out previously) why they do not. The rates in the BQ are not retail prices for which the contractor is offering to execute individual pieces of work in isolation, but are merely a notional breakdown of the total offer. However, even if it were possible to compel the contractor to show true cost estimates, and to enforce standard practice in defining and allocating overhead costs, there would still be the difficulty of setting production costs against square metres of design elements, as so many of these costs (e.g. the tower crane) are not directly related to element unit quantity.

Even if these limitations could be overcome, the whole system suffers because under normal conditions there is no contractual commitment to the cost plan. The QS has no responsibility for the tender amounts and the contractor has no interest in the cost plan. The exercise thus has a great deal in common with weather forecasting. The cost planner is able to draw on a considerable body of past experience and can consider present trends, but is in no position to guarantee the result. The forecast will often be right, but if any unusual circumstances manifest themselves it can still be badly wrong.

The comparatively crude methods which have been set out in this and preceding chapters are usually adequate for forecasting within the wide limits of market pricing; any attempt to get much closer in these conditions is unlikely to be a worthwhile use of resources. Therefore if the cost planner cannot offer any kind of guarantee to the client, the latter is unlikely to agree to expenditure on expensive and sophisticated methods of control which are not in fact controlling anything but are just hopeful forecasts.

If a higher standard of cost control is required than is provided by the reasonable development of these traditional methods, then it will be necessary to sacrifice some measure of market freedom to obtain it, for example by involving the contractor at an early stage. There is just no way round this.

17.8   Key points

- Elemental cost planning should ensure that the tender amount is close to the first estimate, or alternatively that any likely difference between the two is anticipated and is acceptable.
- Elemental cost planning should ensure that the money available for the project is allocated consciously and economically to the various components and finishes.
- Elemental cost planning does not mean minimum standards and a cheap job; it aims to achieve good value at the desired level of expenditure.
- Elemental cost planning always involves the measurement and pricing of approximate quantities at some stage of the cost plan or cost check.

Further reading

Cartlidge, D. (2006) *New Aspects of Quantity Surveying Practice*. Butterworth-Heinemann, Oxford.

Jaggar, D., Ross, A., Smith, J. and Love, P. (2002) *Building Design Cost Management*. Blackwell Science Ltd, Oxford.

Smith, J. and Jaggar, D. (2006) *Building Cost Planning for the Design Team*. Butterworth-Heinemann, Oxford.

Phase 3

Chapter 18
Cost Control (2): Real Time

18.1 Why real-time cost control?

The methods of cost planning and control that have been covered so far are concerned the with planning and monitoring costs during the:

- investigation stage;
- planning stage;
- design stage;

and finishes at the point where tenders are received, or a contract entered into. During these stages nothing is irrevocable:

- Drawings can be revised.
- The scheme can be reduced in size or its configuration altered.
- The whole thing can be started again from scratch or postponed.
- The scheme can even be totally abandoned.

The cost of these will be (from the point of view of the total project cost) a negligible cost for abortive professional work. Even if the project has got as far as the submission of tenders by contractors, the scheme can still be radically changed or abandoned without any commitment to the tenderers or any need to recompense them for their trouble.

However, while the accurate forecasting of a tender amount and the signing of a contract for that sum (or an amended one where the forecast has not quite worked out) is of considerable importance, the matter does not end there. The final agreed cost, after both the project and the various financial negotiations arising from it have been completed, is rarely time same as the original contract amount. It is this final figure that the client has to pay and that is the building cost for which the finance will have had to be raised and serviced. On profit-oriented projects in particular, it is the figure on which the success or failure of the project will be judged.

In many cases the contract sum may be merely an intermediate stage in arriving at this final cost; a system of cost planning and control which stops at this halfway point will only be of limited use (and possibly not worth the money it costs). It is important, therefore, that the cost control process continues to the point of completion and handover of the project. This continuation is often termed post-contract cost control. It differs from design cost planning in many ways, and these will depend on the exact contractual arrangements, but two differences will be inescapable:

- The interests of a contractor or contractors now form an important addition to the considerations involved.
- Major expenditure is currently being incurred and future options proportionately reduced.

It becomes increasingly difficult to make major alterations of any kind at the stroke of a pen, once drawings and specifications are being translated into an expensive organisation of resources and finished work. Even delay caused by hesitation or reconsideration may involve the client in heavy costs.

To emphasise this similarity with real-time computer control systems, and because in some circumstances there may not even be a contract, the more appropriate term 'real-time' rather than 'post-contract' will be used here to describe this type of control.

The basis of real-time cost control is reporting at regular (often monthly) intervals or on a special occasion when a major decision has to be made. This cost control report will set out the client's likely final cost commitment in some detail and also the cost consequences of any remaining major options. In some instances this report may be made to the architect or other professional project controller, but it is preferable that it should be made direct to the client.

This service should be envisaged when the QS's fee is being negotiated, and should be written into the agreement. Its purpose is to:

- enable the client to budget for the likely expenditure;
- enable the cost effect of any major changes to be seen in the context of the project as a whole;
- enable avoiding action to be taken if the total cost appears to be escalating unduly.

One of the problems attached to this service is the extent to which sums of money should be kept in hidden reserve by:

- underestimation of savings;
- overestimation, or at least the pessimistic evaluation, of additional costs.

There is a natural tendency towards this type of caution. In its most extreme forms this may be taken to the extent of inflating the projections sufficiently to cover without disclosure any mistakes which may be made by the architect or QS, and which will still leave the client with a pleasant surprise when the final account is inevitably settled at below the forecast amount. Such conduct would be as unprofessional and unhelpful as inflating the quantities in the BQ for similar reasons.

Nevertheless, caution dictates that some allowance is made for inevitable minor additions which are not immediately apparent, or for things going slightly wrong. This particularly applies where final costs will be subject to negotiation or where the full consequences of decisions or events cannot be exactly foreseen. The reasonable judgement of these allowances, avoiding both excessive pessimism or optimism, is one of the major factors in making this a truly professional task. In particular, the QS who is successful in this role will develop a feeling for the average project, where not everything is going to go wrong, and for the occasional project where it would be wise to assume that it might.

18.2 The problem of information

In carrying out real-time cost control it is essential for the QS always to be fully informed as to what is happening and what is intended. Ensuring this may well be one of the more difficult parts of the operation. Ideally the QS should be part of the decision-making process, in which case the difficulty should not arise, but this is not usually the situation. However, merely sitting in the office waiting for information to come in is not good enough. Most contracts contain no requirement for communication to and from the contractor to flow through the QS's office. The QS is likely to be informed eventually of everything necessary for the final settlement of accounts, but this information will be far too late to be of any use for real-time cost control purposes.

It must be remembered that cost control requires a record not just of costs incurred to date but also of likely eventual cost commitments arising from:

- current proposals for variations;
- other decisions that have been taken by the design team and/or the client which will create variations and/or cause delay or difficulties in working;
- failure by the design team to meet deadlines for supply of information or for appointing nominated sub-contractors etc., which will have the same effect.

The QS must, therefore, instigate current awareness procedures, and in particular must:

- insist on seeing immediate copies of all official orders, drawings, and letters;
- attend all site meetings and generally look around the site to see what is going on;
- use this opportunity to find out what verbal instructions might have been given.

It has often been alleged that on many projects the official documentation is merely trying to catch up with the real but informal site communications system. The preparation of interim valuations for payment purposes provides an excellent opportunity to monitor what is actually happening and should always be used for this purpose. The maintenance and updating of real-time cost control records is an ideal within e-procurement systems, whether these are based on simple spreadsheets or specialist control systems.

Although, as shown above, many of the factors in real-time cost control are common to any type of contractual situation, it will be most convenient to look at the detail in the context of each of the different types.

18.3   Real-time cost control of lump sum contracts based on BQs

In the real-time cost control system for this type of project, the starting point will be the contract sum and the provisions the contract makes for amending it. These provisions will usually cover:

- The adjustment of provisional sums of money or quantities of work, which are embodied in the contract, in the light of the actual cost or quantity of work carried out. This type of provision is often made for work such as foundation excavation which cannot be foreseen exactly.
- Payment for variations and for additions to, and omissions from, the work.
- The adjustment of amounts included in the contract for work to be carried out by nominated sub-contractors and others, in the light of actual cost.
- The correction of errors in BQs and other contract documents.
- Adjustment for the effects of inflation (if applicable).
- Adjustment for the effects of new legislation.
- Compensation to the contractor for the cost of delays in the work caused by circumstances for which the client is responsible (or, in more generous contracts, by circumstances for which the contractor is not responsible). These circumstances may include delays caused by the instruction of variations or by delay in the provision of drawings, etc., or where the more generous provisions apply, delay caused by bad weather or industrial action.

In addition to the likely revised contract amount the QS's cost projections should include:

- updated additional amounts for professional fees;
- furnishings (unless it has been agreed to exclude these);
- any other items which the client wishes to consider as part of the total cost.

Phase 3

18.3.1   Adjusting the contingency sum

An item which can give rise to controversy is the adjustment of the provisional contingency sum, which is an amount included in the contract sum by the design team to cover unforeseen expenditure. The controversy centres around whether this sum is at the architect's disposal to cover design development (or more bluntly, mistakes) or is at the disposal of the client to spend on extra items. Some public authorities hold the second view so strongly as to demand the omission of the contingency sum as the first variation on the contract, so as to effectively take it out of the architect's control (while of course retaining it as a buffer in their own calculations).

Except where the client holds such extreme views it is probably best to show the contingency sum separately in the cost control report, deducting any incidental extras of the appropriate kind and carrying the balance forward to cover further similar extras which can occur at any stage of the project. In any case a provision of some kind must be made for contingencies during the remainder of the project. This sum is sometimes split into two, in order to make separate provision for:

- design development and unforeseen circumstances;
- additional items (i.e. extras).

18.3.2   Variations

A general difficulty facing the QS is that although the contractor is required by most contracts to give written notice to the architect of circumstances having arisen which will give rise to an extra cost, this notice is usually only required to be given within a reasonable time of the occurrence. In contractual terms this has to be interpreted fairly generously, and may well be too late for cost control purposes. The remarks already made about current awareness therefore apply strongly. A further difficulty under the JCT Standard Building Contract (although not always overseas) is that there is no requirement for the financial effect to be stated or agreed at the same time, and in fact this is positively discouraged both by the contract clauses and by traditional practice.

The QS's estimates, therefore, have to bear in mind that the figures may be subject to negotiation, and if the contractor also suggests a figure at this stage (although not legally bound to do so) the QS will have to guess the point between them at which a deal is likely to be made. For this reason the QS's real-time cost projections on this type of contract should never be shown to the contractor, since the QS may have prudently assumed a higher figure for extras than would at that time be conceded.

Although the contract usually contains a similar provision for notification of reduced or omitted work by the contractor, there is obviously less incentive for this to be done, and the duty is easily forgotten. The QS again must be in a position to see that omissions are duly notified and adjusted.

18.3.3   Nominated and named sub-contracts

The growing tendency for main contractors to attempt to transfer blame for project variations on to nominated sub-contractors (i.e. for delays) and thereby escape responsibility, has led to a sharp decline in the use of nominated sub-contractors. Naming has become far more common; this is where the employer provides the principal contractor with a list of firms to send tenders out to – this is the naming process.

Where a problem occurs, the first adjustment will be done when a nominated sub-contractor's tender is accepted for the work; this tender is likely to contain similar types of provision for further adjustment as the main contract does, and similar subsequent changes in cost are likely to occur which will have to be reflected in the cost-control report.

18.3.4   Inflation

On fluctuation contracts, the adjustment of the cost control report for inflation is a difficult one, particularly at an early stage of the project. It involves:

- a forecast of rises in building costs over the contract period;
- an estimate of the likely progress of the work in cost terms so that the forecast increases can be applied to an appropriate proportion of the cost.

There are three different figures to be calculated for inflation, and this will apply whether the contract adjustment is on the basis of the difference between ascertained actual costs of resources and costs at the time of tender, or whether it is done on an index-linked basis:

- increased cost incurred to date;
- estimated additional cost for remainder of contract of increases announced to date;
- allowances for effect of future increases.

The figure that the client will require for budgeting purposes is the total of the three, although the total should be split into the three categories to emphasise the progressively more conjectural nature of the estimates. Although it is theoretically possible for there to be a decrease in cost through the operation of this clause of a contract, there have been no known instances of a reduction in the contract sum from this cause in the UK during the 60 years prior to 1999.

Phase 3

18.3.5   Budgetary control versus best deal

A most important point regarding cost control of lump sum contracts is the perennial conflict between tight budgetary control and getting the best deal from market forces. The best budgetary control is achieved if a firm contractual commitment can be obtained at the earliest possible moment – in the case of extra work, for instance, before the order is given to go ahead. Even the JCT Standard Form of Contract makes some provision for this in its use of the words 'unless otherwise agreed' in its set of procedures for valuing variations. However, as previously mentioned, the contractor is under no obligation to agree a firm figure beforehand, with its attendant risk of underestimation, and may prefer to leave negotiation until the final costs are available.

It is up to the QS to advise the client on the advantages and disadvantages of agreeing extras at an early stage in the light of issues such as the client's particular budgetary problems and the apparent willingness or otherwise of the contractor to do a reasonable deal on this basis.

18.3.6   Cost reports

As an example of a system of cost reporting, two consecutive monthly cost reports on the early stages of a job are shown in Figures 18.1 and 18.2. In order to illustrate the difficulties involved, a fluctuations contract has been assumed. The references A.I.1, A.I.3, etc., refer to sequentially numbered architect's instructions.

There are many possible ways of setting out such reports, and readers may be able to devise what they consider to be a better format. What is most important is that any changes since the previous report should be clearly highlighted, either by the use of asterisks and notes, as in the example, or by simply giving the previous figures as totals. Using the latter system, the figures in the second example for 'net additions' could simply have been shown thus:

Net additions	£
As before	53,100
Extra work to entrance vestibule and staircase (A.I.6)	6,000
	59,100

Points where other people's tastes might differ include:

- The rounding off of estimated figures – a figure of £18,577 implies a standard of accuracy which £18,600 does not.
- The item 'sundry minor variations', which could have been shown in detail.

The view taken here is that this document is prepared to assist budgetary control, not to start a premature witch-hunt against the architect by highlighting minor problems which have arisen. Consideration of such matters

OFFICE BLOCK MIDTOWN

Cost Report No 2 10 Jun 06			**Contract sum**	£1,910,200

Net omissions   £

Partial substitution of Conglint panels for				
Portland stone (A.I.3)			27,050	
*Savings in foundations and piling			7,000	
			34,050	34,050
				1,876,150

Net additions

Decorations and partitions in offices (A.I.1)			44,000	
Carpeting in offices in lieu of wood block flooring			9,100	
			53,100	53,100
				1,929,250

Adjustment of prime cost sums   **Omit**   **Add**
£   £

Heating installation (Midland Heating Company Ltd)		7,500	
Electrical installation (Sparks and Co.)		975	
*Curtain walling (The Curtain Walling Co.)	2,065		
	2,065	**8,475**	
		2,065	
		6,410	
Profit and attendance		500	
Net addition		**6,910**	6,910
			1,936,160

**Adjustment for fluctuations (inflation)
including sub-contractors**

*(i) increase on work to date	2,000	
*(ii) allowance for effect of current increases on		
remainder of work	60,000	
*(iii) allowance for possible future increases (say)	50,000	
	112,000	112,000

Total estimated final cost 2,048,160

Amount included for contingencies

Contingency sum in contract	30,000
*Sundry minor variations (net extra)	1,000
Balance remaining	**29,000**

*Indicates changed or additional item since previous cost report
VAT and professional fees not included in above figures

Figure 18.1   Real-time cost report no. 2.

Phase 3

OFFICE BLOCK MIDTOWN

Cost Report No 3 10-Jul-06		**Contract sum**	£1,910,200
Net omissions		£	

Partial substitution of Conglint panels for		
Portland stone (A.I.3)	27,050	
*Savings in foundations and piling	6,652	
	33,702	33,702
		1,876,498

Net additions

Decorations and partitions in offices (A.I.1)	44,000	
Carpeting in offices in lieu of wood block flooring	9,100	
*Extra work to entrance vestibule and staircase (A.1.6)	6,000	
	59,100	59,100
		1,935,598

Adjustment of prime cost sums	**Omit**	**Add**	
	£	£	
Heating installation (Midland Heating Company Ltd)		7,500	
Electrical installation (Sparks and Co.)		975	
*Curtain walling (The Curtain Walling Co.)	2,065		
*Lifts (Ascenseurs Ltd)	1,995		
	4,060	**8,475**	
		4,060	
		4,415	
Profit and attendance		450	
Net addition		**4,865**	4,865
			1,940,463

**Adjustment for fluctuations (inflation)
including sub-contractors**

*(i) increase on work to date	6,500	
*(ii) allowance for effect of current increases on		
remainder of work	65,000	
*(iii) allowance for possible future increases (say)	40,000	
	111,500	111,500

	Total estimated final cost	**2,051,963**
Amount included for contingencies		
Contingency sum in contract		30,000
*Sundry minor variations (net extra)	2,200	
*Likely claim for delay in connection with GF		
walling layout (site meeting 05.07.2006)	12,000	
	14,200	14,200
	Balance remaining	**15,800**

Indicates changed or additional item since previous cost report
VAT and professional fees not included in above figures

Figure 18.2   Real-time cost report no. 3.

can usually be better left until the final account, when these problems can be seen in the context of the finished project.

It will be noted that the report takes account of variations ordered, and quotations accepted, for work that is not yet done or even started. The possible claim for delay has been charged against the contingency fund, which will not be able to stand many such demands upon it.

Finally, it must again be emphasised that under normal contractual arrangements, the contractor has no duty to participate in any cost control scheme or cash flow control scheme of the client, and there are sound commercial reasons for refusing to do so. Any such requirement, to be binding, must be written into the form of contract with legal advice and cannot simply be stated in the preliminaries of the BQ.

18.4   Real-time cost control of negotiated contracts

A negotiated contract provides excellent opportunities for cost control. Indeed, this is one of its principal advantages for which, as usual, some measure of market-force benefit may have to be sacrificed. From the moment of involvement in the scheme, the contractor should be committed by the architect and QS to participate in the cost control process both during the design stage and subsequently. This requirement should be made clear during the early stages of negotiation and duly written into the contractual arrangements, together with a provision for prior negotiation and agreement of extra costs.

18.5   Real-time cost control of cost reimbursement contracts

The situation here is quite different. There is no contract sum as a starting-off point and no possibility of contractual commitment to estimates of original or extra costs, so that real-time cost control by the QS becomes of fundamental importance. On the other hand, there is no price mechanism to cloud the issue and thus no valid commercial reason why the contractor's own estimates and costings should not be made fully available to the client's representatives on a so-called open book basis.

It is preferable that cost control of the project should be carried out by the QS co-operatively with the contractor, since two heads will certainly be better than one and full advantage can be taken of the absence of commercial secrecy – one of the main benefits of this type of contract. At any point in the project the QS's cost control report is likely to be based on:

■ the QS's original estimate of cost, which should have incorporated an allowance for inflation;

Phase 3

- the QS's estimate of cost of variations, which also should have incorporated an allowance for inflation;
- any adjustment of estimated cost for actual cost of work completed or expenditure committed, where this is different to what was envisaged;
- any adjustment to the estimated cost of current or future work in the light of this experience.

Towards the end of the project the QS will probably switch to a simpler system involving the ascertained cost of work completed plus the estimated cost of the remaining work required.

The QS's original estimate is likely to be based on approximate quantities, and so are the QS's estimates of the likely cost of variations. However, in both cases and especially the latter, a resource-based approach may be adopted in co-operation with the contractor.

Even where approximate quantities are used, however, all estimates should be discussed with the contractor before being given to the client. The purpose of this co-operation is:

- to ensure that the QS has not made any false assumptions;
- to enable a positive contribution to be made where the QS thinks that the methods or equipment which the contractor is proposing to use are unnecessarily expensive or inefficient.

It must be remembered that the builder's usual incentive to cost cutting will be absent, since it is the client's money, not the contractor's, which is being spent. In such an event there may have to be a meeting between the architect, QS and contractor to decide what is to be done; this could in some instances involve minor redesign to enable something to be built more efficiently. This aspect of cost control is an important part of the QS's contribution.

A most difficult part of the whole task will be the replacement of estimated costs by actual ascertained costs. Under some forms used for cost reimbursement contracts, the contractor's responsibility for costs is limited to providing a periodic statement of total labour and plant costs to date, and a list of invoices received. Such a document is totally useless for cost control purposes both in format and in timing, since invoices are often not available until weeks or even months after the cost commitment has been incurred. The QS must, therefore, ensure that the contractual arrangements provide for:

- facilities for the client's representatives to participate in the contractor's estimating and work planning procedures and in the appointment and control of sub-contractors;
- the use of the contractor's own detailed costing system for the benefit of the client's representatives, or alternatively the installation of a suitable system where the contractor's system is unsuitable or non-existent;
- early disclosure of all papers concerned with ordering, costing, delivery or pricing.

It must be made clear prior to the contractor's appointment that these requirements exist, and they must be incorporated into the contract. The arrangements must not rely on goodwill, since at the least they are inconvenient to the contractor and are certain to involve some expense. This will need to be reflected in the contractor's fee.

The costing system will have to separate the costs in respect of specific parts of the work which correspond to identifiable portions of the QS's estimate, otherwise any form of reconciliation between costs and estimate will be impossible until the whole job is complete. This means in turn that the QS's estimate will need to be structured suitably with this need in mind, an elemental basis being quite a good one. However, work executed by different trades at different times should not be telescoped into a single priced item (for example roof carcassing and roof tiling).

A further point to watch is delay in reporting; costs of plant and materials may take some time to work through the contractor's office system and a method has to be devised for overcoming this delay. For example, materials could be priced from delivery dockets – which should be instantly available. Reconciliation of delivered materials with measurements of major items of work should also be carried out to ensure that there are no unexplained differences up or down due to theft, excessive wastage or clerical error.

Reconciliation of allowances for inflation poses some further problems, since these will often not be separately identified in the ascertained costs of resources. In most cases some form of approximation will be adequate in showing how much of the cost of some particular operation should be set against the allowance for inflation, since expensive and complicated bookkeeping arrangements which do not affect the total to be paid are unlikely to yield any commensurate benefit to the client. Under this type of contract the client is always going to have to meet the cost of inflation. Allowances for future inflation may need to be revised from time to time, as was the case with cost control of fluctuating lump sum contracts.

18.5.1 *Cost control of sub-contractors on cost reimbursement contracts*

It is necessary to watch closely the arrangements for sub-contractors. The builder, in collaboration with the QS, will be seeking separate tenders from all the specialist sub-contractors, and each of these has to be dealt with separately. It might seem a good idea to try and obtain the most important tenders at more or less the same time, since it would then be possible to consider their total effect on the cost of the project. However, this is rarely advantageous in practice since cost reimbursement is most likely to be used on projects where design and construction proceed more or less simultaneously, with the design of later stages of the work going on while earlier stages are being executed.

Specialist work should not be tendered for until the design work on that particular specialism has been completed, otherwise unnecessary variations are likely (and this will be no help whatever to cost control). So it is more usual to carry out the cost control of such projects with the budgets for the specialist works contractors being turned into firm prices one by one. This means that it may often be possible to appoint the specialists on a price basis rather than a cost basis, as the work may be more clearly defined by the time they are involved. For example, the structural items to which finishes are to be applied may actually have been built.

Competitive tendering for sub-contracts might well be possible; again if the QS does not look after this, nobody else will. Without such control it is much easier for the builder just to ring up a friendly sub-contractor and make the necessary arrangements.

Leaving the calling of tenders for specialist work until they are needed also has the advantage that it is possible to amend the budgets (and specifications) for later specialists if there are cost overruns on the earlier ones, so that tenders can be sought on the revised basis. This gives better cost control than having to amend works contracts to obtain savings, as may happen if tenders have been called too early.

18.5.2   Difficulties with combining price-based work and cost reimbursed work on the one project

Great care is necessary when work on a price basis and work on a cost reimbursement basis are included in the same contract. This is only likely to be satisfactory where one or more of the following circumstances exists:

- The cost reimbursement work is only a negligible proportion of the whole (e.g. incidental dayworks on a lump sum contract).
- The two types of payment apply to two separate organisations (e.g. the main contractor and a sub-contractor).
- The cost reimbursement work is carried out by a separate gang of people using special materials.
- The two types of work are carried out at different periods or at different sites.

Unless at least one of these conditions is fulfilled, there will be a tendency for labour and/or materials used in the price-based work to be charged to the cost reimbursement work, thus being paid for twice. This is almost impossible to check without continuous monitoring. An example of the type of project where this could happen would be an extension to an existing building where the extension was to be paid for as a lump sum but the alterations in the existing building were done on a cost reimbursement basis. It would be very difficult to ensure that operatives' time was charged to the correct part of the work if it was all proceeding simultaneously, and even more difficult to check afterwards that this had been done. Such hybrid arrangements should therefore be avoided.

It will have been seen that the real-time cost control of cost reimbursement contracts involves the skilful combination of contractors' costing systems and client budgeting and cost strategies, and is likely to be expensive. There is no halfway house, however; cost control of this type of project can only be done properly or not at all. Clients must make up their minds about this at the beginning, under appropriate advice. The internal cost control of a package deal contract would have many features in common with the above procedures.

18.6   Real-time cost control of management contracts

In many ways the procedures for management contracts would resemble those on a cost reimbursement contract, but probably at a more strategic level since the work will be undertaken using a number of different contracts, any of which in turn may require its own cost control procedures of the appropriate type already discussed. The remarks made above concerning sub-contracts on cost reimbursement work will apply to all the works contracts.

On a really large project it is not necessary that the whole of a particular work package must be let at once, and there would be advantages in letting a package in several parts at different times if only some areas of the project have been fully designed. Nevertheless, the general pattern set out in sections 18.3 and 18.4 will also apply to the project as a whole.

18.7   A spreadsheet cost report on a management contract

Figure 18.3 shows a financial report (No. 5) on a sports training centre. Where reports such as this are being generated by computer, they are likely to be updated as changes occur rather than on a monthly or other periodic basis. Spreadsheet output should always be checked arithmetically, as a matter of good practice. Although computers do not make mistakes in carrying out the calculations they have been instructed to do, it is always possible to give them the wrong instructions in compiling a spreadsheet.

Some points to note in this example (Figure 18.3):

- The figures exclude VAT and professional fees.
- The 'Estimated value' column represents the cost plan, against which the client's budget has been set. The client may be holding additional project contingencies, not displayed here, as on the open book basis the contractor has input to, and receives, this summary.

Phase 3

Trade package	Package reference	Estimated value cost plan (A)	Transfers within cost plan	Adjusted cost plan (A2)	Package sum (B)	Confirmed instructions (C)	Anticipated instructions (D)	Anticipated final cost E = (B+C+D)	Difference with adjusted cost plan (E−A2)	Package procurement status
Substructure and groundworks	1,100	282,000.00	81,313.00	363,313.00	370,086.07	3,200.00	5,300.00	378,586.07	15,273.07	Contract let and on site
Steelwork	1,200	275,000.00		275,000.00	258,921.00	1,000.00	15,000.00	274,921.00	−79.00	Contract let and on site
Pre-cast floors and stairs	1,300	44,500.00		44,500.00	42,194.00			42,194.00	−2,306.00	Contract let
Roof system and rainwater goods	1,400	199,000.00		199,000.00	192,545.00			192,545.00	−6,455.00	Out to tender
Rooflights, patent glazing, windows and curtain walling	1,500	110,000.00		110,000.00	110,000.00			110,000.00	0.00	Out to tender
Masonry	1,600	109,000.00	13,600.00	122,600.00	122,600.00			122,600.00	0.00	Out to tender
Internal partitions	1,700	35,000.00		35,000.00	35,000.00			35,000.00	0.00	
Joinery, doors and ironmongery	2,000	100,000.00	12,000.00	112,000.00	112,000.00			112,000.00	0.00	
Plastering/screeds	2,100	60,000.00		60,000.00	60,000.00			60,000.00	0.00	
Decorations/painting	2,200	55,000.00		55,000.00	55,000.00			55,000.00	0.00	
Floor finishes	2,500	145,000.00		145,000.00	145,000.00			145,000.00	0.00	
Ceilings	2,700	45,000.00		45,000.00	45,000.00			45,000.00	0.00	
Mechanical and electrical installations	3,000	800,000.00	−81,313.00	718,687.00	718,687.00			718,687.00	0.00	Out to tender
Builder's work in connection with services	3,500	30,000.00		30,000.00	30,000.00			30,000.00	0.00	
Toilet cubicles, back panels and vanity units	4,000	40,000.00		40,000.00	40,000.00			40,000.00	0.00	
Loose fittings, furniture and equipment	5,000	270,000.00		270,000.00	270,000.00			270,000.00	0.00	
External works and landscaping	6,000	220,000.00		220,000.00	220,000.00			220,000.00	0.00	
Sub total		**2,819,500.00**	**25,600.00**	**2,845,100.00**	**2,827,033.07**	**4,200.00**	**20,300.00**	**2,851,533.07**	**6,433.07**	
Contractor's percentage for overheads and profit @ 4%		**112,780.00**	**1,024.00**	**113,804.00**	**113,081.32**	**168.00**	**812.00**	**114,061.32**	**257.32**	
Contractor's preliminaries		291,556.16		291,556.16	291,556.16		20,000.00	311,556.16	20,000.00	
Package interface (contingencies)		189,900.84		163,276.84	182,066.45		−45,480.00	136,586.45	−26,690.39	
Total project cost (£)		**3,413,737.00**	**26,624.00**	**3,413,737.00**	**3,413,737.00**	**4,368.00**	**−4,368.00**	**3,413,737.00**	**0.00**	

Figure 18.3 Bloggsville United Training Centre Financial Summary No. 5.

- Figures in the 'Transfers within cost plan' column indicate where amounts have been transferred between packages in the cost plan, for example the transfer of drainage from 'Mechanical and electrical installations' to the 'Groundworks' package. These changes result in the adjusted cost plan, which should always total the same as the cost plan, although constituent parts will vary.
- In the 'Package sum (B)' column, the figures are the same as in the adjusted cost plan unless tenders have been received and sub-contractors appointed, when the revised figures are inserted (in italics, to make clear that this has been done).
- Columns C and D represent variations to the package values, whether formally instructed or merely anticipated.
- The most important column is the 'Anticipated final cost' (B + C + D) column, which indicates where the project is heading financially, both on a package-by-package basis and overall. It will be seen that this total is at the moment the same as for the 'Adjusted cost plan', but note that this has been achieved by a reduction in the 'Package interface (contingencies)' figure. Zero in the 'Difference with adjusted cost plan' column confirms that the anticipated final cost does not exceed the budget.
- The main contractor's (construction manager's) costs and the preliminaries costs are included as a percentage of the package costs. (These were determined by competitive tender.)
- Note that a sum of £20,000 is included for estimated preliminaries costs associated with an anticipated two-week extension of time relating to problems encountered with the substructure and foundations.

18.8 Control of cash flow

So far in this chapter we have ignored the phasing of expenditure, except in so far as the effect of inflation on costs is concerned. In practice, client organisations will be concerned about the sums of money they will be called upon to pay at any given time, and may require a cash flow estimate from the QS before signing the contract, to ensure that they will be able to meet these demands. They will need this in order to make adequate arrangements to raise the money, irrespective of whether they are operating in the profit sector or in the social sector. Quite clearly they are unlikely to be able to place the whole contract amount in their bank trading accounts at the beginning of the project, ready to be drawn on as required over the contract period. Even if they were theoretically in a position to do this, they would still have more profitable uses for the money in the interim, which would probably mean that it would not be instantly available.

The cash flow estimate will require to be revised by the QS from time to time, preferably at the same intervals as the real-time cost control reports.

Phase 3

These revisions will take into account not only any changes in total cost but also the extent to which some aspects of the programme are running faster or slower than anticipated, or indeed whether the whole programme has slipped. In many cases, a periodic report of this kind may be all that is required, giving the client an updated warning of cash requirements. As well as performing this function, such a report may also be very useful in monitoring programme slippage, because the level of cash flow required to complete on time may be seen to be quite impossible in the light of experience to date.

However, in some instances a client will need not merely a report of what is likely to happen but some positive form of control. In the case of a cost reimbursement contract it should be possible to order an increased or decreased tempo of work, and few problems arise. In management projects, also, the placing of contracts can be deferred or work accelerated, but serious difficulties occur in the case of lump sum BQ contracts on the standard form. Once a contract has been signed using the British JCT Standard Building Contract, the contractor is not answerable for the phasing of expenditure. The contracting firm does not have to provide an estimate of client's cash flow, and even if they agree to do so as a favour they cannot be held to perform in accordance with it. Their only duty is to proceed 'regularly and diligently' and to complete by the appointed date or a later date that may be agreed on because of delays.

It is in the contractor's interest to obtain high levels of payment in the early stages of the project in order to finance the later stages and so the prices for earlier parts of the work may be inflated, and early deliveries of materials arranged, to achieve this. Therefore if a positive form of cash flow control is required, a suitable provision will have to be made in the actual contract (with legal advice), and not merely in the BQ or in correspondence. The contract might state a month-by-month cash flow programme with provisions that the contractor would not be reimbursed ahead of this programme if the work was carried out faster than scheduled, and if expenditure fell behind the programme, there could either be a provision for damages to be ascertained or, more constructively, the difference between the programmed amount and the actual expenditure could be paid into a trust fund. It might be thought that there would be no damage to the client's interests if expenditure fell behind the programme (provided the job was finally completed on time), but in the public sector and in some large private corporations, construction is often funded on an annual basis and amounts unspent at the end of the financial year may be permanently lost to the authority or department concerned.

In the profit sector, however, quite apart from their general wish to reduce the cost of financing their cash flow, clients may find it difficult to raise the total cost of construction until nearer the time when the building will actually be producing an income. They might therefore arrange for the contractor to bear part of the building cost during construction or even to take a share in the risk of the development. In such cases it would be essential for the arrangements for funding during the progress of the works to be formalised in the contract.

18.9   Cash flow control of major development schemes

Major development schemes may extend over many years and may include a large number of building contracts entered into at different times. Cash flow control is necessary in the running of public development schemes to ensure that money is available to meet outgoings. If a development scheme is funded on an accrual basis (that is, unspent funds which have been allocated to it can be carried forward from year to year), and if it is not possible to lend out money at a profit (perhaps because the authority does not have the necessary powers), then nothing more than a passive control of cash flow will be required. The authority will simply need to order its financial affairs to suit what is happening.

However, public funding is often on a yearly basis, in which unspent funds in one sector are not carried forward for the benefit of the project or authority for which they were allocated, but are used to meet overspending in other sectors or applied to a reduction in government expenditure. In addition, most major building projects involve an unavoidable commitment to expenditure for some years ahead, although the authority's budget for these years may be unconfirmed when the commitment is entered into.

Trouble may then occur if major construction projects fall behind schedule, so that much of the expenditure which should have taken place in the current year becomes a commitment in future years, and the cash balance which is left at the end of the current year will be lost to the authority with no certainty of picking it up again when it is needed. It is, therefore, necessary to manipulate cash flow in the light of changes so that:

- The year's allocation of funds is completely but productively spent and a realistic revision can be made of cash requirements for ensuing years.
- The ongoing programme can be hastily amended if funds are reduced (or increased!) in any particular year.

This could be called active or positive control of cash flow.

In any authority's development budget the expenditure will fall into three categories, and in each of these categories the discretionary factor in expenditure is progressively increased.

Old projects

These are projects which have been completed but for which full payment has not yet been made (possibly depending on the outcome of litigation, arbitration, or negotiation). Once payments are due they will have to be met.

Phase 3

Current projects

It may be possible, and worthwhile, for an authority to order a speed-up or slow-down of current projects in order to change its cash commitment during the year, even at the expense of an eventual increase in total cost. What more usually happens, however, is that circumstances beyond the authority's control cause such changes, and the cash flow budgeting has to be revised to suit. Almost invariably, expenditure tends to fall behind estimate because of delays in progress, although this may sometimes be counterbalanced to some extent by extras and by inflation.

It might be thought that if an authority had a large number of projects on hand (such as a programme of ten schools), the differences would tend to average out. But this tends not to happen in practice because the causes of delay are often national or regional in character – weather, industrial disputes, labour shortages, etc.

New projects

It is these which offer the greatest scope for control. If the start of a two-year project worth £10,000,000 is brought forward or put back by one month, the effect is likely to be an increase or decrease of £400,000 in the current year's cash flow, and correspondingly greater for longer periods than one month. There are two major difficulties however:

- The starting of a major project may have a comparatively small effect in the year in which the go-ahead is given (but it will then join the ranks of the current projects and will be an inescapable and major commitment for the following two years or more).
- Expenditure on projects tends to follow the well-known S-curve where the rate of spending is at its greatest in the middle period of the project, with a comparatively slow build up in the early stages.

In many ways the best way to control the cash flow situation is by having a number of smaller short-term projects ready to roll. These can be largely completed during the financial year in question, even if the order is not given until the year is quite well advanced.

An obvious method of disbursing funds on a project that is falling behind programme is to pay for the work, or for materials, ahead of the legal liability to do so. There are obvious dangers in doing this and a public auditor would be unlikely to accept a straightforward attempt to do it. It might, however, be possible to pay money into a joint trust fund.

18.10   Control of short-term cash flow

Up to now we have been considering cash flow as though we are simply concerned with the year as a whole, but this is rarely the case. Funds are not normally paid across to an authority in a single sum at the beginning of the year, nor are they paid as each cash commitment arises, but are usually paid in quarterly instalments or something of the kind. Clearly therefore, the authority must avoid too much cash expenditure in the early part of the year since funds will not be available; if it is empowered to lend surplus short-term money profitably, it may be able to arrange matters so that it can do this. With quarterly funding, even where there is a smooth monthly outgoing throughout the year, the money for the third month of any quarter will be available for alternative use during the first two months.

18.11   Payment delays on profit projects will be to the client's advantage

Although the above section has been written in the context of public or social development, much of it is equally applicable to profit schemes, except that the problem of spending all the money by the end of the year is not so likely to occur. Any pressure is likely to be in the opposite direction; the question of interest on money borrowed becomes of paramount importance so that any pushing back of payment dates will be welcomed, so long as final completion is not held up. It is not unknown for this to happen in the public domain.

An overseas regional government, which was anxious to minimise the effects of a slump in the local building industry but had spent all its funds for the year, let contracts on the basis that no payment would be made until the following financial year, so that for the first six months or more the contractors were totally financing the programme. However, as would be expected, the tenders reflected this additional burden.

18.12   Cost control on a resource basis

As with real-time cost control on any other basis, cost control on a resource basis fundamentally depends on swift and reliable reporting of what is actually happening, and the rapid reconciliation of this with the control document.

The cost controller will be concerned with two different aspects. The first is the overall time for the project. The preparation of the network will have involved a large number of assumptions about the time to be taken by the various activities, and revisions must be made in the light of actual progress. In this context it is only the activities on the critical or near-critical paths which have to be considered, but any substantial change (for better or worse) may cause a change in the critical path itself and in the duration of the works. Decisions will then have to be made as to whether a different duration is acceptable, or whether the time for remaining activities must be adjusted somehow. In both cases cost will be substantially affected.

Phase 3

The second aspect is the cost of individual activities. A costing system has three objectives:

- to measure actual expenditure against estimated cost;
- to indicate whether performance needs to be improved;
- to provide feedback which will improve future estimating performance.

Costs are recorded under cost centres. These can be anything from the whole project to a tiny part of it, but to meet the above objectives costs need to be recorded under headings which can be identified with sections of the estimate. The estimate in turn needs to be prepared in a suitable format to achieve this.

There is a need to establish a level of breakdown which is neither so fine as to make the recording and allocation of labour and material costs unduly detailed, nor so crude that the cost centres are too broad to mean anything. Individual BQ items are normally at too fine a level; a whole trade section would be too crude. A single activity, or a group of activities, from the network is usually a suitable basis.

Costing of site work has two main difficulties that distinguish it from normal production costing as practised in factories and as described in most textbooks on costing. Both difficulties spring from the one-off nature of most of the work in terms of the finished product and of the conditions under which it is produced.

The first problem is that the usual costing system of standard costs does not really apply. A standard cost is an ascertained cost of carrying out an operation, which can then be built into an estimating system and which should normally be achieved in future – any substantial deviation requiring to be investigated. Machining operations in a woodworking shop can be costed on this basis: once a standard sequence of events has been carried out on a machine a number of times (inserting the work, operating on it, withdrawing it), there is no reason why the same operation should not always cost exactly the same in the future.

Site operations are not only less standardised than this, but the position of the work, weather conditions, etc. vary continuously. It is very difficult to get a run of identical work sufficient to establish standard costs and still have enough similar work ahead to make it worthwhile to apply them. Certainly it is not usually possible to apply such data from job to job, as a joinery works or engineering works would do.

The second difficulty is that of actually recording and processing costs of labour and materials. When operatives are dispersed about the site, it is almost impossible to record accurately what they are doing except in fairly general terms, unless a quite unacceptably expensive staff is engaged for the purpose. Many sophisticated costing systems have failed through the time sheets being filled in from memory on a Friday afternoon in the foreman's office.

Phase 3

Materials to some extent are an equal problem, as nothing like a factory control system is operated on most sites when materials are drawn from stock. The invoice is often the main weapon in materials cost control, so it is convenient if the cost centres are sufficiently large and identifiable for particular invoices to be identified with them. However, some contractors' systems do not bother overmuch about materials. As already seen, these are much less liable than labour and plant costs to deviate from estimate, given a reasonable level of management. This is another advantage, from the contractor's point of view, of not using all-in labour and material rates.

All this may sound very defeatist – difficulties exist to be overcome, not to be excused. They can indeed be overcome, but except in some limited circumstances (such as repetitive house building) nobody has found that the cost and trouble of doing this have produced economically worthwhile results, because of the already mentioned difficulty of deriving standard costs from the data for reuse. All one is likely to get is an expensive post-mortem report. A further point to remember is that some of the people on a site, particularly labourers, are doing work that has to be done but which may not have been identified as an operation in the network – sweeping up, carrying odd materials, attending on the clerk of works.

It is, therefore, preferable to use fairly broad cost centres – 'fixing formwork to second storey' – and devote one's attention to getting the feedback rapidly and with a fair degree of reliability. It should then be possible to measure actual expenditure against estimated cost with reasonable accuracy, and also to a somewhat lesser extent fulfil the other two objectives of a costing system.

In the reconciliation of costs and estimate, the difficulties concerning inflation and alterations to the scope of the work will again be encountered. However, the inflation problem can be minimised by working in operative-hours and machine-hours rather than money, for comparison purposes. A contractor's costing system will in any case need to highlight the cost of alterations, because very often these will form an extra to be charged to the client. It will also be necessary to ensure that the cost performance of sub-contractors is kept up to date in order to arrive at the projected total for the project, complicated by the fact that these may often be on a price, rather than a cost reimbursement, basis.

18.13 Key points

■ Real-time or post-contract cost control differs from design cost planning in many ways, and these will depend on the exact contractual arrangements, but two differences will be inescapable:
   □ The interests of a contractor or contractors now form an important addition to the considerations involved.

Phase 3

- ☐ Major expenditure is currently being incurred and future options proportionately reduced.
- ■ However, today most clients are even more interested in the final figure than in the amount of the tender. Real-time cost control depends above all on the QS/cost planner being kept informed as to what is going on. If necessary, requirements to this end must be included in the contract arrangements.
- ■ Regular reports on the financial progress of the job must be submitted to the client during the progress of the work, so that any necessary steps to prevent cost overruns can be taken before it is too late.

Chapter 19
Cost Planning of Renovation and Maintenance Work

19.1 Introduction

The sustainability paradigm has irrevocably changed the way that clients procure work and the response to this from the construction industry. This, in combination with the heavy demand for development opportunities and increasing land costs, has led to unprecedented growth in refurbishment work, particularly in the residential sector. Liverpool, for example, before becoming European Capital of Culture in 2008, has experienced unprecedented levels of work on complex, conservation-based refurbishment projects including façade retention and complete structural renovation.

The archetypal exemplar of this is the refurbishment/construction of St Pancras International, the new Eurostar rail terminal in London. Prior to the redevelopment, St Pancras comprised two of the most important Victorian era structures: the train shed (completed 1868 and the largest single-span structure of that time) by engineer William Henry Barlow, and the St Pancras Chambers, formerly the Midland Grand Hotel (constructed between 1868–76) by the renowned architect George Gilbert Scott. The project involved the complete renovation of the train shed and the hotel buildings, plus the construction of new platforms and a permanent way to link the station with other National Rail networks. It is truly one of the most ambitious schemes to grace the UK and is overdue, given that the Midland Hotel has stood empty since the demise of the London Midland and Scottish railway company prior to the nationalisation of British Rail.

Historically, this approach has not been the norm; 40 years or more earlier it was perhaps habitual to think of building schemes in terms of greenfield projects, that is to say projects erected on an open site (one which had never been built on previously and had none of the issues associated with this, such as contaminated land). The classic cost planning techniques that emerged in this era tended to be oriented towards such projects as the norm. This was rather a convenient assumption of course; it avoided many complications but it is doubtful whether it was ever completely appropriate. Today it is clearly untrue, and a very substantial part of the current UK construction programme comprises:

- refurbishment;
- renewal work;
- life-cycle replacement, major maintenance and repair work programmes.

As the last two categories nearly always include some element of improvement, and as refurbishment will usually involve renewal and repair, they can all be considered together for the purposes of cost planning.

Some of the points considered here will also apply to a large proportion of the so-called new work being undertaken, which involves building on very restricted urban sites between existing buildings with complicated access and close proximity to the general public.

19.2   The appropriateness of lump sum competitive tenders

The uncertainties associated with many major refurbishment projects usually dictate that it is unsuitable and perhaps inadvisable to undertake the scheme on the traditional basis of lump sum competitive tenders for the whole of the works. Lump sum competitive tenders usually:

- require that the work to be done can be accurately foreseen;
- formalise an adversarial relationship between contractor and client's representatives, which is wholly inappropriate for this type of work.

There are exceptions to this, of course, but not very many. The major exception is where a series of similar buildings are being refurbished (on local authority housing estates for example), and experience with the first few projects by both design team and contractors will enable a schedule of rates, or even a BQ, to be prepared and priced. But normally collaborative methods of procurement strategy will be applied, either on a cost-plus basis or perhaps through management contracting, partnering or a bespoke approach negotiated between the parties.

As part of the Constructing Excellence demonstration projects, Oldham Metropolitan Borough Council Housing Department entered into a partnering agreement for the delivery of £10 million of external refurbishment work over a three-year period. The works were procured in accordance with EU procurement directives and through a two-stage evaluation method based on an 80/20 quality/price matrix[1]. At Benwell in Newcastle upon Tyne, another demonstration project involved the refurbishment of 163 homes (to include homes and flats) together with a community centre and associated environmental works. This project was procured using a three-stage selection procedure to obtain partnering contractors and sub-contractors[2].

19.3   Elemental cost planning inappropriate

Whilst it is possible to set out any estimates in an elemental format, the normal process of elemental cost planning is not really appropriate. Principally, it will not be practicable to produce the element-by-element cost comparisons with other projects, which lie at the heart of this technique. The costs will depend on the state of the individual building in relation to what is proposed within the refurbishment works; thus comparisons with other projects on an elemental basis would effectively be futile. In essence, the cost planning and control of refurbishment work should start from first principles.

19.4   Conflict of objectives

The most important initial step is to recognise the conflict of objectives, which is inherent in refurbishment work. There is of course a conflict between cost, time, quality and size in carrying out straightforward new work, but on such projects the client's interest is simply to optimise the scheme in these terms, and elemental cost planning allows this to be done. However, in refurbishment work these basic objectives are usually supplemented, or even outweighed, by major secondary objectives.

Occasionally a single objective, such as speed of completion, may be strongly dominant and this is little different to new work. But more often, two or more conflicting objectives are perceived as dominant; for instance, continuing occupation required by a user-client may hinder the achievement of speedy completion, which is the priority of the owner-client. Even the emergence of a single dominant secondary objective, such as safety of the structure or of the public, may have consequences that cut across normal procedures, particularly procedures for efficient construction methods or financial control.

The likely extent of conflict can only be identified after a thorough assessment of risks and objectives, and the trade-offs between them. Frequent reviews of the relative benefits compared with the likely cost, time, and quality implications will probably be required.

19.5   Risk and uncertainty

Uncertainty is a major characteristic of refurbishment work. Generally there is a high level of uncertainty, not only in the client objectives but also in

Phase 3

available physical data. This is generally due to the rather bespoke nature that refurbishment work assumes. Such uncertainties will probably extend into the construction period, with a high likelihood that unforeseen events will occur. This is likely to have a consequence on cost, so careful management is essential. Problems due to uncertainty may be mitigated by:

- allowing a longer lead-in time;
- assembling the work into discrete packages;
- seeking advice both of the principal contractor (or construction manager) and any specialist sub-contractors and involving them within the project as early as is possible.

Even so there will remain a need for client, designer and constructor to respond quickly to the discovery of previously unknown features of, or defects in, the existing building.

19.6 Safety

Safety is perhaps the most dominant feature of refurbishment work, and the problems in this area are intensified by the general uncertainty already referred to. Safety affects:

- operatives;
- the users of the building;
- the general public;

and may relate either to the safety of the structure or to the safety of the operations.

Clearly, where safety and cost are in conflict, safety must be paramount, so it is vital that safety issues are identified at an early stage if estimates are to be relied on. However, safety is not discretionary so there are no real choices to be made as regards the level of safety, only the means of ensuring it.

19.7 Occupation and/or relocation costs

Another major issue does, however, involve fundamental choice, and that is the extent to which (if at all) the building should remain in use during the works – often referred to as partial possession[3]. Here we can be dogmatic. Just as real estate agents are supposed to believe that the three most important things about a property are location, location and location, so the three most important decisions in connection with refurbishment are undoubtedly occupation, occupation and occupation!

The effects on costs, programme and safety if the building remains in use are so enormous that a client who wishes to do this should be encouraged to think again – and again – to see if some other alternative can be found. It is incumbent on the cost planner to keep pointing this out. A problem in this regard is that, seen in advance, the difficulties of moving out seem much greater than the difficulties of staying put. Staying put surely only means having to put up with a little dirt, noise and inconvenience for a year or so, whereas moving out will involve:

- finding alternative premises;
- moving into them, and out again at completion;
- getting customers and staff to come to terms with the new location;
- reprinting stationery and literature, etc.

Seen in advance, moving out appears to be far more of a hassle. This is an illusion. Staying put involves much more than slight inconvenience.

In 2006 Network Rail assumed a possession[4] of the West Coast main line south of Birmingham; where possible, trains were diverted to other routes but on the whole passengers had to travel by replacement bus services. The possession was in order to install new permanent way, overhead line equipment and upgrade the signalling. The project could certainly have been done piecemeal without such drastic action, but the time, cost and safety aspects would all have been badly affected and there would have been just as much inconvenience, only spread over a longer timescale. Exactly the same situation applies to buildings.

If relocation is decided on, it may be worthwhile to make the timing of the whole project dependent on the availability of suitable alternative accommodation, rather than deciding on the programme first and then trying to see if anything is available at that time. If relocation does prove to be impossible or unacceptable, then a very realistic allowance must be made for the additional cost and time involved. However, this will not represent the total penalty imposed by the decision; the health and wellbeing of staff and inconvenience to customers or residents, caused by dust and dirt, continuous noise of plant, temporary access arrangements, etc., will not appear on the project balance sheet but will show itself in staffing difficulties and reduced patronage.

Even if most of the work is carried out at weekends and at night, the dust problem is difficult to overcome. Finding stock, papers, furniture, etc. covered in dust every morning is one of the greatest irritants to occupants and tends to lead to complaints about everything else. These problems are bad enough for workers and casual users in public and commercial buildings, but are even worse if people's homes are concerned. Here, unless it is possible to move them out, occupiers will be subjected to the effects of noise, vibration and dust on a more-or-less permanent basis, and the option of mitigating the nuisance by working at night or at weekends will not of course be available. Householders are likely to be a good deal more militant than office or shop workers when subjected to these annoyances, and the presence of children on the site will give rise to problems of their safety on the one hand and of security of the works and vandalism on the other.

Phase 3

19.8   The need for effective liaison and oversight

The problems connected with refurbishment of buildings in occupation under-line the importance of the client having a very senior manager employed full-time to liaise between the building team and the occupants, with ultimate authority to override either party. The client's buildings officer, who is likely to be the first person thought of for this role, is unlikely to be suitable. In the case of the refurbishment of commercial or public buildings, this person will not be senior enough to be able to dictate to departmental managers, and is probably more used to being ordered around by them. And in the case of housing refurbishment, the role of liaison between residents and contractors will be a particularly demanding one, requiring a person with exceptional interpersonal skills.

This cost of the time of a senior person – who in a commercial organisation would need to be at board level – must be allowed for. Allowance will also have to be made for the cost consequences of the decisions which this liaison manager will have to make in order to keep the client's organisation running as smoothly as possible, or in order to satisfy the reasonable demands of tenants.

It is almost impossible, within the bounds of reason, to overestimate the total costs of refurbishing a building that is in occupation.

19.9   Costs excluding occupation

Even without occupation the costs of refurbishment are difficult to estimate, because of the factors associated with uncertainty already mentioned. But before any figures are given, the position of the building with regard to listed status must be investigated, since the planning authorities may impose extremely expensive requirements on the scheme. The cost of obtaining, and working with, obsolete materials, if this is either required or dictated, is likely to be considerable.

A further point to notice is that the cost of temporary works – scaffolding, shoring, etc. – is almost certain to play a much greater part than in the case of new works. Where a major gutting of a building's interior is taking place within retained external walls, such works are likely to be very sophisticated and expensive indeed. This is a type of work where the average cost planner's knowledge of both technology and cost is sometimes weaker than it is with new permanent works. It is therefore essential that a structural engineer is involved in the estimates at an early stage if any major structural work is contemplated.

The advantages of working with companies with an established compet-ence in refurbishment work cannot be overestimated; this applies as much to design and management consultants as it does to construction staff and con-tractors. Such firms are likely to reflect this within their costs. One method of identifying such organisations within the UK (for public sector clients) is through the DTI Constructionline database. A response to the Rethinking Construction initiative, it is the UK's register of pre-qualified construction contractors and consultants. Constructionline is designed to streamline pre-qualification procedures and is thus ideal for this type of work. The database was revised and updated for 2006, and at the time of writing, over 12,500 contractors and consultants are registered and have achieved the pre-qualification requirements supported by the DTI. Constructionline is used regularly by large central government departments and local authorities, universities, and NHS trusts, where refurbishment works are required. Constructionline is currently working with e-procurement solution providers (see Chapter 2) to help public sector buyers migrate towards e-procurement whilst continuing to access accredited construction suppliers at no extra cost.

This really leads to the conclusion that if there is a choice between new-build and refurbishment for the client's premises, the problems must be faced squarely at the start; tight, over-optimistic estimating has absolutely no place here. Cost overruns are not preferable at any time, but if the implication is that with hindsight the incorrect decision had been taken by the client (with professional advice), then the consequences are likely to be more than usually serious.

The cost control process should follow the usual real-time methods set out in Chapter 18. A particular point to observe, however, concerns the letting of work packages to trade or specialist contractors. On new-build work it is usually considered beneficial to secure this fairly early in the project, so that actual quotations can go into the estimated cost in place of the cost planner's estimates. However, it is often better in refurbishment work to wait until the efforts of other trades have reached a point where the specialist's work package can be properly defined and a firm price obtained (perhaps even in competition). In the case of a large project the inconvenience of letting, say, the plastering work in several separate packages could well be justified by the ability to get firm prices for each stage of the work.

Finally, it is essential to be aware, and keep the client aware also, of the current cost situation as the project develops. It is crucial on this type of project, since the probability of cost escalation is higher than it would be on other types of project.

19.10 Key points

Uncertainty is a major characteristic of refurbishment work, and this means that it will usually be inadvisable to undertake the project on the traditional basis of lump sum competitive tenders for the whole of the works. Other more collaborative methods of procurement have to be used – either cost-plus or some form of management contracting. And although it is obviously possible to set out any estimates in an elemental format, the normal process of elemental cost planning is not really applicable and cost planning must start from first principles.

The most important initial step is to recognise the conflict of objectives, which is inherent in such work. The effect on costs, programme and safety if the building remains in use are so enormous that a client who wishes to do this should be encouraged to see if some other alternative can be found. If not, it is important for the client to have a very senior manager employed full-time to liaise between the building team and the occupants, with ultimate authority to override either party.

Notes

(1) Client: First Choice Homes (Oldham Metropolitan Borough Council), Construction value £10 million, Refurbishment works, November 2000, Collaborative Procurement (Partnering) agreement using PPC 2000.

(2) Client: The Guinness Trust, Construction value £1 million, Refurbishment works, November 2000, Collaborative Procurement (Partnering) agreement using PPC 2000.

(3) It should be noted that partial possession is an important concept in contractual issues (such as in the JCT 2005 Standard Building Contract) and occurs when the employer takes possession of part of the works before the project has been fully completed.

(4) A possession, in railway parlance, is the occupation of a section of permanent way by a contractor in order to complete a project. This requires that no trains use the 'possessed' section during the period set aside for the works.

Further reading

BCIS (2007) *Wessex Alterations and Refurbishment Estimating Price Book 2007*, 12th Edition. Building Cost Information Service, London.

BRE (2006) *Sustainable Refurbishment of Victorian Housing*, IHS BRE Press, Bracknell.

Highfield, D. (2000) *Refurbishment and Upgrading of Buildings*. Spon Press, Abingdon.

RICS Building Maintenance Panel (2000) *Building Maintenance: Strategy, Planning and Procurement*. RICS Books, London.

Appendix A
BCIS Elemental Cost Analysis

The BCIS elements are described in full in the Standard Form of Cost Analysis (SFCA). The SFCA was in the process of being updated in March 2007 when this publication went to press. The element lists and definitions presented here are based on a draft of the updated SFCA, which was still subject to consultation. The revised version of the SFCA should be published in summer 2007.

Standard Elements

1 SUBSTRUCTURE

1A SUBSTRUCTURE

Definition: All work below underside of screed or, where no screed exists, to underside of lowest floor finishes including damp-proof membrane, together with relevant excavations and foundations (includes walls to basements designed as retaining walls).

Measurement: Area of lowest floor measured to the internal face of the external wall (as for gross internal floor area) (m^2).

1A1 *Standard Foundations*: All standard foundations up to and including the damp-proof course.

1A2 *Special Foundations*: All special foundations up to and including the damp-proof course.

1A3 *Lowest Floor Bed/Slab*: the entire lowest floor assembly below the underside of screed or lowest floor finish.

1A4 *Basement Excavation*: all work to basement excavation.

1A5 *Basement Retaining Walls*: up to and including the damp-proof course.

2 SUPERSTRUCTURE

2A   FRAME

Definition: Loadbearing framework. Main floor and roof beams, ties and roof trusses of framed buildings. Casing to stanchions and beams for structural or protective purposes.

Measurement: Area of floors related to the frame measured to internal face of external walls (as for gross internal floor area) (m^2).

2B   UPPER FLOORS

Definition: Upper floors including suspended floors over or in basements, service floors, balconies, sloping floors and walkways.

Measurement: Total area of upper floor measured to the internal face of the external wall (as for gross internal floor area) (m^2).

Notes:

- Sloping surfaces such as galleries, tiered terraces and the like should be measured flat on plan.
- Where balconies are included, the sum of the upper and lowest floors will exceed the gross internal floor area.

2C   ROOF

Definition: Roof Structure, roof coverings, roof drainage, rooflights and roof features.

Measurement: Area on plan measured to the inside face of the external wall (m^2).

2C1   *Roof Structure*: All components of the roof.
2C2   *Roof Coverings*: Protective covering to roof.
2C3   *Roof Drainage*: Rain water disposal systems to roof.
2C4   *Roof Lights and Openings*: Roof lights and openings to roof.
2C5   *Roof Features*: Roof features not forming part of the main structure.

2D   STAIRS

Definition: Construction of ramps, stairs, ladders, etc connecting floors at different levels.

Measurement: Number of storey flights (Nr) ie the number of staircases multiplied by the number of floors served (excluding the lowest floor served in each case).

2D1   *Stair structure*: Construction of ramps, stairs and landings. Note: if the cost of the staircase structure is included in the elements frame or upper floor this should be stated.

2D2   *Stair Finishes*: Finishes to stairs, ramps and landings.

2D3   *Stair Balustrades and Handrails*: Balustrades and handrails to stairs, ramps and landings.

2E   EXTERNAL WALLS

Definition: External enclosing walls including walls to basements but excluding walls to basement designed as retaining walls and items included with *2C Roof Structure* and *2F Windows and Doors*.

Measurement: Area of external walls measured on the inner face (excluding openings measured as for *2H Windows and External Doors*) (m^2). NB: the total of the area of *2E External Walls* and *2F Windows and External Doors* should equal the area of the vertical enclosure.

2E1   *External Enclosing Walls*.

2E2   *Solar/Rain Screening*: Cladding systems etc. attached to the exterior of the building to protect the external walls.

2E3   *Basement Walls*.

2E4   *Façade Access*.

2F   WINDOWS AND EXTERNAL DOORS

Definition: Windows, doors and openings in external walls.

Measurement: Total area of windows and external doors measured over frames (m^2). NB: the total of the area of *2E External Walls* and *2F Windows and External Doors* should equal the area of the vertical enclosure.

2F1   *External Windows*: Windows and openings in external walls for ventilation and light.

2F2   *External Doors*: Doors and openings in external walls for physical movement.

2G   INTERNAL WALLS AND PARTITIONS

Definition: Internal walls, partitions, balustrades, moveable room dividers, cubicles and the like.

Measurement: Total area of internal walls and partitions excluding openings (m^2).

2G1   *Walls/Partitions*: Internal walls and fixed partitions.

2G2 *Balustrades and Handrails*: Internal balustrades, handrails and other fixed non storey height divisions.

2G3 *Moveable Room Dividers*: Moveable partitions intended to divide rooms into smaller spaces.

2G4 *Cubicles*: Proprietary cubicle partitions and doors.

2H   INTERNAL DOORS

Definition: Doors, hatches and other openings in internal walls and partitions.

Measurement: Number of doors (Nr).

3   FINISHES

3A   WALL FINISHES

Definition: Preparatory work and finishes to surfaces of walls and other vertical surfaces internally.

Measurement: Total area of finished walls (m²) ie the area of wall to which finishes are applied.

3B   FLOOR FINISHES

Definition: Preparatory work and finishes to internal floor surfaces.

Measurement: Total area of floor finishes (m²).

3B1 *Finishes to Floors*: Preparatory work finishes applied to floor surfaces.

3B2 *Raised Access Floors*: Construction and finishes of raised access floors.

3C   CEILING FINISHES

Definition: Preparatory work and finishes to internal ceiling surfaces.

Measurement: Total area of ceiling finishes (m²).

3C1 *Finishes to Ceilings*: Preparatory work finishes applied to ceiling surfaces.

3C2 *Suspended Ceilings*: Construction and finishes of suspended ceilings.

4   FITTINGS AND FURNISHINGS

4A   FITTINGS AND FURNISHINGS

Definition: Fittings, fixtures, furniture, works of art and non-mechanical and electrical equipment.

Measurement: Gross internal floor area (m²).

4A1   Fittings, Fixtures and Furniture.
4A2   Soft Furnishings.
4A3   Works of Art.
4A4   Equipment.

5   SERVICES

5A   SANITARY APPLIANCES

Definition: Baths, basins, sinks, WCs and the like.
Measurement: Number of fittings (Nr).
  Notes on measurement:

- Count sanitary fittings only, exclude ancillary items.
- Include the designed number of occupants for continuous urinals and the like.

5A1   Sanitaryware: Toilet-roll holders, towel rails, etc. Traps, waste fittings, overflows and taps as appropriate.
5A2   Pods: Bathroom, toilet and shower pods supplied as completed units manufactured off site.

State if pods replace any of the building structure e.g. internal or external walls, floors, etc.

5B   SERVICES EQUIPMENT

Definition: Mechanical and electrical equipment.
Measurement: Number of fittings (Nr). Note: Count fittings only exclude ancillary items.

5C   DISPOSAL INSTALLATIONS

Definition: Internal drainage, refuse disposal and chemical and industrial liquid waste disposal.
Measurement: Number of fittings serviced ie total of items listed below (Nr).
  Costs and measurement should be shown separately for:

- Internal drainage: Number of sanitary appliances (5A) and services equipment fittings (5B) (Nr).
- Refuse disposal:
    - Number of self contained fittings (Nr)

- ☐ Number of entry points to rubbish chutes etc (Nr)
- ☐ Number of entry points for chemical and industrial waste (Nr).
- ■ Chemical and industrial liquid waste appliances (Nr).

5C1   *Internal Drainage*: Waste pipes to sanitary appliances, services equipment, etc.

5C2   *Refuse Disposal*: Refuse chutes, local incinerators and the like.

5C3   *Chemical and Industrial Liquid Waste Disposal*.

5D   WATER INSTALLATIONS

Definition: Mains supply, hot and cold water services, steam and condensate services.

Measurement: Number of draw off points (Nr).

5D1   *Mains supply*: Incoming water main from external face of external wall at point of entry into buildings.

5D2   *Cold Water Services*: Cold water supply from storage tanks to appliances and equipment.

5D3   *Hot Water Services*: Hot water and/or mixed water supply from, and including, storage cylinders etc. to appliances and equipment.

5D4   *Steam and Condensate*: Steam distribution and condensate return pipework to and from services equipment within the building.

5E   HEAT SOURCE

Definition: Boilers and other sources of heat production for heating, hot water and power generation, including combined heat and power and ancillary installations.

Measurement: Rating in kilowatts (kW)

5F   SPACE HEATING AND AIR CONDITIONING

Definition: Heating, cooling and air conditioning systems and fixed equipment.

Measurement: Treated floor area (m^2). Treated volume (m^3) ie the floor area of treated spaces multiplied by their storey height should also be stated (Note: Atriums etc should be measured full height).

5F1   *Central Heating*: Systems where heating is generated centrally.

5F2   *Local Heating*: Systems where heating is generated in or adjacent to the space to be treated.

5F3   *Central Cooling*: Systems where cooling is performed at a central point and distributed to the space being treated

5F4   *Local Cooling*: Systems where cooling is performed in or adjacent to the space to be treated.

5F5   *Central Heating and Cooling*: Combined systems where heating and cooling are performed at a central point and distributed to the space being treated

5F6   *Local Heating and Cooling*: Combined systems where heating and cooling are performed in or adjacent to the space to be treated.

5F7   *Central Air Conditioning*: Combined systems where heating, cooling, dehumidification and other air treatments are performed at a central point and ducted to the space being treated

State the air treatments included in the systems e.g.

- Heating
- Cooling
- Humidification/dehumidification
- Filtration
- Pressurisation

5F8   *Local air conditioning*: Combined systems where heating, cooling dehumidification and other air treatments are performed in or adjacent to the space to be treated.

State the air treatments included in the systems e.g.

- Heating
- Cooling
- Humidification/dehumidification
- Filtration
- Pressurisation

5F9   *Instrumentation and Controls for More than One System*: Combined instrumentation and controls serving more than one heating, cooling or air conditioning system.

5G   VENTILATION SYSTEMS

Definition: Ventilating system not incorporating heating or cooling installations.
Measurement: Treated volume (m^3) ie the floor area of treated spaces multiplied by their storey height. Treated floor area should also be stated (m^2).

5G1   *Central Ventilation.*
5G2   *Smoke Ventilation.*
5G3   *Local and Special Ventilation.*

5H   ELECTRICAL INSTALLATIONS

Definition: Electric source and mains, power distribution, electric lighting distribution and fittings.

Measurement: Total number of power outlet points and light fittings

5H1   *Electric Source and Mains*: All work from external face of building (the supplier's meter) up to and including local distribution boards. Note: show high voltage and low voltage supplies separately.

5H2   *Electric Power Supplies*: (Small power) General purpose power supplies and supply to other services installations.

5H3   *Electric Lighting*: Power supply to lighting.

5H4   *Electric Light Fittings*: Light fittings including fixing.

5H5   *Specialist Lighting*: Specialist lighting installations e.g. display lighting, stage lighting, studio lighting, operating theatre lighting and the like.

5H6   *Local Electric Supply*: Local electric generation, emergency power supplies, etc.

5H7   *Earthing Systems*: Separate earthing systems.

5I   FUEL INSTALLATIONS

Definition: Fuel services from meter or from point of entry to appliances and equipment.
Measurement: Number of draw-off points (Nr).
Different types of supply should be shown separately.

5J   LIFT AND CONVEYOR INSTALLATIONS

Definition: Lifts, hoists, escalators, travelators, conveyors and the like.
Measurement: Number of stops (Nr).
Also show:

- Passenger lifts: Number of stops (Nr). Wall climbing and conventional lifts should be shown separately.
- Goods lifts: (in enclosed shafts): Number of stops (Nr).
- Escalators: Number of storey flights (Nr) ie the number of escalators multiplied by the number of floors served.
- Travelators: length of travel (m).
- Hoists, cranes, etc (not enclosed in shafts): Total rise (m).
- Conveyors: length of travel (m). People conveyors and goods conveying systems should be shown separately.
- Dock levellers, scissor lifts: Total rise (m).

5J1   *Lifts and Enclosed Hoists*: Passenger and goods lifts and hoists enclosed in shafts.

5J2   *Escalators*: Escalators, travelators, stair lifts, etc for the movement of people.

5J3   *Conveyors*: Conveyors etc for movement of materials and goods.

5J4   *Dock Levellers and Scissor Lifts*: Localised lifting systems for goods and people.

5J5   *Cranes and Unenclosed Hoists*: Movable lifting systems for goods.

5J6   *Car Lifts, Turntables and the Like*: Vehicle lifting and moving systems.

5K   FIRE AND LIGHTNING PROTECTION

Definition: Fire suppression systems, fire-fighting and lightning protection installation.
Measurement: Gross internal floor area (m²).

5K1   *Automatic Fire Suppression Systems*: sprinkler, dry chemical, foam and inert gas extinguishing installations and the like.

5K2   *Fire-fighting installations*: Services and equipment for manual fire-fighting.

5K3   *Lightning Protection*: Lightning protection installation.

5L   COMMUNICATION AND SECURITY INSTALLATIONS

Definition: Warning, communication and access control installations.
Measurement: Gross internal floor area (m²).

5L1   *Warning Installations*: Fire, theft and other emergency warning installations.

5L2   *Visual, Audio and Data Installations*: Communications systems.

5L3   *Security*: Observation and access control installations and the like.

5M   SPECIAL INSTALLATIONS

Definition: All other mechanical and/or electrical installations (separately identifiable) which have not been included elsewhere.
Measurement: Costs and measurement should be shown separately for each installation

5M1   *Mechanical and Electrical Systems*.

5M2   *Building Management Control Installations*: Central control and management systems for mechanical, electrical, and other building services systems.

5N   BUILDER'S WORK IN CONNECTION WITH SERVICES

Definition: Work carried out solely to facilitate the provision of services installation not provided by other elements.

Measurement: Gross internal floor area (m^2).
Costs and measurement should be shown separately for each installation.

5O   MANAGEMENT OF WHOLE BUILDING SERVICES COMMISSIONING

6   EXTERNAL WORKS

6A   SITE WORKS

Definition: Site preparation and external works.
Measurement: Area of external works (m^2) ie total site area excluding the building footprint. Exclude any areas used temporarily for the works that do not form part of the delivered project.

6A1   *Site Preparation*: Preparatory earth works to form new contours.
6A2   *Site Remediation and Decontamination*: Preparing site by removal or containment of decontamination.
6A3   *Surface Treatments*: Preparation and completion of unenclosed surfaces within the site.
6A4   *Site Enclosure and Division*: Fences, walls, etc.
6A5   *Fittings and Furniture*: Street furniture, site signage, etc.

6B   DRAINAGE

Definition: Drainage from the building and the site, on-site waste water treatment, etc.
Measurement: Area of external works (m^2) ie total site area excluding the building footprint. Exclude any areas used temporarily for the works that do not form part of the delivered project.

6B1   *Drainage Under the Building.*
6B2   *Drainage Outside the Building.*

6C   EXTERNAL SERVICES

Definition: Service supplies to the building and services to external works.
Measurement: Area of external works (m^2) ie total site area excluding the building footprint. Exclude any areas used temporarily for the works that do not form part of the delivered project.
Costs and measurement should be shown separately for each installation.

6C1   *Service Mains*: Mains from existing supply up to external face of building. Costs and measurement should be shown separately for each service.

6C2   *Site Lighting*: Lighting to external surfaces.
6C7   *Other Site Services*: Services to external surfaces (each shown separately).
6C8   *Builder's Work in Connection with External Services.*

6D   MINOR BUILDING WORK

Definition: Ancillary buildings, alterations to existing buildings, other buildings and work included in the contract.
Measurement: Gross internal floor area of the buildings (m^2).
  Costs and measurement should be shown separately for each building.

6D1   *Ancillary Buildings*: Separate minor building.
6D2   *Alterations to Existing Buildings* – Alterations and work to existing buildings.
6D3   *Other Buildings and Work Included in the Contract.*

6E   DEMOLITIONS AND WORK OUTSIDE THE SITE

Definition: Major demolitions and building works required outside the site boundary.

6E1   *Demolition.*
6E2   *Work outside the site*: Any work required to be carried out as part of the contract.

7   PRELIMINARIES

Definition: Priced items in Preliminaries Bill and Summary but excluding contractor's price adjustments.
  Lump sum adjustments which should be allocated to the builders work.

8   EMPLOYER'S CONTINGENCIES

Definition: Employer's contingencies included in the contract.

9   DESIGN FEES (on Design and Build Schemes)

Definition: Design fees included in the contract.

Appendix B
Discount Rate Tables

Please refer to Chapter 6 of this text book for the relevant formulae associated with these tables.

Table B.1 Compound Interest.

Value at end of each period of £1 invested at the beginning of period 1 and accumulating at compound interest from 1% to 30% per period.

Period	1% £	1.5% £	2% £	2.5% £	3% £	3.5% £	4% £	5% £	6% £	7% £	8% £	9% £	10% £	11% £	12% £	13% £	14% £	15% £	20% £	25% £	30% £
1	1.01	1.02	1.02	1.03	1.03	1.04	1.04	1.05	1.06	1.07	1.08	1.09	1.10	1.11	1.12	1.13	1.14	1.15	1.20	1.25	1.30
2	1.02	1.03	1.04	1.05	1.06	1.07	1.08	1.10	1.12	1.14	1.17	1.19	1.21	1.23	1.25	1.28	1.30	1.32	1.44	1.56	1.69
3	1.03	1.05	1.06	1.08	1.09	1.11	1.12	1.16	1.19	1.23	1.26	1.30	1.33	1.37	1.40	1.44	1.48	1.52	1.73	1.95	2.20
4	1.04	1.06	1.08	1.10	1.13	1.15	1.17	1.22	1.26	1.31	1.36	1.41	1.46	1.52	1.57	1.63	1.69	1.75	2.07	2.44	2.86
5	1.05	1.08	1.10	1.13	1.16	1.19	1.22	1.27	1.34	1.40	1.47	1.54	1.61	1.69	1.76	1.84	1.93	2.01	2.49	3.05	3.71
6	1.06	1.09	1.13	1.16	1.19	1.23	1.27	1.34	1.42	1.50	1.59	1.68	1.77	1.87	1.97	2.08	2.19	2.31	2.99	3.81	4.83
7	1.07	1.11	1.15	1.19	1.23	1.27	1.32	1.41	1.50	1.61	1.71	1.83	1.95	2.08	2.21	2.35	2.50	2.66	3.58	4.77	6.27
8	1.08	1.13	1.17	1.22	1.27	1.32	1.37	1.50	1.59	1.72	1.85	1.99	2.14	2.30	2.48	2.66	2.85	3.06	4.30	5.96	8.16
9	1.09	1.14	1.20	1.25	1.30	1.36	1.42	1.59	1.69	1.84	1.99	2.17	2.36	2.56	2.77	3.00	3.25	3.52	5.16	7.45	10.60
10	1.10	1.16	1.22	1.28	1.34	1.41	1.48	1.63	1.79	1.97	2.16	2.37	2.59	2.84	3.11	3.39	3.71	4.05	6.19	9.31	13.79
11	1.12	1.18	1.24	1.31	1.38	1.46	1.54	1.71	1.90	2.10	2.33	2.58	2.85	3.15	3.48	3.84	4.23	4.65	7.43	11.64	17.92
12	1.13	1.20	1.27	1.34	1.43	1.51	1.60	1.80	2.01	2.25	2.52	2.81	3.14	3.50	3.90	4.33	4.82	5.35	8.92	14.55	23.30
13	1.14	1.21	1.29	1.38	1.47	1.56	1.67	1.89	2.13	2.41	2.72	3.07	3.45	3.88	4.36	4.90	5.49	6.15	10.70	18.19	30.29
14	1.15	1.23	1.32	1.41	1.51	1.62	1.73	1.98	2.26	2.58	2.94	3.34	3.80	4.31	4.89	5.53	6.26	7.08	12.84	22.74	39.37
15	1.16	1.25	1.35	1.45	1.56	1.68	1.80	2.08	2.40	2.76	3.17	3.64	4.18	4.78	5.47	6.25	7.14	8.14	15.41	28.42	51.19
16	1.17	1.27	1.37	1.48	1.60	1.73	1.87	2.18	2.54	2.95	3.43	3.97	4.59	5.31	6.13	7.07	8.14	9.36	18.49	35.53	66.54
17	1.18	1.29	1.40	1.52	1.65	1.79	1.95	2.29	2.69	3.16	3.70	4.33	5.05	5.90	6.87	7.99	9.28	10.76	22.19	44.41	86.50
18	1.20	1.31	1.43	1.56	1.70	1.86	2.03	2.41	2.85	3.38	4.00	4.72	5.56	6.54	7.69	9.02	10.58	12.38	26.62	55.51	112.46
19	1.21	1.33	1.46	1.60	1.75	1.92	2.11	2.53	3.03	3.62	4.32	5.14	6.12	7.26	8.61	10.20	12.06	14.23	31.95	69.39	146.19
20	1.22	1.35	1.49	1.64	1.81	1.99	2.19	2.65	3.21	3.87	4.66	5.60	6.73	8.06	9.65	11.52	13.74	16.37	38.34	86.74	190.05
21	1.23	1.37	1.52	1.68	1.86	2.06	2.28	2.79	3.40	4.14	5.03	6.11	7.40	8.95	10.80	13.02	15.67	18.82	46.01	108.42	247.06
22	1.24	1.39	1.55	1.72	1.92	2.13	2.37	2.93	3.60	4.43	5.44	6.66	8.14	9.93	12.10	14.71	17.86	21.64	55.21	135.53	321.18
23	1.26	1.41	1.58	1.76	1.97	2.21	2.46	3.07	3.82	4.74	5.87	7.26	8.95	11.03	13.55	16.63	20.36	24.89	66.25	169.41	417.54
24	1.27	1.43	1.61	1.81	2.03	2.28	2.56	3.23	4.05	5.07	6.34	7.91	9.85	12.24	15.18	18.79	23.21	28.63	79.50	211.76	542.80
25	1.28	1.45	1.64	1.85	2.09	2.36	2.67	3.39	4.29	5.43	6.85	8.62	10.83	13.59	17.00	21.23	26.46	32.92	95.40	264.70	705.64
26	1.30	1.47	1.67	1.90	2.16	2.45	2.77	3.56	4.55	5.81	7.40	9.40	11.92	15.08	19.04	23.99	30.17	37.86	114.48	330.87	917.33
27	1.31	1.49	1.71	1.95	2.22	2.53	2.88	3.73	4.82	6.21	7.99	10.25	13.11	16.74	21.32	27.11	34.39	43.54	137.37	413.59	1192.53
28	1.32	1.52	1.74	2.00	2.29	2.62	3.00	3.92	5.11	6.65	8.63	11.17	14.42	18.58	23.88	30.63	39.20	50.07	164.84	516.99	1550.29
29	1.33	1.54	1.78	2.05	2.36	2.71	3.12	4.12	5.42	7.11	9.32	12.17	15.86	20.62	26.75	34.62	44.69	57.58	197.81	646.23	2015.38
30	1.35	1.56	1.81	2.10	2.43	2.81	3.24	4.32	5.74	7.61	10.06	13.27	17.45	22.89	29.96	39.12	50.95	66.21	237.38	807.79	2620.00
31	1.36	1.59	1.85	2.15	2.50	2.91	3.37	4.54	6.09	8.15	10.87	14.46	19.19	25.41	33.56	44.20	58.08	76.14	284.85	1009.74	3405.99
32	1.37	1.61	1.88	2.20	2.58	3.01	3.51	4.76	6.45	8.72	11.74	15.76	21.11	28.21	37.58	49.95	66.21	87.57	341.82	1262.18	4427.79
33	1.39	1.63	1.92	2.26	2.65	3.11	3.65	5.00	6.84	9.33	12.68	17.18	23.23	31.31	42.09	56.44	75.48	100.70	410.19	1577.72	5756.13
34	1.40	1.66	1.96	2.32	2.73	3.22	3.79	5.25	7.25	9.98	13.69	18.73	25.55	34.75	47.14	63.78	86.05	115.80	492.22	1972.15	7482.97
35	1.42	1.68	2.00	2.37	2.81	3.33	3.95	5.52	7.69	10.68	14.79	20.41	28.10	38.57	52.80	72.07	98.10	133.18	590.67	2465.19	9727.86
36	1.43	1.71	2.04	2.43	2.90	3.45	4.10	5.79	8.15	11.42	15.97	22.25	30.91	42.82	59.14	81.44	111.83	153.15	708.80	3081.49	12646.22
37	1.45	1.73	2.08	2.49	2.99	3.57	4.27	6.08	8.64	12.22	17.25	24.25	34.00	47.53	66.23	92.02	127.49	176.12	850.56	3851.86	16440.08
38	1.46	1.76	2.12	2.56	3.07	3.70	4.44	6.39	9.15	13.08	18.63	26.44	37.40	52.76	74.18	103.99	145.34	202.54	1020.67	4814.82	21372.11
39	1.47	1.79	2.16	2.62	3.17	3.83	4.62	6.70	9.70	13.99	20.12	28.82	41.14	58.56	83.08	117.51	165.69	232.92	1224.81	6018.53	27783.74
40	1.49	1.81	2.21	2.69	3.26	3.96	4.80	7.04	10.29	14.97	21.72	31.41	45.26	65.00	93.05	132.78	188.88	267.86	1469.77	7523.16	36118.86

Table B.2 Future Value of £1 invested at regular intervals.

Value of £1 invested regularly at end of each period (i.e. weekly, monthly, annually) accumulating at compound interest.

Period	1% £	1.5% £	2% £	2.5% £	3% £	3.5% £	4% £	5% £	6% £	7% £	8% £	9% £	10% £	11% £	12% £	13% £	14% £	15% £	20% £	25% £	30% £
1	1.00	1.00	1.00	1.00	1.00	1.00	1.00	1.00	1.00	1.00	1.00	1.00	1.00	1.00	1.00	1.00	1.00	1.00	1.00	1.00	1.00
2	2.01	2.01	2.02	2.03	2.03	2.04	2.04	2.05	2.06	2.07	2.08	2.09	2.10	2.11	2.12	2.13	2.14	2.15	2.20	2.20	2.30
3	3.03	3.05	3.06	3.08	3.09	3.11	3.12	3.15	3.18	3.21	3.25	3.28	3.31	3.34	3.37	3.41	3.44	3.47	3.64	3.81	3.99
4	4.06	4.09	4.12	4.15	4.18	4.21	4.25	4.31	4.37	4.44	4.51	4.57	4.64	4.71	4.78	4.85	4.92	4.99	5.37	5.77	6.19
5	5.10	5.15	5.20	5.26	5.31	5.36	5.42	5.53	5.64	5.75	5.87	5.98	6.11	6.23	6.35	6.48	6.61	6.74	7.44	8.21	9.04
6	6.15	6.23	6.31	6.39	6.47	6.55	6.63	6.80	6.98	7.15	7.34	7.52	7.72	7.91	8.12	8.32	8.54	8.75	9.93	11.26	12.76
7	7.21	7.32	7.43	7.55	7.66	7.78	7.90	8.14	8.39	8.65	8.92	9.20	9.49	9.78	10.09	10.40	10.73	11.07	12.92	15.07	17.58
8	8.29	8.43	8.58	8.74	8.89	9.05	9.21	9.55	9.90	10.26	10.64	11.03	11.44	11.86	12.30	12.76	13.23	13.73	16.50	19.84	23.86
9	9.37	9.56	9.75	9.95	10.16	10.37	10.58	11.03	11.49	11.98	12.49	13.02	13.58	14.16	14.78	15.42	16.09	16.79	20.80	25.80	32.01
10	10.46	10.70	10.95	11.20	11.46	11.73	12.01	12.58	13.18	13.82	14.49	15.19	15.94	16.72	17.55	18.42	19.34	20.30	25.96	33.25	42.62
11	11.57	11.86	12.17	12.48	12.81	13.14	13.49	14.21	14.97	15.78	16.65	17.56	18.53	19.56	20.65	21.81	23.04	24.35	32.15	42.57	56.41
12	12.68	13.04	13.41	13.80	14.19	14.60	15.03	15.92	16.87	17.89	18.98	20.14	21.38	22.71	24.13	25.65	27.27	29.00	39.58	54.21	74.33
13	13.81	14.24	14.68	15.14	15.62	16.11	16.63	17.71	18.88	20.14	21.50	22.95	24.52	26.21	28.03	29.98	32.09	34.35	48.50	68.76	97.63
14	14.95	15.45	15.97	16.52	17.09	17.68	18.29	19.60	21.02	22.55	24.21	26.02	27.97	30.09	32.39	34.88	37.58	40.50	59.20	86.95	127.91
15	16.10	16.68	17.29	17.93	18.60	19.30	20.02	21.58	23.28	25.13	27.15	29.36	31.77	34.41	37.28	40.42	43.84	47.58	72.04	109.69	167.29
16	17.26	17.93	18.64	19.38	20.16	20.97	21.82	23.66	25.67	27.89	30.32	33.00	35.95	39.19	42.75	46.67	50.98	55.72	87.44	138.11	218.47
17	18.43	19.20	20.01	20.86	21.76	22.70	23.70	25.84	28.21	30.84	33.75	36.97	40.54	44.50	48.88	53.74	59.12	65.08	105.93	173.64	285.01
18	19.61	20.49	21.41	22.39	23.41	24.50	25.65	28.13	30.91	34.00	37.45	41.30	45.60	50.40	55.75	61.73	68.39	75.84	128.12	218.04	371.52
19	20.81	21.80	22.84	23.95	25.12	26.36	27.67	30.54	33.76	37.38	41.45	46.02	51.16	56.94	63.44	70.75	78.97	88.21	154.74	273.56	483.97
20	22.02	23.12	24.30	25.54	26.87	28.28	29.78	33.07	36.79	41.00	45.76	51.16	57.27	64.20	72.05	80.95	91.02	102.44	186.69	342.94	630.17
21	23.24	24.47	25.78	27.18	28.68	30.27	31.97	35.72	39.99	44.87	50.42	56.76	64.00	72.27	81.70	92.47	104.77	118.81	225.03	429.68	820.22
22	24.47	25.84	27.30	28.86	30.54	32.33	34.25	38.51	43.39	49.01	55.46	62.87	71.40	81.21	92.50	105.49	120.44	137.63	271.03	538.10	1067.28
23	25.72	27.23	28.84	30.58	32.45	34.46	36.62	41.43	47.00	53.44	60.89	69.53	79.54	91.15	104.60	120.20	138.30	159.28	326.24	673.63	1388.46
24	26.97	28.63	30.42	32.35	34.43	36.67	39.08	44.50	50.82	58.18	66.76	76.79	88.50	102.17	118.16	136.83	158.66	184.17	392.48	843.03	1806.00
25	28.24	30.06	32.03	34.16	36.46	38.95	41.65	47.73	54.86	63.25	73.11	84.70	98.35	114.41	133.33	155.62	181.87	212.79	471.98	1054.79	2348.80
26	29.53	31.51	33.67	36.01	38.55	41.31	44.31	51.11	59.16	68.68	79.95	93.32	109.18	128.00	150.33	176.85	208.33	245.71	567.38	1319.49	3054.44
27	30.82	32.99	35.34	37.91	40.71	43.76	47.08	54.67	63.71	74.48	87.35	102.72	121.10	143.08	169.37	200.84	238.50	283.57	681.85	1650.36	3971.78
28	32.13	34.48	37.05	39.86	42.93	46.29	49.97	58.40	68.53	80.70	95.34	112.97	134.21	159.82	190.70	227.95	272.89	327.10	819.22	2063.95	5164.31
29	33.45	36.00	38.79	41.86	45.22	48.91	52.97	62.32	73.64	87.35	103.97	124.14	148.63	178.40	214.58	258.58	312.09	377.17	984.07	2580.94	6714.60
30	34.78	37.54	40.57	43.90	47.58	51.62	56.08	66.44	79.06	94.46	113.28	136.31	164.49	199.02	241.33	293.20	356.79	434.75	1181.88	3227.17	8729.99
31	36.13	39.10	42.38	46.00	50.00	54.43	59.33	70.76	84.80	102.07	123.35	149.58	181.94	221.91	271.29	332.32	407.74	500.96	1419.26	4034.97	11349.98
32	37.49	40.69	44.23	48.15	52.50	57.33	62.70	75.30	90.89	110.22	134.21	164.04	201.14	247.32	304.85	376.52	465.82	577.10	1704.11	5044.71	14755.98
33	38.87	42.30	46.11	50.35	55.08	60.34	66.21	80.06	97.34	118.93	145.95	179.80	222.25	275.53	342.43	426.46	532.04	664.67	2045.95	6306.89	19183.77
34	40.26	43.93	48.03	52.61	57.73	63.45	69.86	85.07	104.18	128.26	158.63	196.98	245.48	306.84	384.52	482.90	607.52	765.37	2456.12	7884.61	24939.90
35	41.66	45.59	49.99	54.93	60.46	66.67	73.65	90.32	111.43	138.24	172.32	215.71	271.02	341.59	431.66	546.68	693.57	881.17	2948.34	9856.76	32422.87
36	43.08	47.28	51.99	57.30	63.28	70.01	77.60	95.84	119.12	148.91	187.10	236.12	299.13	380.16	484.46	618.75	791.67	1014.35	3539.01	12321.95	42150.73
37	44.51	48.99	54.03	59.73	66.17	73.46	81.70	101.63	127.27	160.34	203.07	258.38	330.04	422.98	543.60	700.19	903.51	1167.50	4247.81	15403.44	54796.95
38	45.95	50.72	56.11	62.23	69.16	77.03	85.97	107.71	135.90	172.56	220.32	282.63	364.04	470.51	609.83	792.21	1031.00	1343.62	5098.37	19255.30	71237.03
39	47.41	52.48	58.24	64.78	72.23	80.72	90.41	114.10	145.06	185.64	238.94	309.07	401.45	523.27	684.01	896.20	1176.34	1546.17	6119.05	24070.12	92609.14
40	48.89	54.27	60.40	67.40	75.40	84.55	95.03	120.80	154.76	199.64	259.06	337.88	442.59	581.83	767.09	1013.70	1342.03	1779.09	7343.86	30088.66	120392.88

Table B.3 Present Value of £1

Present Value of £1 payable (or receivable) at the end of any period 1 to 40, discounted at interest rates from 1% to 30% per period.

Period	1% £	1.5% £	2% £	2.5% £	3% £	3.5% £	4% £	5% £	6% £	7% £	8% £	9% £	10% £	11% £	12% £	13% £	14% £	15% £	20% £	25% £	30% £
1	0.990	0.985	0.980	0.976	0.971	0.966	0.962	0.952	0.943	0.935	0.926	0.917	0.909	0.901	0.893	0.885	0.877	0.870	0.833	0.800	0.769
2	0.980	0.971	0.961	0.952	0.943	0.934	0.925	0.907	0.890	0.873	0.857	0.842	0.826	0.812	0.797	0.783	0.769	0.756	0.694	0.640	0.592
3	0.971	0.956	0.942	0.929	0.915	0.902	0.889	0.864	0.840	0.816	0.794	0.772	0.751	0.731	0.712	0.693	0.675	0.658	0.579	0.512	0.455
4	0.961	0.942	0.924	0.906	0.888	0.871	0.855	0.823	0.792	0.763	0.735	0.708	0.683	0.659	0.636	0.613	0.592	0.572	0.482	0.410	0.350
5	0.951	0.928	0.906	0.884	0.863	0.842	0.822	0.784	0.747	0.713	0.681	0.650	0.621	0.593	0.567	0.543	0.519	0.497	0.402	0.328	0.269
6	0.942	0.915	0.888	0.862	0.837	0.814	0.790	0.746	0.705	0.666	0.630	0.596	0.564	0.535	0.507	0.480	0.456	0.432	0.335	0.262	0.207
7	0.933	0.901	0.871	0.841	0.813	0.786	0.760	0.711	0.665	0.623	0.583	0.547	0.513	0.482	0.452	0.425	0.400	0.376	0.279	0.210	0.159
8	0.923	0.888	0.853	0.821	0.789	0.759	0.731	0.677	0.627	0.582	0.540	0.502	0.467	0.434	0.404	0.376	0.351	0.327	0.233	0.168	0.123
9	0.914	0.875	0.837	0.801	0.766	0.734	0.703	0.645	0.592	0.544	0.500	0.460	0.424	0.391	0.361	0.333	0.308	0.284	0.194	0.134	0.094
10	0.905	0.862	0.820	0.781	0.744	0.709	0.676	0.614	0.558	0.508	0.463	0.422	0.386	0.352	0.322	0.295	0.270	0.247	0.162	0.107	0.073
11	0.896	0.849	0.804	0.762	0.722	0.685	0.650	0.585	0.527	0.475	0.429	0.388	0.350	0.317	0.287	0.261	0.237	0.215	0.135	0.086	0.056
12	0.887	0.836	0.788	0.744	0.701	0.662	0.625	0.557	0.497	0.444	0.397	0.356	0.319	0.286	0.257	0.231	0.208	0.187	0.112	0.069	0.043
13	0.879	0.824	0.773	0.725	0.681	0.639	0.601	0.530	0.469	0.415	0.368	0.326	0.290	0.258	0.229	0.204	0.182	0.163	0.093	0.055	0.033
14	0.870	0.812	0.758	0.708	0.661	0.618	0.577	0.505	0.442	0.388	0.340	0.299	0.263	0.232	0.205	0.181	0.160	0.141	0.078	0.044	0.025
15	0.861	0.800	0.743	0.690	0.642	0.597	0.555	0.481	0.417	0.362	0.315	0.275	0.239	0.209	0.183	0.160	0.140	0.123	0.065	0.035	0.020
16	0.853	0.788	0.728	0.674	0.623	0.577	0.534	0.458	0.394	0.339	0.292	0.252	0.218	0.188	0.163	0.141	0.123	0.107	0.054	0.028	0.015
17	0.844	0.776	0.714	0.657	0.605	0.557	0.513	0.436	0.371	0.317	0.270	0.231	0.198	0.170	0.146	0.125	0.108	0.093	0.045	0.023	0.012
18	0.836	0.765	0.700	0.641	0.587	0.538	0.494	0.416	0.350	0.296	0.250	0.212	0.180	0.153	0.130	0.111	0.095	0.081	0.038	0.018	0.009
19	0.828	0.754	0.686	0.626	0.570	0.520	0.475	0.396	0.331	0.277	0.232	0.194	0.164	0.138	0.116	0.098	0.083	0.070	0.031	0.014	0.007
20	0.820	0.742	0.673	0.610	0.554	0.503	0.456	0.377	0.312	0.258	0.215	0.178	0.149	0.124	0.104	0.087	0.073	0.061	0.026	0.012	0.005
21	0.811	0.731	0.660	0.595	0.538	0.486	0.439	0.359	0.294	0.242	0.199	0.164	0.135	0.112	0.093	0.077	0.064	0.053	0.022	0.009	0.004
22	0.803	0.721	0.647	0.581	0.522	0.469	0.422	0.342	0.278	0.226	0.184	0.150	0.123	0.101	0.083	0.068	0.056	0.046	0.018	0.007	0.003
23	0.795	0.710	0.634	0.567	0.507	0.453	0.406	0.326	0.262	0.211	0.170	0.138	0.112	0.091	0.074	0.060	0.049	0.040	0.015	0.006	0.002
24	0.788	0.700	0.622	0.553	0.492	0.438	0.390	0.310	0.247	0.197	0.158	0.126	0.102	0.082	0.066	0.053	0.043	0.035	0.013	0.005	0.002
25	0.780	0.689	0.610	0.539	0.478	0.423	0.375	0.295	0.233	0.184	0.146	0.116	0.092	0.074	0.059	0.047	0.038	0.030	0.010	0.004	0.001
26	0.772	0.679	0.598	0.526	0.464	0.409	0.361	0.281	0.220	0.172	0.135	0.106	0.084	0.066	0.053	0.042	0.033	0.026	0.009	0.003	0.001
27	0.764	0.669	0.586	0.513	0.450	0.395	0.347	0.268	0.207	0.161	0.125	0.098	0.076	0.060	0.047	0.037	0.029	0.023	0.007	0.002	0.001
28	0.757	0.659	0.574	0.501	0.437	0.382	0.333	0.255	0.196	0.150	0.116	0.090	0.069	0.054	0.042	0.033	0.026	0.020	0.006	0.002	0.001
29	0.749	0.649	0.563	0.489	0.424	0.369	0.321	0.243	0.185	0.141	0.107	0.082	0.063	0.048	0.037	0.029	0.022	0.017	0.005	0.002	0.000
30	0.742	0.640	0.552	0.477	0.412	0.356	0.308	0.231	0.174	0.131	0.099	0.075	0.057	0.044	0.033	0.026	0.020	0.015	0.004	0.001	0.000
31	0.735	0.630	0.541	0.465	0.400	0.344	0.296	0.220	0.164	0.123	0.092	0.069	0.052	0.039	0.030	0.023	0.017	0.013	0.004	0.001	0.000
32	0.727	0.621	0.531	0.454	0.388	0.333	0.285	0.210	0.155	0.115	0.085	0.063	0.047	0.035	0.027	0.020	0.015	0.011	0.003	0.001	0.000
33	0.720	0.612	0.520	0.443	0.377	0.321	0.274	0.200	0.146	0.107	0.079	0.058	0.043	0.032	0.024	0.018	0.013	0.010	0.002	0.001	0.000
34	0.713	0.603	0.510	0.432	0.366	0.310	0.264	0.190	0.138	0.100	0.073	0.053	0.039	0.029	0.021	0.016	0.012	0.009	0.002	0.001	0.000
35	0.706	0.594	0.500	0.421	0.355	0.300	0.253	0.181	0.130	0.094	0.068	0.049	0.036	0.026	0.019	0.014	0.010	0.008	0.002	0.000	0.000
36	0.699	0.585	0.490	0.411	0.345	0.290	0.244	0.173	0.123	0.088	0.063	0.045	0.032	0.023	0.017	0.012	0.009	0.007	0.001	0.000	0.000
37	0.692	0.576	0.481	0.401	0.335	0.280	0.234	0.164	0.116	0.082	0.058	0.041	0.029	0.021	0.015	0.011	0.008	0.006	0.001	0.000	0.000
38	0.685	0.568	0.471	0.391	0.325	0.271	0.225	0.157	0.109	0.076	0.054	0.038	0.027	0.019	0.013	0.010	0.007	0.005	0.001	0.000	0.000
39	0.678	0.560	0.462	0.382	0.316	0.261	0.217	0.149	0.103	0.071	0.050	0.035	0.024	0.017	0.012	0.009	0.006	0.004	0.001	0.000	0.000
40	0.672	0.551	0.453	0.372	0.307	0.253	0.208	0.142	0.097	0.067	0.046	0.032	0.022	0.015	0.011	0.008	0.005	0.004	0.001	0.000	0.000

Table B.4 Present Value of £1 payable at regular intervals ('year's purchase').

Present Value of £1 payable (or receivable) at the end of any period (i.e. weekly, monthly, annually) discounted at interest rates from 1% to 30% per period.

Period	1% £	1.5% £	2% £	2.5% £	3% £	3.5% £	4% £	5% £	6% £	7% £	8% £	9% £	10% £	11% £	12% £	13% £	14% £	15% £	20% £	25% £	30% £
1	0.99	0.99	0.98	0.97	0.96	0.96	0.95	0.94	0.93	0.93	0.93	0.92	0.91	0.90	0.89	0.89	0.88	0.88	0.87	0.87	0.86
2	1.97	1.96	1.93	1.91	1.89	1.87	1.85	1.83	1.81	1.79	1.77	1.76	1.73	1.72	1.70	1.69	1.67	1.65	1.63	1.62	1.60
3	2.94	2.91	2.86	2.83	2.78	2.75	2.70	2.67	2.63	2.60	2.56	2.53	2.49	2.47	2.43	2.40	2.36	2.34	2.31	2.29	2.25
4	3.90	3.85	3.76	3.72	3.63	3.59	3.51	3.47	3.39	3.36	3.28	3.25	3.18	3.15	3.08	3.05	2.99	2.96	2.91	2.88	2.82
5	4.85	4.78	4.65	4.58	4.46	4.40	4.28	4.22	4.12	4.06	3.96	3.91	3.82	3.77	3.69	3.64	3.56	3.52	3.44	3.40	3.33
6	5.80	5.70	5.51	5.42	5.25	5.17	5.01	4.94	4.80	4.73	4.59	4.53	4.41	4.35	4.23	4.18	4.07	4.02	3.92	3.87	3.78
7	6.73	6.60	6.35	6.24	6.02	5.91	5.71	5.62	5.44	5.35	5.18	5.10	4.95	4.88	4.74	4.67	4.54	4.48	4.36	4.30	4.19
8	7.65	7.49	7.18	7.03	6.76	6.63	6.38	6.26	6.04	5.94	5.74	5.64	5.46	5.37	5.20	5.12	4.97	4.89	4.75	4.68	4.55
9	8.57	8.36	7.98	7.80	7.47	7.31	7.02	6.88	6.61	6.49	6.25	6.14	5.93	5.82	5.63	5.54	5.36	5.27	5.11	5.03	4.89
10	9.47	9.22	8.77	8.55	8.16	7.97	7.63	7.46	7.16	7.01	6.74	6.61	6.36	6.24	6.03	5.92	5.72	5.62	5.44	5.35	5.19
11	10.37	10.07	9.54	9.28	8.82	8.61	8.21	8.02	7.67	7.50	7.19	7.04	6.77	6.64	6.39	6.27	6.05	5.95	5.74	5.65	5.47
12	11.26	10.91	10.29	9.99	9.47	9.22	8.76	8.55	8.15	7.97	7.62	7.46	7.15	7.01	6.73	6.60	6.36	6.24	6.02	5.92	5.72
13	12.13	11.73	11.02	10.68	10.09	9.80	9.30	9.05	8.62	8.41	8.03	7.84	7.51	7.35	7.05	6.91	6.65	6.52	6.28	6.16	5.95
14	13.00	12.54	11.73	11.36	10.69	10.37	9.81	9.54	9.06	8.83	8.41	8.21	7.85	7.67	7.35	7.19	6.91	6.77	6.52	6.39	6.17
15	13.87	13.34	12.43	12.01	11.27	10.92	10.30	10.00	9.47	9.22	8.77	8.55	8.16	7.97	7.63	7.46	7.16	7.01	6.74	6.61	6.36
16	14.72	14.13	13.12	12.65	11.83	11.44	10.77	10.44	9.87	9.60	9.11	8.88	8.46	8.26	7.89	7.71	7.39	7.23	6.94	6.80	6.55
17	15.56	14.91	13.79	13.27	12.37	11.95	11.22	10.87	10.25	9.96	9.44	9.19	8.74	8.52	8.13	7.94	7.60	7.44	7.14	6.99	6.72
18	16.40	15.67	14.44	13.87	12.90	12.44	11.65	11.27	10.61	10.30	9.75	9.48	9.00	8.77	8.36	8.16	7.81	7.63	7.31	7.16	6.88
19	17.23	16.43	15.08	14.46	13.41	12.92	12.06	11.66	10.96	10.63	10.04	9.76	9.25	9.01	8.58	8.37	8.00	7.81	7.48	7.32	7.03
20	18.05	17.17	15.71	15.04	13.90	13.37	12.46	12.04	11.29	10.94	10.32	10.02	9.49	9.24	8.79	8.57	8.17	7.98	7.64	7.47	7.17
21	18.86	17.90	16.32	15.60	14.38	13.81	12.85	12.39	11.61	11.23	10.58	10.27	9.72	9.45	8.98	8.75	8.34	8.14	7.79	7.61	7.30
22	19.66	18.62	16.92	16.14	14.84	14.24	13.22	12.74	11.91	11.52	10.83	10.50	9.93	9.65	9.16	8.92	8.50	8.29	7.93	7.74	7.42
23	20.46	19.33	17.50	16.67	15.29	14.66	13.58	13.07	12.20	11.79	11.07	10.73	10.13	9.84	9.34	9.09	8.65	8.44	8.06	7.87	7.54
24	21.24	20.03	18.08	17.19	15.73	15.05	13.92	13.39	12.48	12.05	11.30	10.95	10.33	10.03	9.50	9.24	8.79	8.57	8.18	7.99	7.65
25	22.02	20.72	18.64	17.70	16.15	15.44	14.25	13.69	12.75	12.29	11.52	11.15	10.51	10.20	9.66	9.39	8.93	8.70	8.30	8.10	7.75
26	22.80	21.40	19.19	18.19	16.57	15.82	14.57	13.99	13.00	12.53	11.73	11.35	10.69	10.36	9.81	9.53	9.06	8.82	8.41	8.20	7.85
27	23.56	22.07	19.72	18.67	16.96	16.18	14.88	14.27	13.25	12.76	11.93	11.54	10.85	10.52	9.95	9.67	9.18	8.94	8.51	8.30	7.94
28	24.32	22.73	20.25	19.14	17.35	16.53	15.18	14.54	13.48	12.98	12.13	11.71	11.01	10.67	10.08	9.79	9.29	9.04	8.61	8.40	8.02
29	25.07	23.38	20.76	19.60	17.73	16.87	15.47	14.81	13.71	13.19	12.31	11.89	11.17	10.81	10.21	9.91	9.40	9.15	8.71	8.49	8.11
30	25.81	24.02	21.27	20.04	18.09	17.20	15.74	15.06	13.93	13.39	12.49	12.05	11.31	10.95	10.33	10.03	9.51	9.25	8.80	8.57	8.18
31	26.54	24.65	21.76	20.48	18.45	17.52	16.01	15.31	14.14	13.59	12.66	12.21	11.45	11.08	10.45	10.14	9.61	9.34	8.88	8.66	8.26
32	27.27	25.27	22.25	20.90	18.80	17.83	16.27	15.54	14.35	13.77	12.82	12.36	11.58	11.21	10.56	10.24	9.70	9.43	8.97	8.73	8.33
33	27.99	25.88	22.72	21.32	19.13	18.13	16.53	15.77	14.54	13.95	12.98	12.50	11.71	11.32	10.67	10.34	9.79	9.52	9.04	8.81	8.40
34	28.70	26.48	23.18	21.72	19.46	18.42	16.77	15.99	14.73	14.12	13.13	12.64	11.84	11.44	10.77	10.44	9.88	9.60	9.12	8.88	8.46
35	29.41	27.08	23.64	22.12	19.78	18.71	17.01	16.21	14.91	14.29	13.27	12.78	11.95	11.55	10.87	10.53	9.96	9.68	9.19	8.94	8.52
36	30.11	27.66	24.08	22.50	20.09	18.98	17.23	16.41	15.09	14.45	13.41	12.91	12.07	11.65	10.96	10.62	10.04	9.75	9.26	9.01	8.58
37	30.80	28.24	24.52	22.88	20.39	19.25	17.46	16.61	15.26	14.61	13.55	13.03	12.18	11.76	11.05	10.70	10.12	9.82	9.32	9.07	8.64
38	31.48	28.81	24.95	23.25	20.68	19.51	17.67	16.80	15.42	14.76	13.68	13.15	12.28	11.85	11.14	10.79	10.19	9.89	9.38	9.13	8.69
39	32.16	29.36	25.37	23.61	20.97	19.76	17.88	16.99	15.58	14.90	13.80	13.26	12.38	11.95	11.23	10.86	10.26	9.96	9.44	9.18	8.75
40	32.83	29.92	25.78	23.96	21.25	20.00	18.08	17.17	15.73	15.04	13.92	13.37	12.48	12.04	11.31	10.94	10.33	10.02	9.50	9.24	8.79

Index